I0055959

Essential Theory and Concepts of Signal Processing

Volume I

Essential Theory and Concepts of Signal Processing

Volume I

Edited by **George Pilato**

New Jersey

Published by Clanrye International,
55 Van Reypen Street,
Jersey City, NJ 07306, USA
www.clanryeinternational.com

Essential Theory and Concepts of Signal Processing: Volume I
Edited by George Pilato

International Standard Book Number: 978-1-63240-225-7 (Hardback)

Printed in the United States of America.

Contents

Preface

Signal processing is a part of the fields of electrical engineering, systems engineering and applied mathematics that deals with analysis of or operations on analog as well as digitized signals which represent spatially varying or time-varying physical quantities. Signal Processing can also be said to be the art and science of adapting obtained time-series data for the purpose of breakdown, analysis and augmentation. Signals of interest usually include sound, images, electromagnetic radiation and sensor readings. Such signals can be lifted from various machines and instances in the real world, for example, from biological measurements such as electrocardiograms, telecommunication transmission signals, control system signals and many others. Signal processing has various applications in the real world such as audio signal processing for electrical signals representing sound; image processing in digital cameras, computers and various imaging systems; wireless communication via filtering, waveform generation, equalization and demodulation; array processing for processing signals from arrays of sensors; video processing for interpreting moving pictures; quality improvement of signal like image enhancement, noise reduction and echo cancellation; and even control systems and seismology. Such applications for this field in our daily lives underline the importance of signal processing in the industry today. Therefore the rising demand for skilled graduates in this arena is keeping up its emergence in the global industry.

This book is an attempt to compile and collate all available research on signal processing under one aegis. I am grateful to those who put their hard work, effort and expertise into these researches as well as those who were supportive in this endeavour.

Editor

1

Sparse-Coding-Based Computed Tomography Image Reconstruction

Sang Min Yoon[1] **and Gang-Joon Yoon**[2]

[1] *School of Computer Science, Kookmin University, 77 Jeongneung-ro, Sungbuk-gu, Seoul 136-702, Republic of Korea*
[2] *National Institute for Mathematical Science, KT Daeduk 2 Research Center, 463-1 Jeonmin-dong, Yuseong-gu, Daejeon 305-811, Republic of Korea*

Correspondence should be addressed to Gang-Joon Yoon; ykj@nims.re.kr

Academic Editors: O. Hadar and E. P. Ong

Computed tomography (CT) is a popular type of medical imaging that generates images of the internal structure of an object based on projection scans of the object from several angles. There are numerous methods to reconstruct the original shape of the target object from scans, but they are still dependent on the number of angles and iterations. To overcome the drawbacks of iterative reconstruction approaches like the algebraic reconstruction technique (ART), while the recovery is slightly impacted from a random noise (small amount of ℓ_2 norm error) and projection scans (small amount of ℓ_1 norm error) as well, we propose a medical image reconstruction methodology using the properties of sparse coding. It is a very powerful matrix factorization method which each pixel point is represented as a linear combination of a small number of basis vectors.

1. Introduction

The challenge of continually improving the technology for generating noninvasive medical images of internal anatomy is an area of great interest in the area of radiology. In particular, computed tomography (CT) is one of the most common modalities used for noninvasive medical diagnostic imaging. Cross-sectional CT images are reconstructed from projection data, which is generated by passing X-rays through the target object and measuring the resulting attenuation of those rays. For some noninvasive imaging modalities, the measurements made (i.e., the projection data) can be converted into samples of a Radon transform [1] in order to be reconstructed. In CT, dividing the measured photon counts by the incident photon counts and taking the negative logarithm yields the linear attenuation map of each projection, which, when noted at each projection angle, can be used to determine the Radon transform of the object of interest.

In the fields of accelerator physics, one expects that the relatively simple charged particle beam distributions can be reconstructed from a small number of projections. Tomographic imaging involves the reconstruction of an image from its projections. The reconstruction problem belongs to the class of inverse problems, and it is characterized by the fact that the information of interest is not directly available for measurement. If $\mathbf{f}$ denotes the unknown distribution and $\mathbf{b}$ the quantity measured by the imaging device, then the measurement system is written as $\mathbf{b} = \mathbf{Rf}$ for the Radon transform $\mathbf{R}$ [1]. The discrete problem, which is based on expanding $\mathbf{f}$ in a finite series of basis function, can be written as $\mathbf{b} = \mathbf{Af}$. The matrix $\mathbf{A}$, typically large and sparse, is a discretization of the Radon transform. An approximate solution to this linear system could be computed by iterative methods, which only require matrix-vector products and hence do not alter the structure of $\mathbf{A}$.

There are numerous approaches to reconstruct the shape of the target object from numerous angular projections. The iterative reconstruction algorithms are known to generate higher quality CT images than the filtered backprojection (FBP) algorithm. For example, the algebraic reconstruction technique (ART) [2] efficiently solves the inverse problem. Yet, although ART is effective, it still produces noise within a reconstructed image. To reduce the noise from reconstructed images using ART, the postprocessing is

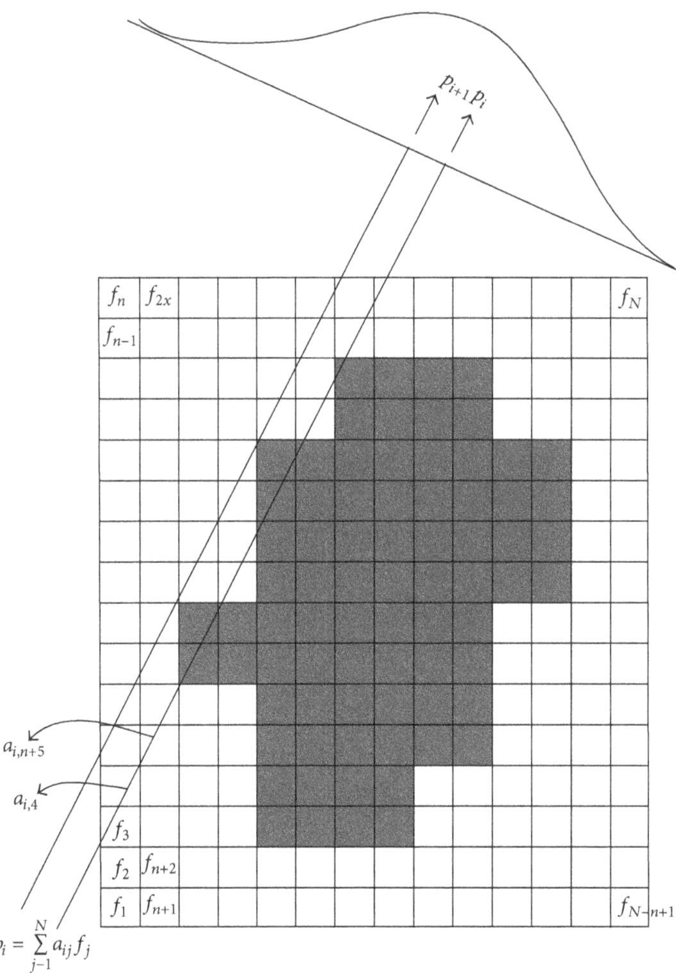

FIGURE 1: Our proposed CT image reconstruction method and its mathematical notations: b_i is the line integral of $\mathbf{f}$ along the ray p_i, and a_{ij} is the length of the intersection of the ray p_i and the pixel I_j.

required. Unfortunately, it is possible to lose important data while removing the noise. Here, we propose a sparse-coding-based reconstruction method to overcome the drawbacks of iterative reconstruction approaches like ART and to minimize the influence of the deficiency of the projection measurements that are very sparse, while the recovery is slightly impacted from random noises. The recovery is achieved by finding an approximation vector with small ℓ_1 and ℓ_2 norm errors, which provides a good fit locally and globally. Our contribution in this paper is to solve the inverse problem using sparse coding by reducing both the random noise and the dependency on projection scans, which is less dependent on the number of angular projections than previous iterative reconstruction methods.

2. Our Proposed Approach

To reconstruct the original shape from projected measurements, we regard an original image as a discrete function $f(x, y)$ defined on a domain $\Omega = \{(x, y) \in \mathbb{R}^2 : 0 \le x \le n, 0 \le y \le m\}$ such that $f(x, y)$ is constant on each cell $I_{i,j} = [i-1, i] \times [j-1, j]$ for $i = 1, \ldots, n$ and $j = 1, \ldots, m$. We number the cells in lexicographical order so that $I_{(i-1)m+j} = I_{i,j}$ for $i = 1, \ldots, n$ and $j = 1, \ldots, m$. Let f_i be the constant in the ith cell. For each angle θ_k given in degree, $k = 1, \ldots, q$, we project the image along parallel rays p with width w as follows:

$$b_i = \sum_{j=1}^{N} a_{ij} f_j, \quad i = 1, \ldots, P \ (N := nm), \tag{1}$$

where P is the total number of projections, and the coefficient a_{ij} is defined to be the length (or area) of the ith ray through the jth cell and $a_{ij} = 0$ if the ray does not touch the cell; that is, the value b_i is the ray sum measured along the ith ray and a_{ij} is the weighting factor representing the contribution of the jth cell to the ray sum. The tomography problem is then related to the reconstruction of an unknown image $\mathbf{f} = (f_1, \ldots, f_N)^T$ from the observed vector $\mathbf{b} = (b_1, \ldots, b_P)^T$ given as

$$\mathbf{Af} = \mathbf{b}, \tag{2}$$

Here, $\mathbf{A} = (a_{ij})_{i=1,j=1}^{P,N}$, and $\cdot^T$ denotes the transpose of a vector/matrix. In general, $P < N$, and the problem is

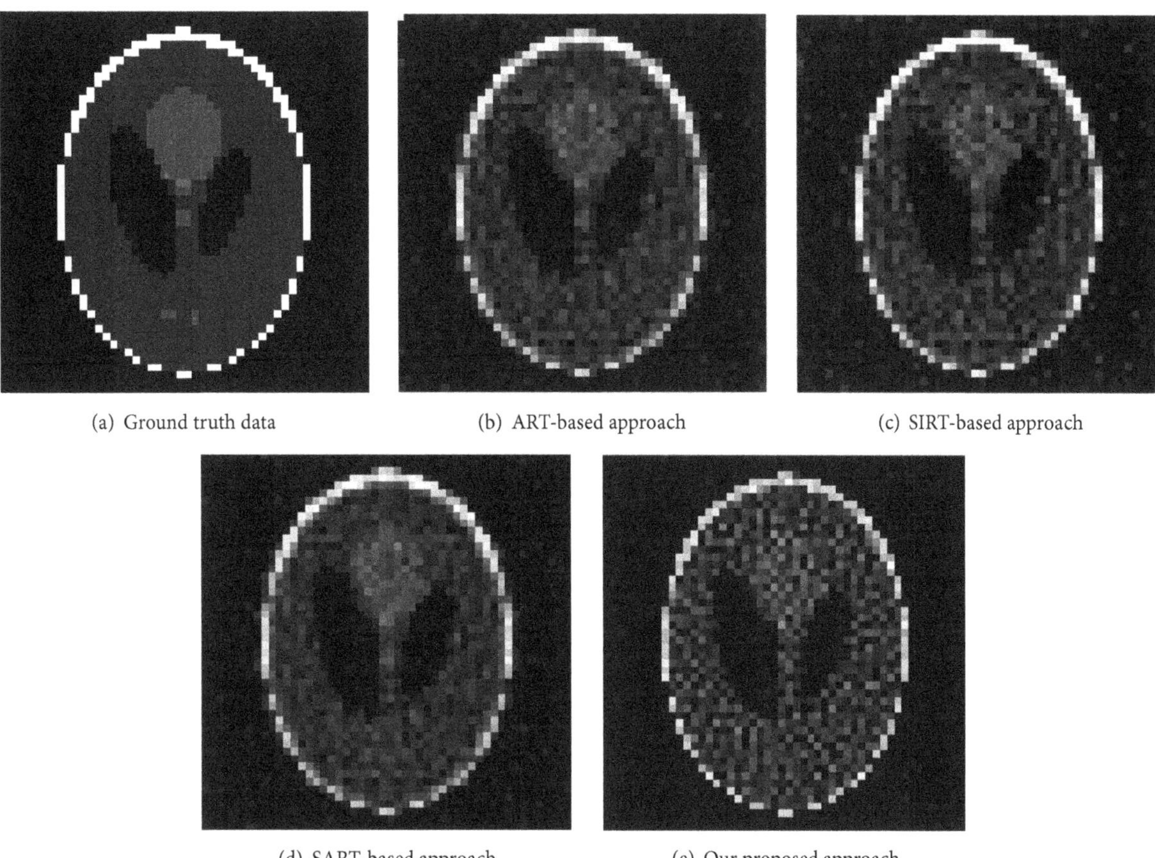

(a) Ground truth data (b) ART-based approach (c) SIRT-based approach

(d) SART-based approach (e) Our proposed approach

FIGURE 2: Comparison of image reconstruction from 2250 angular projections around the target object.

underdetermined and so is ill-posed. Figure 1 shows our proposed CT image reconstruction approach and mathematical notations in (1) and (2).

This inverse problem emerges in areas of study involving 2D and 3D imaging such as medical imaging, geophysics, and material science [3–5]. To solve the inverse problem (2), many methods such as the least squares method, ART [6, 7], SARTs (simultaneous algebraic reconstruction techniques [8]), and SIRT (simultaneous iterative reconstruction techniques [9]) have been proposed and studied (see [10, 11] and references therein). We observe that almost all entries of $\mathbf{A}$ are zeros, and the vector $\mathbf{f}$ from CT images also has sparse pixel values. Based on this observation, we propose a method to solve the inverse problem by applying the compressive sampling (CS) technique. CS pertains to the recovery of $\mathbf{x} \in \mathbb{R}^K$ from rather a small amount of measurements $\mathbf{y} \in \mathbb{R}^L$ with $L < K$. Given an underdetermined systems

$$\mathbf{y} = \Phi\mathbf{x}, \tag{3}$$

where Φ is an $L \times K$ matrix, CS enables us to perform exact recovery when $\mathbf{x}$ is sparse, and the matrix Φ satisfies the restricted isometry property [12, 13] (also see [14, 15]). The recovery performs through the so-called basis pursuit optimization:

$$\arg\min_{\mathbf{x}\in\mathbb{R}^K} \|\mathbf{x}\|_1 \text{ subject to } \mathbf{y} = \Phi\mathbf{x}. \tag{4}$$

Using $\|\mathbf{v}\|_p$ $(p = 1, 2)$, we denote the ℓ_p norm of a vector $\mathbf{v} = (v_1, \dots, v_K)^T \in \mathbb{R}^K$ defined by $\|\mathbf{v}\|_p = (|v_1|^p + \cdots + |v_K|^p)^{1/p}$. In the paper, we also consider the case where noise is added to the projected image, $\mathbf{f}$, so that we can model such an imaging system as follows:

$$\mathbf{Af} = \mathbf{b} + \eta. \tag{5}$$

In order to recover the sparse image from the inverse problem (5) while preserving intrinsic geometric structures such as edges, we try to find an image $\tilde{\mathbf{f}}$ which provides a good fit to the original image $\mathbf{f}$ globally (small amount of $\|\mathbf{f} - \tilde{\mathbf{f}}\|_2$) and locally (small amount of $\|\mathbf{f} - \tilde{\mathbf{f}}\|_1$) as well. So, we propose the following elastic net optimization:

$$\tilde{\mathbf{f}} = \arg\min_{\mathbf{f}\in\mathbb{R}^N} \frac{1}{2}\|\mathbf{Af} - \mathbf{b}\|_2^2 + \beta\|\mathbf{f}\|_2^2 + \gamma\|\mathbf{f}\|_1. \tag{6}$$

The model in (6) is a combination of the basis pursuit denoising ($\beta = 0$) and the ridge regression ($\gamma = 0$). In (6), the first term measures the fitting, and the third regularization term is added to recover a sparse data where γ controls the trade-off between sparsity and reconstruction fidelity. The quadratic term of the regularization gives rise to the grouping effect of highly correlated variables and removes the limitation on the number of selected variables [16]. The basis

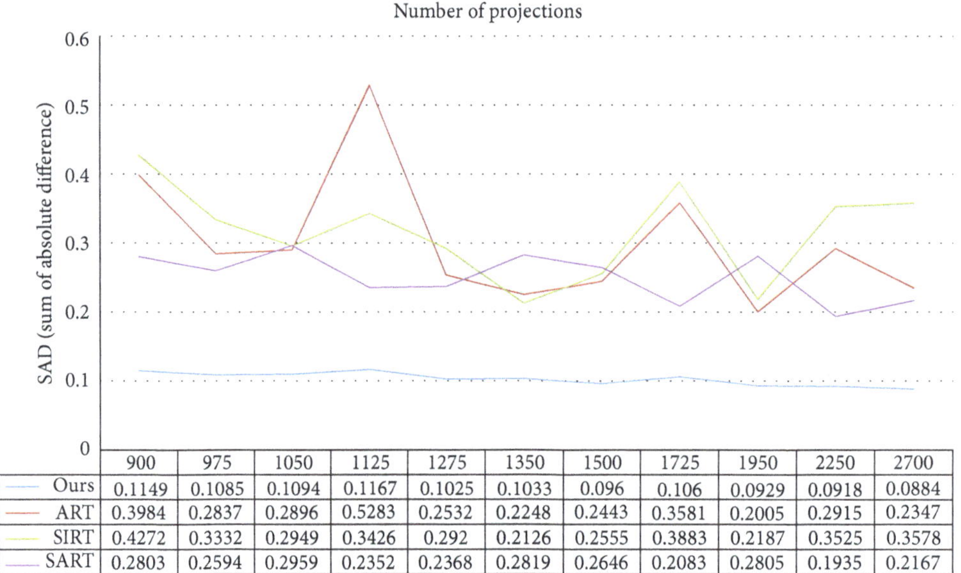

	900	975	1050	1125	1275	1350	1500	1725	1950	2250	2700
Ours	0.1149	0.1085	0.1094	0.1167	0.1025	0.1033	0.096	0.106	0.0929	0.0918	0.0884
ART	0.3984	0.2837	0.2896	0.5283	0.2532	0.2248	0.2443	0.3581	0.2005	0.2915	0.2347
SIRT	0.4272	0.3332	0.2949	0.3426	0.292	0.2126	0.2555	0.3883	0.2187	0.3525	0.3578
SART	0.2803	0.2594	0.2959	0.2352	0.2368	0.2819	0.2646	0.2083	0.2805	0.1935	0.2167

(a) Quantitative measurement using SAD

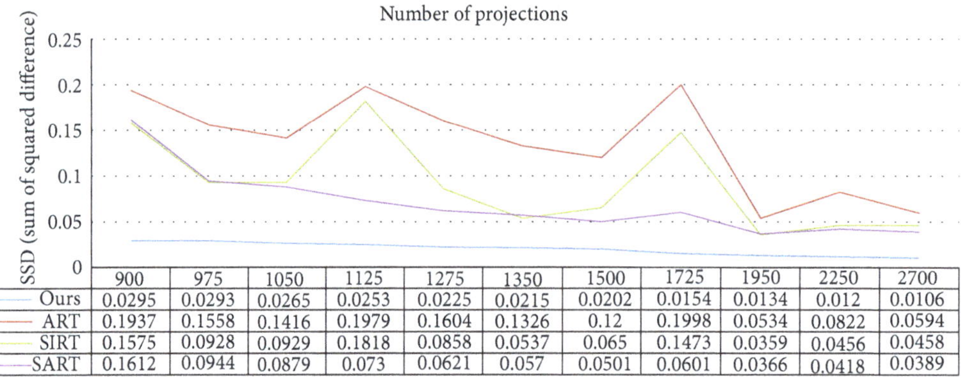

	900	975	1050	1125	1275	1350	1500	1725	1950	2250	2700
Ours	0.0295	0.0293	0.0265	0.0253	0.0225	0.0215	0.0202	0.0154	0.0134	0.012	0.0106
ART	0.1937	0.1558	0.1416	0.1979	0.1604	0.1326	0.12	0.1998	0.0534	0.0822	0.0594
SIRT	0.1575	0.0928	0.0929	0.1818	0.0858	0.0537	0.065	0.1473	0.0359	0.0456	0.0458
SART	0.1612	0.0944	0.0879	0.073	0.0621	0.057	0.0501	0.0601	0.0366	0.0418	0.0389

(b) Quantitative measurement using SSD

FIGURE 3: Quantitative comparison of previous approaches and our proposed approach using SAD and SSA by changing the number of angular projections from 900 to 2700.

pursuit denoising enables us to reconstruct **f** simply in the ℓ_1 sense, making **b** close to **Af** in the ℓ_2 sense. We solve the optimization (6) using the feature-sign search algorithm [17].

3. Experiments

The schematic setup for sparse-coding-based CT image reconstruction is illustrated in this section. To efficiently show the robustness of our proposed approach, we have compared our sparse-coding-based image reconstruction method to other popular approaches, such as traditional ART, SIRT, and SART in Figure 2. For fair comparison, we reconstructed an image that has 50×50 resolution from 2250 angular projections around the target object. The previous approaches such as ART (Figure 2(b)), SIRT (Figure 2(c)), and SART (Figure 2(d)), require several iterations to robustly reconstruct the target object (100 iterations in our experiment because the reconstructed images have few improvements in resolution after that.) As shown in Figure 2, there is still a noise effect in the background, but our proposed sparse-coding-based approach (Figure 2(e)) dramatically reduces the noise in the background within a reconstructed image.

Image reconstruction using rotated angular projection methodologies is very dependent on the number of projections. To show the robustness of the sparse-coding-based image reconstruction methodology, we have quantitatively compared our approach to previous approaches by changing the number of the projections. For this, we evaluated the ℓ_1-norm-based similarity measure called SAD (sum of absolute difference) and ℓ_2-norm based-similarity measure called SSD (sum of squared difference) between reconstructed images and the original image by changing the number of projections. In Figure 3, we show how we robustly reconstruct the original image by calculating the SAD and SSD from the randomly selected projections from 900 to 2700 around the ground truth data between the reconstructed images using ART, SIRT, SART, and our approach and ground truth data. Figure 3 demonstrates that our proposed approach can reconstruct the original shape of the target object while reducing the effect of the number of projections. SAD-based error comparison as shown in Figure 3(a) provides that our proposed reconstruction method is less dependent on the measurement (b) than other previous approaches, and SSD-based comparison in Figure 3(b) shows that our proposed approach is very efficient to solve the inverse linear problem to recover the original shape of the target object.

4. Discussion

In this paper, we propose a novel reconstruction approach from the projected measurement of the radiation around the target object. This nonparametric approach to the reconstruction of medical CT images using sparse coding shows robustness in the reconstruction of the original shape without prior information of the target object. Sparse coding is a state-of-the-art technique used to reconstruct a sparse data set from a small number of linear measurements. We propose a reconstruction method based on sparse coding, motivated by the fact that medical images are often represented in terms of only a few data pixel values in the spatial domain. Our proposed image reconstruction methodology is less dependent on the number of projections collected around the target object, so the original shape can be recovered with only a few measurements. Based on our experimental evaluation, we argue that the proposed sparse-coding-based CT image reconstruction scheme provides better ℓ_1-norm and ℓ_2-norm error results than previous approaches.

Acknowledgment

S. M. Yoon was supported by the new Faculty Research Program, 2012, of Kookmin University in the Republic of Korea.

References

[1] J. Radon, "On the determination of functions from their integral values along certain manifolds," *IEEE Transactions on Medical Imaging*, vol. 5, no. 4, pp. 170–176, 1986.

[2] R. Gordon and G. T. Herman, "Reconstruction of pictures from their projections," *Communications of the ACM*, vol. 14, pp. 759–768, 1971.

[3] M. Bertero and P. Boccacci, *Introduction to Inverse Problems in Imaging*, IOP publishing Ltd., London, UK, 1998.

[4] H. W. Engl, M. Hanke, and A. Neubauer, *Regularization of Inverse Problems*, Kluwer, Dordrecht, The Netherlands, 1996.

[5] P. C. Hansen, *Discrete Inverse Problems: Insights and Algorithms*, SIAM, Philadelphia, Pa, USA, 2010.

[6] S. Kaczmarz, "Angenäherte auflösung von systemen linearer gleichungen," *Bulletin International de l'Académie Polonaise des Sciences et des Lettres*, vol. A35, pp. 355–357, 1938.

[7] R. Gordon, R. Bender, and G. T. Herman, "Algebraic Reconstruction Techniques (ART) for three-dimensional electron microscopy and X-ray photography," *Journal of Theoretical Biology*, vol. 29, no. 3, pp. 471–481, 1970.

[8] A. H. Andersen and A. C. Kak, "Simultaneous algebraic reconstruction technique (SART): a superior implementation of the art algorithm," *Ultrasonic Imaging*, vol. 6, no. 1, pp. 81–94, 1984.

[9] P. Gilbert, "Iterative methods for the three-dimensional reconstruction of an object from projections," *Journal of Theoretical Biology*, vol. 36, no. 1, pp. 105–117, 1972.

[10] P. C. Hansen and M. S. Hansen, "A MATLAB package of algebraic iterative reconstruction methods," *Journal of Computational and Applied Mathematics*, vol. 236, pp. 2167–2178, 2012.

[11] G. T. Herman, *Fundamentals of Computerized Tomography: Image Reconstruction from Projection*, Springer, New York, NY, USA, 2nd edition, 2009.

[12] E. Candès and T. Tao, "Decoding by linear programming," *IEEE Transactions on Information Theory*, vol. 51, no. 12, pp. 4203–4215, 2005.

[13] E. J. Candès, J. K. Romberg, and T. Tao, "Stable signal recovery from incomplete and inaccurate measurements," *Communications on Pure and Applied Mathematics*, vol. 59, no. 8, pp. 1207–1223, 2006.

[14] E. Candès, "Compressive sampling," in *Proceedings of the International Congress of Mathematics*, pp. 1433–1452, 2006.

[15] D. L. Donoho, "Compressed sensing," *IEEE Transactions on Information Theory*, vol. 52, no. 4, pp. 1289–1306, 2006.

[16] H. Zou and T. Hastie, "Regularization and variable selection via the elastic net," *Journal of the Royal Statistical Society B*, vol. 65, no. 2, pp. 301–320, 2005.

[17] H. Lee, A. Battle, R. Raina, and A. Y. Ng, "Efficient sparse coding algorithms," in *Advanced in Neural Information Processing Systems (NIPS)*, vol. 19, 2007.

2

Recognition of Emotions in Mexican Spanish Speech: An Approach Based on Acoustic Modelling of Emotion-Specific Vowels

Santiago-Omar Caballero-Morales

Technological University of the Mixteca, Road to Acatlima K.m. 2.5, 69000 Huajuapan de León, OAX, Mexico

Correspondence should be addressed to Santiago-Omar Caballero-Morales; scaballero@mixteco.utm.mx

Academic Editors: R. J. Ferrari and S. Wu

An approach for the recognition of emotions in speech is presented. The target language is Mexican Spanish, and for this purpose a speech database was created. The approach consists in the phoneme acoustic modelling of emotion-specific vowels. For this, a standard phoneme-based Automatic Speech Recognition (ASR) system was built with Hidden Markov Models (HMMs), where different phoneme HMMs were built for the consonants and emotion-specific vowels associated with four emotional states (anger, happiness, neutral, sadness). Then, estimation of the emotional state from a spoken sentence is performed by counting the number of emotion-specific vowels found in the ASR's output for the sentence. With this approach, accuracy of 87–100% was achieved for the recognition of emotional state of Mexican Spanish speech.

1. Introduction

Emotion recognition has become an important research subject in human-computer interaction and image and speech processing [1]. Besides human facial expressions, speech has proven as one of the most promising modalities for the automatic recognition of human emotions [2]. Among the different applications of speech emotion recognition the following can be mentioned: psychiatric diagnosis, intelligent toys, lie detection, learning environments, and educational software [3].

Many approaches have been presented to recognize affective states based on specific speech features. Short-term features (formants, formant bandwidth, pitch/fundamental frequency, and log energy) and long-term features (mean of pitch, standard deviations of pitch, time envelopes of pitch, and energy) have been used for this purpose. Short-term features reflect local speech characteristics in a short-time window while long-term features reflect voice characteristics over a whole utterance [4]. Pitch/fundamental frequency (f_0), intensity of the speech signal (energy), and speech rate have been identified as important indicators of emotional status [5–8]. Other works have shown that speech formants, particularly the first and the second, are affected by the emotional states [9, 10].

Acoustic speech features are represented with different methods, most of them related to speech recognition. Linear predictive coefficients (LPCs) have been used to represent the spectral envelope of a digital signal of speech in compressed form, using the information of a linear predictive model [11]. However, a problem faced with the LPCs for the process of formant tracking in emotion recognition is the false identification of the formants [8]. Mel-Frequency Cepstral Coefficients (MFCCs) provide a more reliable representation of the speech signal because they consider the human auditory frequency response [12]. Diverse works have used MFCCs as spectral features with significant results for emotion recognition [1, 3, 7, 13–16]. In [7] an alternative to MFCCs was presented in the form of short-time log frequency power coefficients (LFPCs).

Diverse classification methods are available for the recognition of emotions from the obtained speech features. In [16] high recognition accuracy was obtained with Support Vector

Machines (SVMs) when compared with Naive Bayes and K-Nearest Neighbor. Other works have used Artificial Neural Networks (ANNs) [17–19] and Hidden Markov Models (HMMs) [13, 17, 19] with significant performance. In general, recognition tests with these methods are performed with long-term and short-term features which are obtained from speech corpora utterances with four or six emotions [8].

Most of the emotional speech databases cover the German and English languages (i.e., [19–23]). However, for the Spanish language, just few databases are known as presented in [8]. Particularly for the Mexican Spanish language, no speech database or developments in the field of speech emotion recognition are known.

In this work, the development of a Mexican Spanish emotional speech corpus following guidelines found in the literature for other languages is presented. In addition, an emotional speech recognizer is built with this corpus to test an emotion recognition approach. The approach consists in the phoneme acoustic modelling of vowels associated with each emotional state, considering that an emotional state is reflected as a tone variation of vowels. A standard phoneme-based Automatic Speech Recognition (ASR) system is built with HMMs, where different phoneme HMMs are built for the vowels associated with the considered emotional states. Estimation of the emotional state from a spoken utterance is performed by counting the number of emotion-specific vowels found in the ASR's output for the MFCC-coded utterance. With this approach accuracy of 87–100% was achieved for the recognition of the emotional state of Mexican Spanish speech.

This paper is structured as follows: in Section 2 the details of the Mexican Spanish emotional speech corpus (selection of emotional states, stimuli vocabulary, speech recording, phonetic and orthographic labelling, and acoustic features) are presented. Then, in Section 3 the details of the ASR system are presented while the results are presented, and discussed in Section 4. Finally, in Section 5, the conclusions and future work are presented.

2. Speech Corpus

One important resource for research in the emotion recognition field is the speech databases or corpora. Emotional speech data has been obtained from actors (simulated emotions) as in [22] and from spontaneous (non-acted) speech as in [19]. A more comprehensive list of speech databases with simulated and non-acted emotions is presented in [8]. Speech databases with simulated emotions are widely used for research given the similarities found between "real" and "acted" speech data [5].

In this work simulated emotional speech was obtained from Mexican non-professional actors and volunteers from the Cultural Center of the City of "Huajuapan de Leon" in Oaxaca, Mexico. As in [4], the text contents for the sentences of the corpus were written in a way to stimulate a speaker to speak in the specified emotions. About the number of emotions, in [24] the following 15 basic emotions were proposed: anger, fear, sadness, sensory pleasure, amusement, satisfaction, contentment, excitement, disgust,

Table 1: Emotional stimuli: "enojo" (anger).

No	Sentence
1	Tu siempre llegas tarde (You always arrive late)
2	Voy a matarte (I am going to kill you)
3	No vengas aquí (Do not come here)
4	Ya cállate (Shut up)
5	Si no te gusta hazlo tú (If you do not like it then do it yourself)
6	Qué quieres que te diga (What do you want me to tell you)
7	Eres un sinvergüenza (You are a shameless)
8	Me las vas a pagar María Fernanda (You are going to pay, Maria Fernanda)
9	No te quiero volver a ver nunca más (I do not want to see you ever again)
10	Te voy a poner de patitas en la calle (I will take you out of the house)

contempt, pride, shame, guilt, embarrassment, and relief. However, most of the emotional speech corpora consider four or six emotions [8]. In this work the following emotions were considered as defined in [25–28]: anger, happiness, neutral, and sadness. The details of the corpus sentences for the emotions and the speech acquisition process are presented in the following section.

2.1. Data Preparation. In Tables 1, 2, 3, and 4, the stimuli sentences for anger ("enojo"), happiness ("felicidad"), neutral ("neutro"), and sadness ("tristeza") are presented, respectively. Ten sentences for each emotional state were designed, leading to a total of 40 sentences with 233 words (vocabulary of 140 unique words). The speech data was then obtained from four non-professional actors (two males, two females) from the local Cultural Center of the City of "Huajuapan de León" in Oaxaca, Mexico. Two additional volunteers (one male and one female) took part in the speech data collection. Thus, a total of six speakers (MS1-3, FS1-3) were considered for the emotional Mexican Spanish speech corpus, each one speaking the 40 sentences presented in Tables 1–4. This amount of speakers and speech data is similar to the corpus presented in [7] which considered six speakers for Burmese and Mandarin languages and ten sentences for each emotional state.

In this work, the emotional speech was recorded in WAV format with a sampling rate of 48,000 Hz and two audio channels. The software *WaveSurfer* was used for the orthographic and phonetic labelling of the speech data. Because the target language is the Mexican Spanish, special attention was paid to the pronunciation of the speakers as there are significant differences between the Spanish spoken in the South, Central, and North regions of Mexico.

For the definition of the phonetic repertoire for the labelling of the speech corpus (and development of

Table 2: Emotional stimuli: "felicidad" (happiness).

No	Sentence
1	Hey, pasé el exámen (Hey, I passed the exam)
2	Ya tengo a mi hijo en casa (I finally got my son to come home)
3	Tengo una gran familia (I have a great family)
4	Me encanta que mis hijos me vengan a visitar (I love that my children come to visit me)
5	El canto es mi vidà, me encanta cantar (Singing is my life, I love to sing)
6	Gané un millón de dólares (I won a million dollars)
7	Por fin se pagó la casa (Finally, the house was paid)
8	Mañana es mi fiesta (Tomorrow is my party)
9	Gracias papá por mi ropa nueva (Thank you dad for my new clothes)
10	El bebé nació tan saludable (The baby was born very healthy)

Table 3: Emotional stimuli: "neutro" (neutral).

No	Sentence
1	El cielo está nublado (The sky is cloudy)
2	La casa es rosada (The house is pink)
3	El comedor es para seis personas (The dining room is for six people)
4	Mañana vamos a la escuela (Tomorrow we will go to school)
5	Crucemos la calle (Let's cross the street)
6	Tengo mucha tarea (I have a lot of homework)
7	La guitarra de mi mamá está sucia (My mom's guitar is dirty)
8	Mi niño juega sus juguetes nuevos (My kid plays his new toys)
9	Es enorme tu casa (Your home is huge)
10	Es mejor la leche entera que la deslactosada (Whole milk is better than lactose-free milk)

the classifier) the Mexbet alphabet for the Mexican Spanish language [29] was used. An updated version of the alphabet, proposed by the Master in Hispanic Linguistics Cuetara [30], is shown in Table 5. This alphabet is specific for the Spanish spoken in the City of Mexico (Central Region) and the speakers for the emotional corpus had the associated pronunciation.

In addition to the Mexbet phonemes, Cuetara also proposed the inclusion of the archiphonemes /_D/, /_G/, /_N/ and /_R/ to define the neutralization of the following couples of phonemes: /d/-/t/, /g/-/k/, /n/-/m/, and /ɾ/-/r/ [30]. To represent the pronunciation of the sequence of phonemes /k/ and /s/ (as in "extra"), and the silence, the phonemes /ks/ and /sil/ were added. This led to a final alphabet of 28 phonemes for this work.

Table 4: Emotional stimuli: "tristeza" (sadness).

No	Sentence
1	Me siento tan mal (I feel so bad)
2	Mi perrito murió (My little dog died)
3	Juan Antonio, porqué me haz engañado? (Juan Antonio, why you cheated on me?)
4	Digame que mi hijo está bien doctor (Please doctor, tell me if my son is alright)
5	Los ricos también lloran, pero los pobres lloramos más (The rich also cry, but we, the poor, cry more)
6	Espero que algún dia encontremos a mi hijo (One day I hope to find my son)
7	Mamá, porqué ya no viene mi papá? (Mom, why my father is no longer coming?)
8	Nadie me quiere porque soy fea (Nobody likes me because I am ugly)
9	De qué me sirve el dinero si no la tengo a mi lado? (Money is of no use if I do not have her by my side)
10	Gasté todo en bebida, ahora no me queda más que mi desilusión (I spent all my money in alcohol, now I only have my disappointment)

Table 5: IPA and Mexbet representation of the Mexican Spanish phonemes [30].

Description	IPA	Mexbet
Voiceless bilabial stop	p	p
Voiceless dental stop	t	t
Voiceless velar stop	k	k
Voiced bilabial stop	b	b
Voiced dental stop	d	d
Voiced velar stop	g	g
Voiceless palatal affricate	t͡ʃ	tS
Voiceless labiodental fricative	f	f
Voiceless alveolar sibilant	s	s
Voiceless velar fricative	x	x
Voiced palatal fricative	ʒ	Z
Bilabial nasal	m	m
Palatal nasal	ɲ	ñ
Alveolar nasal	n	n
Alveolar lateral	l	l
Alveolar trill	r	r
Alveolar flap	ɾ	r(
Close front unrounded vowel	i	i
Close-mid front unrounded vowel	e	e
Open front unrounded vowel	a	a
Close-mid back rounded vowel	o	o
Close back rounded vowel	u	u

TABLE 6: Emotion-specific vowels in the emotional stimuli.

Emotion	Vowel/phoneme	Frequency
Anger	a_e	31
	e_e	28
	i_e	11
	o_e	9
	u_e	6
Happiness	a_f	41
	e_f	28
	i_f	18
	o_f	12
	u_f	4
Neutral	a_n	38
	e_n	32
	i_n	7
	o_n	14
	u_n	10
Sadness	a_t	32
	e_t	42
	i_t	26
	o_t	34
	u_t	4

The spectral properties of vowel sounds have been found to be the best indicator of emotions in speech [27]. Also, work presented in [31, 32] reported on significant differences in vowels given the emotion used for their production. In this work for the identification of emotions, the following identifiers were added to the phonemes representing vowels: _e for "enojo" (anger), _f for "felicidad" (happiness), _n for "neutro" (neutral), and _t for "tristeza" (sadness). Thus, the vowels in the sentences for anger had the identifier _e, and the vowels in the sentences for sadness had the identifier _t. In Table 6 the frequency of emotion-specific vowels in the emotional stimuli from Tables 1–4 is presented.

Once the emotional stimuli was recorded with the six speakers and the speech data was orthographically and phonetically labelled, a spectral analysis of the speech segments representing the emotion-specific vowels was performed. The software *WafeSurfer* was used for this task. As presented in Figure 1 a "Spectrum Section Plot" was obtained for each vowel in the speech corpus. The setting for the plot was: FFT Analysis, Hamming Window, Reference: −110.0 dB, Range: 110.0 dB, Order: 40, Pre-emphasis: 0.0, and 512 FFT points. The data points of each plot were saved in a text file by pressing the button "Export".

After all spectrum plots were obtained for all samples of all vowels in the speech corpus, the average spectrum per gender and emotion was computed. In Figure 2 the average spectrum for all emotion-specific vowels across all male speakers is presented. The same concepts are presented in Figure 3 for the female speakers. Note the differences in the spectrum for all vowels depending on the emotion. These results are similar to the ones presented in [28]. Thus, the developed speech corpora is representative of the considered emotions and can be used for classification tasks.

In this work it is considered that by means of acoustic modelling of the emotion-specific vowels, an ASR system built with these models can be used to estimate emotional states. Hence, the state of a spoken sentence can be performed by counting the number of emotion-specific vowels found in the ASR's output. Note that with this approach, if a phoneme lexicon is added, information about the words spoken with a particular emotion can also be estimated. However, in order to perform this development, a suitable feature extraction method must be implemented. This is presented in the following section.

2.2. Feature Extraction. As commented in [3], there are no established analytical methods in the field of voice analysis that can reliably determine the intended emotion carried by the speech signal. However, in this field the spectral features obtained with Mel-frequency cepstral coefficients (MFCCs) have ben used with important results [3, 13]. MFCCs have been widely used in speech recognition because of superior performance over other features. These cepstrum-related spectral features have also been found to be useful in the classification of stress in speech [27, 33].

The Mel-frequency cepstrum is a representation of the short-term power spectrum of a sound, based on a linear cosine transformation of a log power spectrum on a nonlinear Mel scale of frequency [3]. MFCCs are based on the known variation of the human ear's perception to different frequencies, which can be expressed in the Mel-frequency scale [34, 35]. The coding of the emotional speech corpus was performed with the *HCopy* module of the Hidden Markov Model Toolkit (HTK) developed by the Cambridge University Engineering Department in the United Kingdom [34]. Details about the MFCC codification process can be found in [34, 35].

For this work, 12 cepstral coefficients plus energy (E), delta (D), and acceleration (A) coefficients were computed [34]. In Figure 4 some examples of the MFCCs obtained for the vowels of one of the male speakers from the emotional speech database are shown with the parameters used for the codification tool *HCopy*. This corroborates the information presented in Figures 2 and 3, showing that features extracted with MFCCs can be used to identify emotional states given the differences presented in Figure 4.

3. Classification: Recognition Method

The approach of this work is that by means of acoustic modelling, particularly of the vowels, speech emotion recognition can be performed. Among the most common methods for acoustic modelling and classification the following can be mentioned: Vector Quantization (VQ), Gaussian Mixture Density (GMD) Models, Support Vector Machines (SVM), Artificial Neural Networks (ANNs), and Hidden Markov Models (HMMs) [3, 4, 13, 16, 17, 19, 27, 36].

In [4] VQ, ANNs, and GMD were trained with the speech features extracted from different sections of whole emotional sentences, obtaining short- and long-term features. The classification method determined the emotion from a particular sentence. This applied to the work presented in [3, 7, 16]. Because global features were considered by these works,

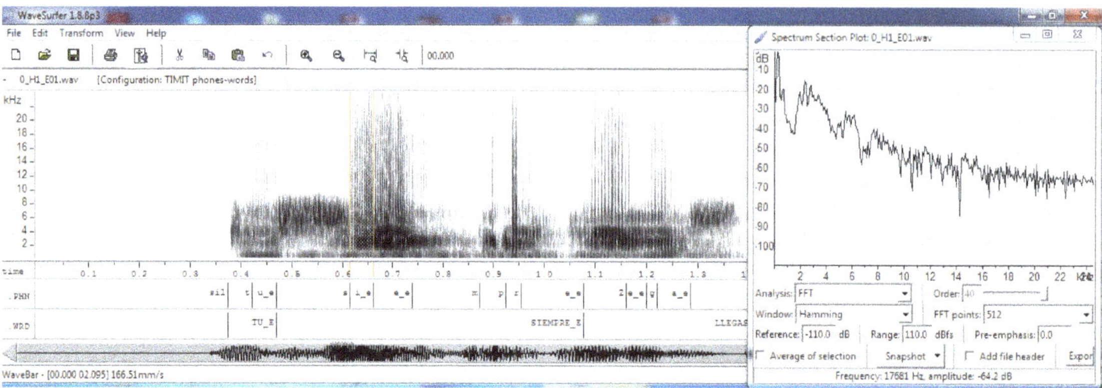

FIGURE 1: Spectrum section plot for the vowel "i_e" (/i/ with anger).

specific features as those of vowels were not fully considered for modelling.

A work that considered this situation was the phoneme-class approach presented in [27]. In that work, two sets of HMM classifiers were built: a generic set of "emotional speech" HMMs (one for each emotion, as in common approaches) and a set of broad phonetic-class-based HMMs for each emotion type considered. Five broad phonetic classes (vowel, glide, nasal, stop, and fricative sounds) were used to explore the effect of emotional "coloring" on different phoneme classes. It was found that spectral properties of vowel sounds were the best indicator of emotions in terms of the classification performance.

Instead of building different HMM classifiers as in [27], with the proposed approach just a single HMM classifier is required. For this classifier, an HMM is built for each of the 28 phonemes in the Mexican Spanish language (23 consonants and 5 vowels) [30]. However, since each vowel can be associated with four emotions, the number of vowels is extended to 20 as presented in Table 6.

An Automatic Speech Recognition (ASR) system built with these HMMs would output phoneme sequences (including the emotion-specific vowels) when tested with emotional speech. A decision about the emotion present in the speech then can be performed by computing the frequency of emotion-specific vowels in the ASR's output. An advantage of using the ASR system is that, by incorporating a phoneme lexicon, word output sequences can be obtained, providing additional information about the sentence spoken with a particular emotion. The proposed HMM ASR system for emotion recognition is presented in Figure 5. Its elements were implemented with the HTK tool [34] and the details are presented in the following sections.

3.1. Acoustic Models. Hidden Markov Models (HMMs) were the method used for the acoustic modelling of the Mexican Spanish phonemes presented in Tables 5 and 6 (which include the emotion-specific vowels). The HMMs had the standard three-state left-to-right structure presented in [34] for acoustic modelling of phonemes with six Gaussian mixture components per state.

3.2. Phoneme Lexicon. The phonetic lexicon was built at the same time as the phonetic labelling of the speech corpus. A lexicon consists of a list of word entries with their respective phonetic transcription based on a given alphabet. For the initial phonetic labelling and creation of the lexicon of the emotional speech corpus, the word transcriptor *TranscribEMex* [37] was used. This tool was developed to phonetically label the DIMEX corpus for Mexican Spanish [37] using the updated Mexbet alphabet [30] presented in Table 5. Then, for the final phonetic labelling and creation of the lexicon, the identifiers _e, _f, _n, and _t were added to the vowel labels obtained with TranscribEMex according to the emotion of the speech.

Because in practice any word can be spoken with any emotion, the words in the system's lexicon also had an identifier associated with the emotion (_E, _F, _N, _T; see Figure 1) and each word was considered to be spoken with all emotions. Thus, for the word CASA (home) the lexicon had the following entries:

(i) CASA_E k a_e s a_e,

(ii) CASA_F k a_f s a_f,

(iii) CASA_N k a_n s a_n,

(iv) CASA_T k a_t s a_t.

This has the possible outcome to recognize the emotion on each word in a sentence. During the ASR process, the lexicon restricts the phoneme sequences decoded by the search algorithm to form valid words. An example of this process is presented in Figure 5. The ASR system would produce the phoneme sequence /t e_e o_e d i_n o_e/ with a phoneme-based language model. When adding the lexicon the phoneme sequence gets restricted to form words. And then, by adding a word based language model, these words get restricted to form phrases. In this case, /t e_e/ = TE and /o_e d i_n o_e/ = ODIO by maximum likelihood.

3.3. Language Models. A language model represents the rules or probabilities that provide information about the valid phoneme/word structures in a language. For this work,

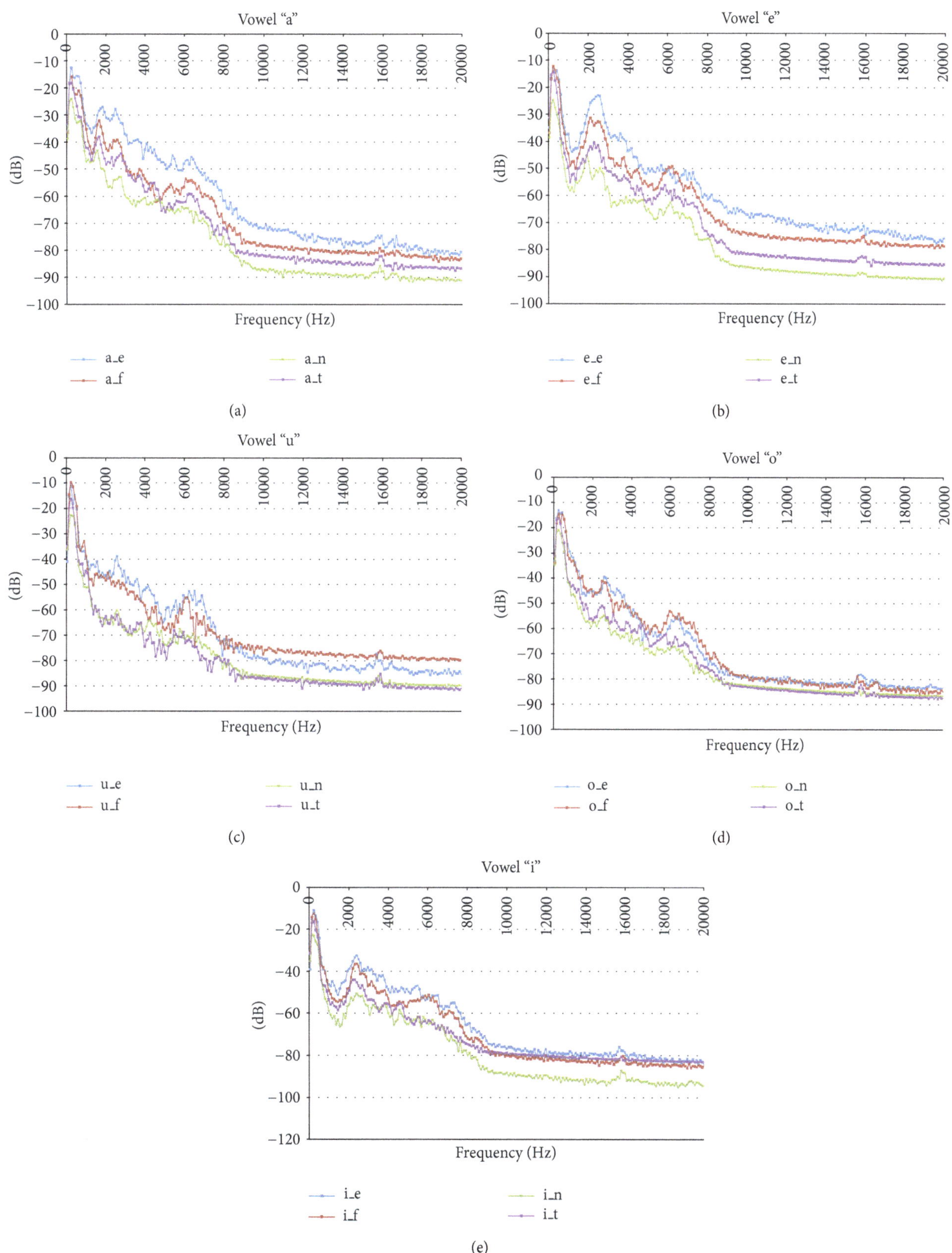

FIGURE 2: Average spectrum for all emotion-specific vowels across all male speakers. Description of identifiers: _e = anger, _f = happiness, _n = neutral, and _t = sadness.

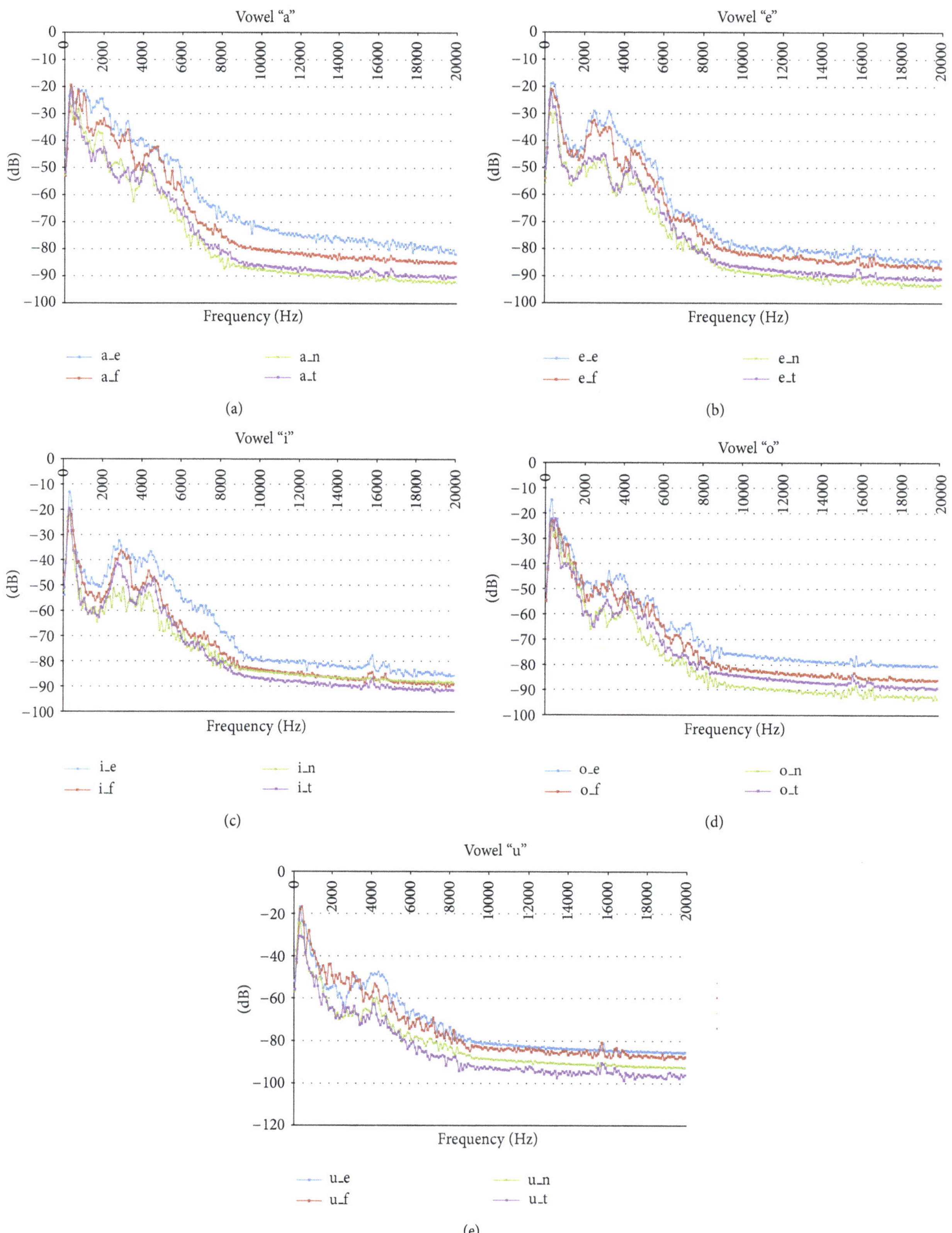

FIGURE 3: Average spectrum for all emotion-specific vowels across all female speakers. Description of identifiers: _e = anger, _f = happiness, _n = neutral, and _t = sadness.

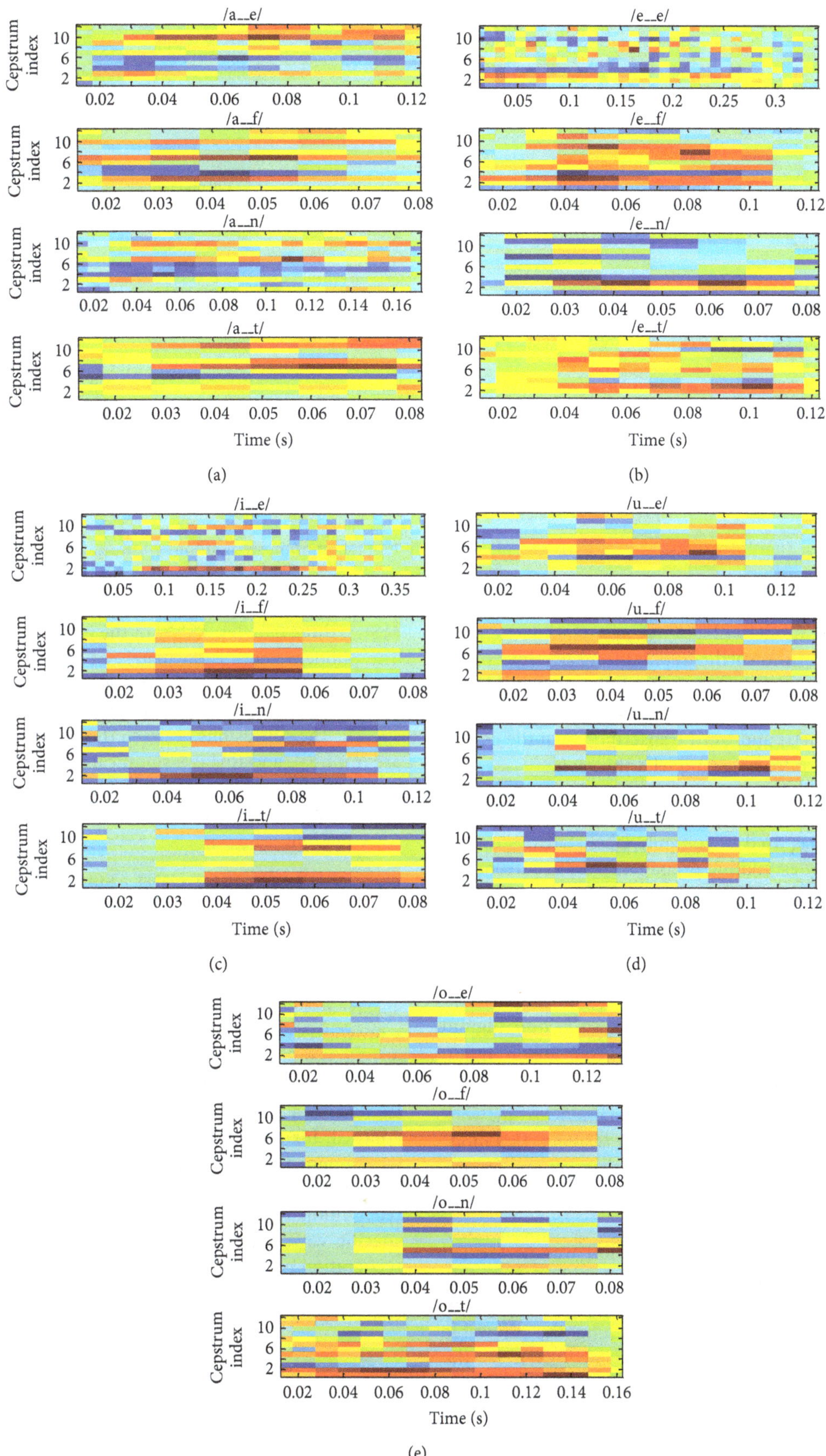

FIGURE 4: MFCCs obtained for some emotion-specific vowels of a male speaker from the built emotional speech database. Description of identifiers: _e = anger, _f = happiness, _n = neutral, and _t = sadness. HTK coding parameters: TARGETKIND = MFCC_E_D_A, WINDOWSIZE = 250000.0, USEHAMMING = T, PREEMCOEF = 0.97, NUMCHANS = 26, CEPLIFTER = 22, and NUMCEPS = 12.

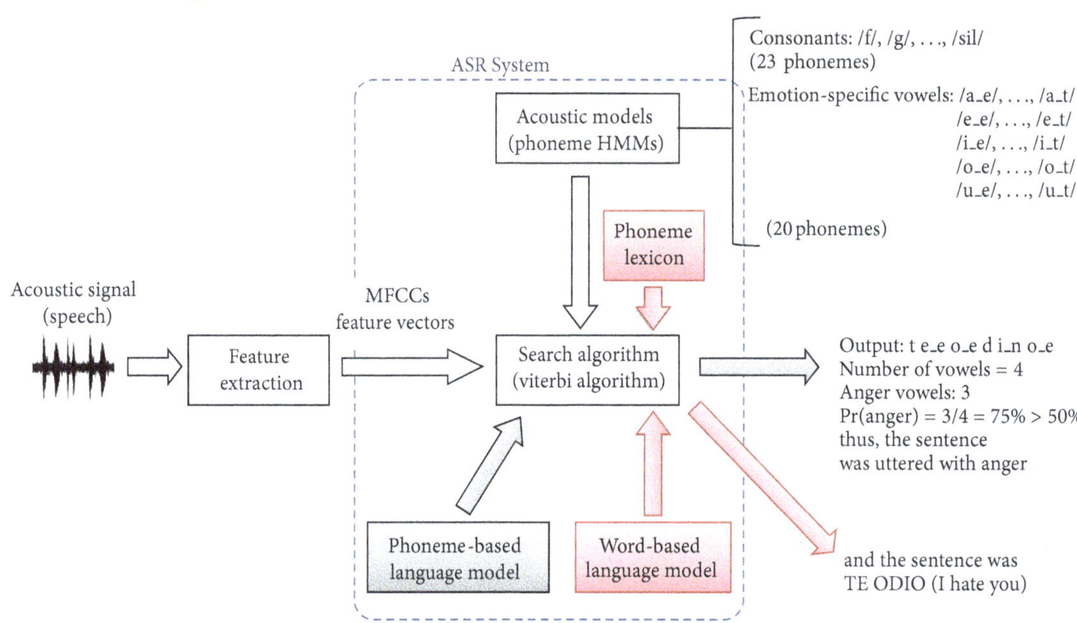

Figure 5: Structure of the ASR system for emotion recognition.

bigram language models (2 grams) were estimated from the phoneme and word (orthographic) transcriptions of the corpus. Because as presented in Section 3.2 each word in the speech corpus was considered to be spoken with all emotions, in total four phonetic and word transcriptions of the corpus were considered for language model estimation.

The following HTK modules were used for this purpose.

(i) *HLStats* was used to compute label statistics for the purpose of estimating probabilities of occurrence of each single phoneme/word in the corpus (unigram probabilities). If configured to estimate bigram probabilities, it provides the associated probabilities of occurrence for the different pairs of phonemes/words found in the phonetic and orthographic transcriptions. For unseen pairs of words, *backed-off* bigram probabilities can be estimated from the unigram probabilities [34].

(ii) *HBuild* was used to build a phoneme/word network with the statistics estimated with *HLStats*. This module generated the statistical language model for the ASR system.

3.4. Search Algorithm. Speech recognition was performed with the Viterbi algorithm implemented with the module *HVite* of HTK. This module takes as input the coded speech to be recognized, the network describing the allowable phoneme/word sequences in the language (given by the language model), the lexicon, and the set of trained HMMs.

4. Performance

Initially the HMM ASR classifier was trained and tested with all the speech samples from the six speakers (three males: MS1-3, three females: FS1-3) of the emotional speech corpus. The phoneme confusion matrix of this test is presented in Figure 6. As observed, there are very few insertion (Ins), deletion (Del), and substitution errors in the recognized speech. The confusions between the emotion-specific vowels are minimal, and this is a significant result because accurate levels of classification of these phonemes are required for emotion recognition. For assessment of recognition accuracy the equation:

$$\text{Accuracy} = \frac{L - \text{Del} - \text{Subst} - \text{Ins}}{L} \times 100, \quad (1)$$

was used, where L is the number of elements (phonemes/words) in the reference transcription of the recognized speech and Subst is the number of substitutions (phonemes/words) present in the ASR's output. For the ASR's phoneme output presented in Figure 6 the following statistics were obtained: $L = 3916$, Del = 152, Ins = 30, and Subst = 148, leading to a phoneme recognition accuracy of 91.57%. In contrast, for the ASR's word output an accuracy of 90.59% was obtained. Both types of performance are normal when testing is performed on training sets. The associated emotion recognition results considering the frequency of emotion-specific vowels are presented in Table 7. For all emotions, recognition performance is higher or equal to 95%.

After the initial test, the system was tested for each speaker in the following way:

(1) a speaker is selected randomly (i.e., MS1);

(2) the HMMs of the ASR system are trained/built with the speech data of the other speakers (i.e., MS2-3, FS1-3);

(3) from the speech data of the selected speaker (i.e., MS1), four randomly selected sentences per emotion

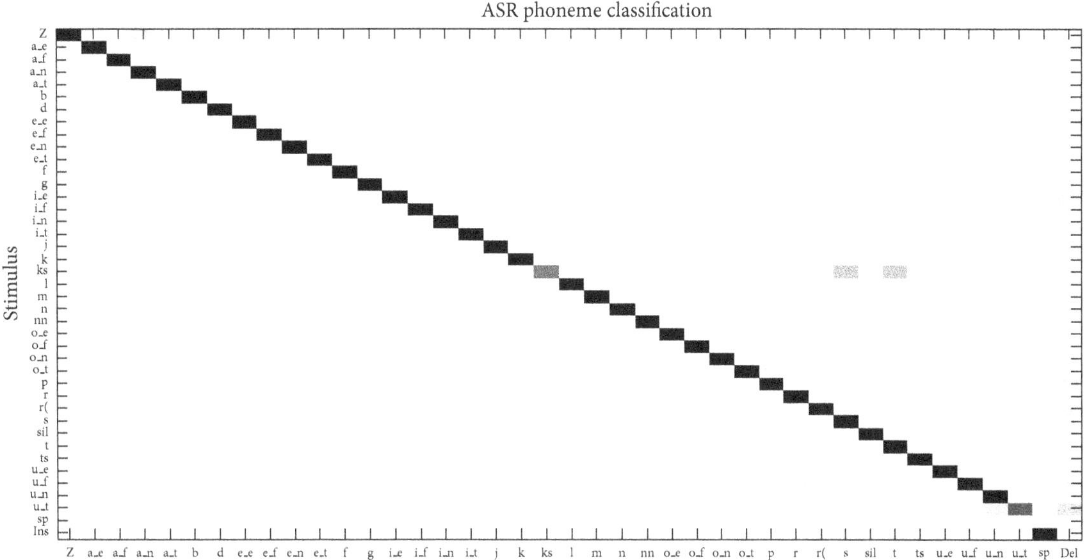

FIGURE 6: Classification performance of the ASR system for speech recognition on the speech corpus (training = testing set).

TABLE 7: Classification performance of the ASR system for emotion recognition on the speech corpus (training = testing set).

	Anger	Happiness	Neutral	Sadness
Anger	95.00	2.50	2.50	0.00
Happiness	0.00	95.24	4.76	0.00
Neutral	0.00	0.00	100.00	0.00
Sadness	0.00	0.00	5.00	95.00

are taken for speaker adaptation. In this case, Maximum Likelihood Linear Regression (MLLR) [34] was used as the adaptation technique;

(4) phoneme-based ASR is performed with the remaining speech data (six sentences per emotion) of the selected speaker (i.e., MS1). With this adaptation/testing scheme more data is available for evaluation of the system in comparison with other works as in [4] where approximately 75% of recorded sentences were used for training and 25% for testing;

(5) vowel and identifier counting is performed on the recognized speech sentences. The identifier with more presence in the vowels found in the ASR's phoneme output (a threshold of 50% was set) determines the dominant emotion in the speech sentence (_e for anger, _f for happiness, _n for neutral, and _t for sadness);

(6) repeat from step (1) until all speakers are selected.

The process described previously was iterated five times in order to obtain different random sets of adaptation and testing sentences per speaker. In Table 8 the details of the emotion recognition results for the individual speakers are presented across the five iterations. Note that for some cases the percentages of word and phoneme recognition accuracy are not as high as the emotion recognition percentages. This is because the accuracy statistics consider all phonemes (see (1)) which consists of vowels and consonants, and emotion is determined based on only vowels. Also, during the word ASR process, for each single word there are four possible choices (see Section 3.2) and thus uncertainty is higher when compared with a standard process.

In Table 9 the average performance and total performance for each speaker (and all speakers) are presented. This data is computed from the results presented in Table 8. As presented, the emotions that are more consistently identified (with a very small standard deviation) are neutral and sadness with 98.89% and 100.00%, respectively. The identification of anger and happiness shows slightly more inconsistencies with a standard deviation of 12.47 and 13.75 although the average recognition is 87.02% and 91.39%, respectively. In this case, significant confusions between these two emotions were found. This situation was observed also in [4, 15, 16, 27]. Nevertheless, for recorded speech data, these results are over the 85% reported by other works with similar number of emotions [16, 27].

5. Conclusions and Future Work

In this paper the development of an emotional speech corpus for Mexican Spanish and an approach for emotion recognition based on acoustic modelling of vowels was presented. HMMs were used for the modelling of consonants and emotion-specific vowels, and these were integrated into an ASR system to generate phoneme and word sequences with emotion identifiers. With this approach the following average recognition results were obtained: 87.02% for anger, 91.39% for happiness, 98.89% for neutral, and 100% for sadness.

Some situations presented by other works were observed in this work. For example, the spectrum differences in vowels give the emotional status [28] and some confusions between anger and happiness [15, 16, 27]. Thus, the speech corpus

TABLE 8: Classification performance of the ASR system for speech and emotion recognition (individual speakers from the speech corpus) across five iterations.

Speaker	Emotion	Iteration 1	Iteration 2	Iteration 3	Iteration 4	Iteration 5
MS1	Anger	100.00	100.00	100.00	100.00	100.00
	Happiness	66.67	66.67	66.67	66.67	66.67
	Neutral	100.00	100.00	100.00	100.00	100.00
	Sadness	100.00	100.00	100.00	100.00	100.00
	Word accuracy	75.00	78.75	80.00	80.00	80.00
	Phoneme accuracy	80.12	82.40	83.46	84.67	83.46
MS2	Anger	71.43	71.43	83.33	83.33	100.00
	Happiness	100.00	83.33	75.00	100.00	75.00
	Neutral	100.00	100.00	100.00	100.00	100.00
	Sadness	100.00	100.00	100.00	100.00	100.00
	Word accuracy	78.26	86.34	88.20	87.58	92.55
	Phoneme accuracy	80.79	88.35	90.02	88.80	93.65
MS3	Anger	83.33	83.33	100.00	83.33	100.00
	Happiness	100.00	100.00	100.00	100.00	75.00
	Neutral	100.00	100.00	100.00	100.00	100.00
	Sadness	100.00	100.00	100.00	100.00	100.00
	Word accuracy	91.30	89.44	98.76	93.79	92.55
	Phoneme accuracy	90.77	91.38	98.49	94.86	93.65
FS1	Anger	100.00	83.33	100.00	100.00	100.00
	Happiness	100.00	100.00	100.00	100.00	100.00
	Neutral	100.00	100.00	66.67	100.00	100.00
	Sadness	100.00	100.00	100.00	100.00	100.00
	Word accuracy	76.25	74.38	68.13	80.63	85.63
	Phoneme accuracy	82.28	81.83	75.83	83.63	88.14
FS2	Anger	62.50	83.33	85.71	66.67	83.33
	Happiness	100.00	100.00	100.00	100.00	100.00
	Neutral	100.00	100.00	100.00	100.00	100.00
	Sadness	100.00	100.00	100.00	100.00	100.00
	Word accuracy	74.38	82.50	85.63	88.13	87.50
	Phoneme accuracy	77.13	87.04	87.80	89.33	88.11
FS3	Anger	85.71	62.50	71.43	83.33	83.33
	Happiness	100.00	100.00	100.00	100.00	100.00
	Neutral	100.00	100.00	100.00	100.00	100.00
	Sadness	100.00	100.00	100.00	100.00	100.00
	Word accuracy	88.75	74.38	92.50	79.38	88.75
	Phoneme accuracy	89.33	79.57	92.23	80.79	90.40

presented similar outcomes as other databases used in the field of emotion recognition.

An advantage of the emotion recognition approach presented in this paper is that it shares the building stages of a general purpose ASR system. Thus, considering that a labelled emotional speech corpus is available, the implementation of the emotion recognizer can be performed quickly. In comparison with other works as [27] only one classifier is required and emotion recognition can be performed over whole sentences or for each word in a sentence.

Also, standard adaptation techniques can be used to make the ASR usable for other speakers for the same purpose of emotion recognition. In this case, the new speaker would be required to produce emotional speech to perform adaptation. For this task, the stimuli presented in Tables 1–4 can be used, and automatic labelling of the adaptation speech can be performed with an emotion-adapted phoneme transcription tool. Online speaker adaptation with the option to add vocabulary to an existing ASR system was presented in [38] with significant results for disordered speech. This can be explored for the case of emotional speech recognition, and ongoing work is focused on the following points:

(i) to test the approach with other emotional speech databases with more emotions;

(ii) to increase the vocabulary and speakers in the Mexican Spanish emotional database. This is important to test emotion recognition performance with a larger vocabulary and a more complex language model;

(iii) to build an ASR with the proposed approach to recognize spontaneous emotional speech. In this

TABLE 9: Average classification performance of the ASR system for speech and emotion recognition (individual speakers from the speech corpus) across five iterations.

Speaker	Statistic	Anger	Happiness	Neutral	Sadness	Word accuracy	Phoneme accuracy
MS1	Average	100.00	66.67	100.00	100.00	78.75	82.82
	Std Dev	0.00	0.00	0.00	0.00	2.17	1.71
MS2	Average	81.90	86.67	100.00	100.00	86.59	88.32
	Std Dev	11.74	12.64	0.00	0.00	5.21	4.70
MS3	Average	90.00	95.00	100.00	100.00	93.17	93.83
	Std Dev	9.13	11.18	0.00	0.00	3.52	3.09
FS1	Average	96.67	100.00	93.33	100.00	77.00	82.34
	Std Dev	7.45	0.00	14.91	0.00	6.59	4.41
FS2	Average	76.31	100.00	100.00	100.00	83.63	85.88
	Std Dev	10.85	0.00	0.00	0.00	5.61	4.96
FS3	Average	77.26	100.00	100.00	100.00	84.75	86.46
	Std Dev	9.96	0.00	0.00	0.00	7.56	5.85
Total (all speakers)	Average	87.02	91.39	98.89	100.00	83.98	86.61
	Std Dev	12.47	13.75	6.09	0.00	7.30	5.55

case, online (automatic) speaker adaptation with live emotional speech is required;

(iv) to compare the performance of the proposed approach when other classification techniques as SVM and ANNs are considered;

(v) to improve current performance.

Acknowledgments

The author thanks Engineer Yara Pérez Maldonado for designing the emotional stimuli and performing the speech recordings of emotional speech for the initial set of four speakers of the speech database and for performing initial phonetic and orthographic labeling on the same speech data.

References

[1] R. Cowie, E. Douglas-Cowie, N. Tsapatsoulis et al., "Emotion recognition in human-computer interaction," *IEEE Signal Processing Magazine*, vol. 18, no. 1, pp. 32–80, 2001.

[2] B. Schuller, G. Rigoll, and M. Lang, "Hidden Markov model-based speech emotion recognition," in *Proceedings of the International Conference on Multimedia and Expo (ICME '03)*, vol. 1, pp. 401–404, 2003.

[3] S. Emerich and E. Lupu, "Improving speech emotion recognition using frequency and time domain acoustic features," in *Proceedings of the Signal Processing and Applied Mathematics for Electronics and Communications (SPAMEC '11)*, pp. 85–88, Cluj-Napoca, Romania, 2011.

[4] Y. Li and Y. Zhao, "Recognizing emotions in speech using short-term and long-term features," in *Proceedings of the International Conference on Spoken Language Processing (ICSLP '98)*, pp. 1–4, 1998.

[5] C. E. Williams and K. N. Stevens, "Emotions and speech: some acoustical correlates," *Journal of the Acoustical Society of America*, vol. 52, no. 4, pp. 1238–1250, 1972.

[6] H. Levin and W. Lord, "Speech pitch frequency as an emotional state indicator," *IEEE Transactions on Systems, Man and Cybernetics*, vol. 5, no. 2, pp. 259–273, 1975.

[7] T. L. Nwe, S. W. Foo, and L. C. De Silva, "Speech emotion recognition using hidden Markov models," *Speech Communication*, vol. 41, no. 4, pp. 603–623, 2003.

[8] D. Ververidis and C. Kotropoulos, "Emotional speech recognition: resources, features, and methods," *Speech Communication*, vol. 48, no. 9, pp. 1162–1181, 2006.

[9] F. J. Tolkmitt and K. R. Scherer, "Effect of experimentally induced stress on vocal parameters," *Journal of Experimental Psychology*, vol. 12, no. 3, pp. 302–313, 1986.

[10] D. J. France, R. G. Shiavi, S. Silverman, M. Silverman, and M. Wilkes, "Acoustical properties of speech as indicators of depression and suicidal risk," *IEEE Transaction Biomedical Engineering*, vol. 7, pp. 829–837, 2000.

[11] L. Deng and D. O'Shaughnessy, *Speech Processing: A Dynamic and Optimization-Oriented Approach*, Marcel Dekker, New York, NY, USA, 2003.

[12] S. B. Davis and P. Mermelstein, "Comparison of parametric representations for monosyllabic word recognition in continuously spoken sentences," *IEEE Transactions on Acoustics, Speech, and Signal Processing*, vol. 28, no. 4, pp. 357–366, 1980.

[13] J. Wagner, T. Vogt, and E. Andr, "A systematic comparison of different HMM designs for emotion recognition from acted and spontaneous speech," in *Affective Computing and Intelligent Interaction*, vol. 4738 of *Lecture Notes in Computer Science*, pp. 114–125, Springer, Berlin, Germany, 2007.

[14] A. B. Kandali, A. Routray, and T. K. Basu, "Emotion recognition from Assamese speeches using MFCC features and GMM classifier," in *Proceedings of the IEEE Region 10 Conference (TENCON '08)*, Hyderabad, India, November 2008.

[15] S. Wu, T. H. Falk, and W.-Y. Chan, "Automatic recognition of speech emotion using long-term spectro-temporal features," in *Proceedings of the 16th International Conference on Digital Signal Processing (DSP '09)*, Santorini-Hellas, Greece, July 2009.

[16] S. Emerich, E. Lupu, and A. Apatean, "Emotions recognition by speech and facial expressions analysis," in *Proceedings of the 17th European Signal Processing Conference (EUSIPCO '09)*, pp. 1617–1621, 2009.

[17] B. D. Womack and J. H. L. Hansen, "Classification of speech under stress using target driven features," *Speech Communication*, vol. 20, no. 1-2, pp. 131–150, 1996.

[18] R. Tato, "Emotional space improves emotion recognition," in *Proceedings of the International Conference on Spoken Language Processing (ICSLP '02)*, vol. 3, pp. 2029–2032, 2002.

[19] R. Fernandez and R. W. Picard, "Modeling drivers' speech under stress," *Speech Communication*, vol. 40, no. 1-2, pp. 145–159, 2003.

[20] K. Alter, E. Rank, and S. A. Kotz, "Accentuation and emotions—two different systems?" in *Proceedings of the ISCA Workshop on Speech and Emotion*, vol. 1, pp. 138–142, 2000.

[21] A. Batliner, C. Hacker, S. Steidl et al., "'you stupid tin box' children interacting with the AIBO robot: a cross-linguistic emotional speech corpus," in *Proceedings of the 4th International Conference of Language Resources and Evaluation (LREC '04)*, pp. 171–174, 2004.

[22] F. Burkhardt, A. Paeschke, M. Rolfes, W. Sendlmeier, and B. Weiss, "A database of German emotional speech," in *Proceedings of the 9th European Conference on Speech Communication and Technology*, pp. 1517–1520, Lisbon, Portugal, September 2005.

[23] M. Grimm, K. Kroschel, and S. Narayanan, "The Vera am Mittag German audio-visual emotional speech database," in *Proceedings of the IEEE International Conference on Multimedia and Expo (ICME '08)*, pp. 865–868, Hannover, Germany, June 2008.

[24] P. Eckman, "An argument for basic emotions," *Cognition and Emotion*, vol. 6, pp. 169–200, 1992.

[25] F. Yu, E. Chang, Y. Q. Xu, and H. Y. Shum, "Emotion detection from speech to enrich multimedia content," in *Proceedings of the IEEE Pacific Rim Conference on Multimedia*, vol. 1, pp. 550–557, Shanghai, China, 2001.

[26] S. Yildirim, M. Bulut, C. M. Lee et al., "An acoustic study of emotions expressed in speech," in *Proceedings of the International Conference on Spoken Language Processing (ICSLP '04)*, vol. 1, pp. 2193–2196, 2004.

[27] C. M. Lee, S. Yildirim, M. Bulut et al., "Emotion recognition based on phoneme classes," in *Proceedings of the International Conference on Spoken Language Processing (ICSLP '04)*, vol. 1, pp. 889–892, 2004.

[28] J. Pribil and A. Pribilov, "Spectral properties and prosodic parameters of emotional speech in Czech and Slovak," in *Speech and Language Technologies*, pp. 175–200, InTech, 2011.

[29] E. Uraga and L. Pineda, "Automatic generation of pronunciation lexicons for Spanish," in *Proceedings of the International Conference on Computational Linguistics and Intelligent Text Processing (CICLing '02)*, A. Gelbukh, Ed., pp. 300–308, Springer, 2002.

[30] J. Cuetara, *Fonetica de la ciudad de Mexico: aporta- ciones desde las tecnologias del habla [MSc. Dissertation]*, National Autonomous University of Mexico (UNAM), Mexico, 2004.

[31] A. Li, Q. Fang, F. Hu, L. Zheng, H. Wang, and J. Dang, "Acoustic and articulatory analysis on mandarin chinese vowels in emotional speech," in *Proceedings of the 7th International Symposium on Chinese Spoken Language Processing (ISCSLP '10)*, pp. 38–43, Tainan, Taiwan, December 2010.

[32] B. Vlasenko, D. Prylipko, D. Philippou-Hiibner, and A. Wendemuth, "Vowels formants analysis allows straight-forward detection of high arousal acted and spontaneous emotions," in *Proceedings of the 12th Annual Conference of the International Speech Communication Association (Interspeech '11)*, pp. 1577–1580, Florence, Italy, 2011.

[33] J. H. L. Hansen and B. D. Womack, "Feature Analysis and Neural Network-Based Classification of Speech under Stress," *IEEE Transactions on Speech and Audio Processing*, vol. 4, no. 4, pp. 307–313, 1996.

[34] S. Young and P. Woodland, *The HTK Book, (for HTK Version 3.4)*, Cambridge University Engineering Department, UK, 2006.

[35] D. Jurafsky and J. H. Martin, *Speech and Language Pro-Cessing*, Pearson Prentice Hall, New Jersey, NJ, USA, 2009.

[36] B. D. Womack and J. H. Hansen, "N-channel hidden Markov models for combined stressed speech classification and recognition," *IEEE Transaction on Speech and Audio Processing*, vol. 7, no. 6, pp. 668–677, 1999.

[37] L. Pineda, L. Villaseñor, J. Cuétara et al., "The corpus DIMEX100: transcription and evaluation," *Language Resources and Evaluation*, vol. 44, pp. 347–370, 2010.

[38] G. Bonilla-Enríquez and S. O. Caballero-Morales, "Communication interface for mexican spanish dysarthric speakers," *Acta Universitaria*, vol. 22, no. NE-1, 98–105 pages, 2012.

3

New Photoplethysmographic Signal Analysis Algorithm for Arterial Stiffness Estimation

Kristjan Pilt,[1] Rain Ferenets,[1] Kalju Meigas,[1] Lars-Göran Lindberg,[2] Kristina Temitski,[1] and Margus Viigimaa[1]

[1] *Department of Biomedical Engineering, Technomedicum, Tallinn University of Technology, Ehitajate tee 5, 19086 Tallinn, Estonia*
[2] *Department of Biomedical Engineering, Linköping Univeristy, 581 85 Linköping, Sweden*

Correspondence should be addressed to Kristjan Pilt; kristjan.pilt@cb.ttu.ee

Academic Editors: N. Bouguila, M. Engin, and T. Yamasaki

The ability to identify premature arterial stiffening is of considerable value in the prevention of cardiovascular diseases. The "ageing index" (*AGI*), which is calculated from the second derivative photoplethysmographic (SDPPG) waveform, has been used as one method for arterial stiffness estimation and the evaluation of cardiovascular ageing. In this study, the new SDPPG analysis algorithm is proposed with optimal filtering and signal normalization in time. The filter parameters were optimized in order to achieve the minimal standard deviation of *AGI*, which gives more effective differentiation between the levels of arterial stiffness. As a result, the optimal low-pass filter edge frequency of 6 Hz and transitionband of 1 Hz were found, which facilitates *AGI* calculation with a standard deviation of 0.06. The study was carried out on 21 healthy subjects and 20 diabetes patients. The linear relationship ($r = 0.91$) between each subject's age and *AGI* was found, and a linear model with regression line was constructed. For diabetes patients, the mean *AGI* value difference from the proposed model y_{AGI} was found to be 0.359. The difference was found between healthy and diabetes patients groups with significance level of $P < 0.0005$.

1. Introduction

There has been an increased interest in the development of innovative noninvasive methods and devices for the diagnosis of cardiovascular diseases [1–3]. Photoplethysmographic (PPG) waveform analysis has been used as one method [4].

PPG is a noninvasive optical technique for measuring changes in blood circulation that is mainly used for monitoring blood perfusion in the skin. The PPG finger sensor consists of a light emitting diode (LED), which is often red or infrared, and a photodetector (PD) [5]. PD and LED are on the opposite side of the finger. The light is emitted from the LED to the skin and a small fraction of light intensity changes is received by the PD, which are related to blood flow, blood volume, blood vessel wall movement, and the orientation of red blood cells in the underlying tissue [6, 7]. The PPG signal consists of different components: DC and AC components and noise, which can be caused by the poor perfusion state and motion artifacts [5]. Noise can be eliminated by using different filtering techniques [8]. The AC component is synchronous with the heart rate and depends on changes in the pulsatile pressure and pulsatile blood volume.

It is apparent that the AC component of the PPG signal changes with age and the waveform transforms from a wavy into a triangular-shaped signal (Figure 1, upper part). Regarding time domain, different methods to analyze the waveform of the PPG signal, measured at the finger, can be used for arterial stiffness estimation and evaluation of cardiovascular aging [9–11].

One option is to use the second derivative of the PPG signal (SDPPG), which was first introduced by Takazawa et al. [12]. The SDPPG is analyzed by using the amplitudes of the distinctive waves "a", "b", "c", "d", and "e", which are situated in the systolic phase of the heart cycle (Figure 1, lower part). The amplitudes of the waves are normalized as follows: b/a, c/a, d/a, and e/a. They found that normalized amplitude b/a increases and c/a, d/a, and e/a decrease in proportion to the increase in the subject's age. As a result an "ageing index" (*AGI*) parameter was proposed according to

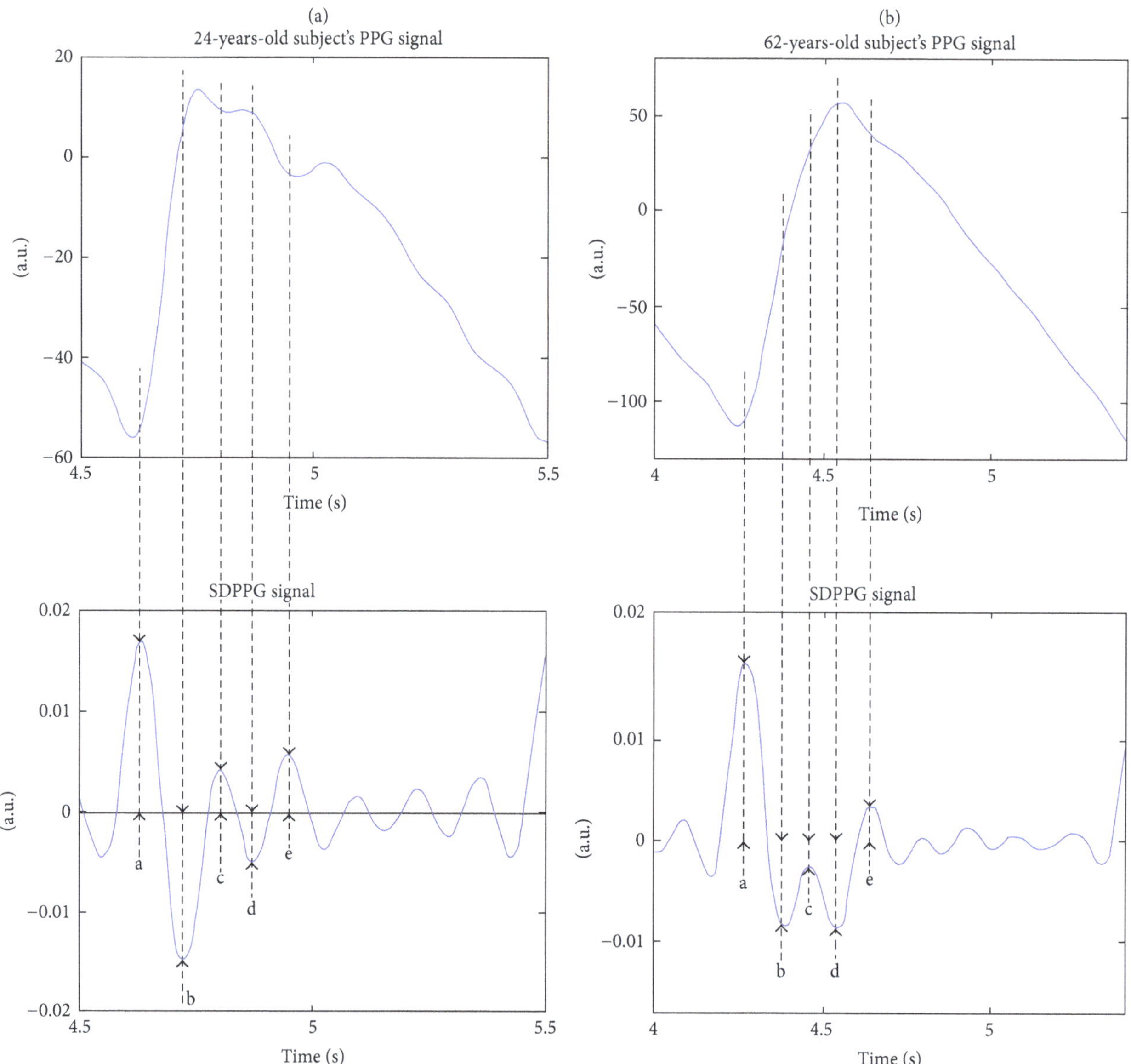

FIGURE 1: The finger PPG signal and its second derivative with distinct waves "a", "b", "c", "d", and "e" of 24-year (a) and 62-year-old (b) subjects.

$AGI = (b-c-d-e)/a$, where the a, b, c, d, and e are the amplitudes of the waves. The *AGI* is used to describe the cardiovascular age of the subject.

In recent publications, the correlation relationship between cardiovascular risk factors and the SDPPG normalized amplitudes values has been analyzed statistically [13–15]. Normalized amplitudes of SDPPG and *AGI* can be good parameters for a screening method to detect increases in the stiffness of the arteries [16].

The sample segment of PPG and SDPPG signal, which has been registered from a 37-year-old healthy subject, with *AGI* values, is shown in Figure 2. The SDPPG signal is processed, and the wave amplitudes are detected according to a study by Millasseau et al. [17]. The similar processing method has been also described in less detail in other studies [12–15]. It is assumed that the cardiovascular system does not change over short periods in cases of healthy subjects. It is visible from Figure 2 that the *AGI* values for the healthy subject are noticeably higher for the first and third periods. The difference between maximal and minimal *AGI* values is 0.47, which constitutes about 39% from the whole scale of *AGI* [12]. Furthermore, the detected peaks in the first and third periods are located to the beginning of systolic phase of the PPG signal compared to the second and fourth periods. As a result the amplitudes of detected peaks in the consecutive periods describe different phase of the PPG waveform and *AGI* values become noticeably different. This leads to higher standard deviation of *AGI* and to faulty interpretation of the results for a single subject. The detection of the peaks from different phases of PPG signal in case of consecutive periods is due to the insufficient suppression of PPG signal higher components and noise.

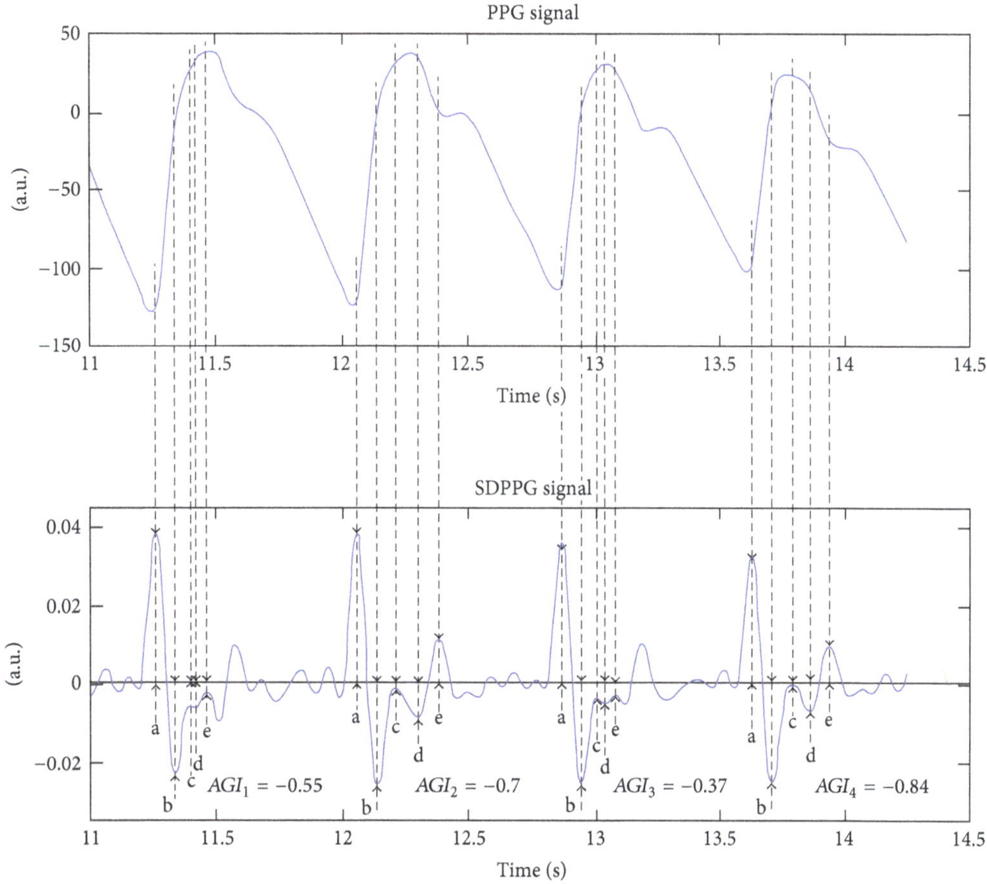

FIGURE 2: The sample segment of the PPG signal (upper part) from a 37-year-old healthy subject and its second derivative (lower part) with detected wave peaks and *AGI* values. The SDPPG signal is processed, and the wave amplitudes are detected according to a study by Millasseau et al. [17].

The *AGI* has to be calculated with low standard deviation in order to differentiate the subjects with increased stiffness from the healthy subjects. In this study, we have improved the SDPPG analysis method in order to obtain the *AGI* values with minimal standard deviation and to detect the waves at the same locations within one period of the PPG signal. The algorithm is tested on group of healthy subjects and a small group of diabetes patients as a pilot study.

2. Methods

2.1. SDPPG Analysis Algorithm. Normalization of the period's length, averaging, filtering, and detection of the waves from the SDPPG signal are illustrated in a block diagram in Figure 3. The PPG signal is filtered with low- and high-pass FIR filters in order to separate DC components and high frequency noise. The cutoff frequencies for the low- and high-pass filters are selected as 30 Hz and 0.5 Hz, respectively. Both filters are designed using the window method, with the Hamming window function where the corresponding filter orders are chosen as 500 for the low-pass after and 4000 for the high-pass filter.

Subsequently the PPG signal is differentiated two and four times (Figure 3). The simplest differentiator calculates the differences between two consecutive samples of the signal, which is also known as the first-difference differentiator. This kind of differentiator works as a high-pass filter, and the high frequencies are amplified as a result. However, the unwanted noise is located at higher frequencies for the PPG signal. Due to the reason outlined previously, the Smooth Noise Robust Differentiator (SNRD) is used.

The SNRD has been developed for different cases that are particularly beneficial for carrying out experiments with noisy data where differentiation is required [18]. This differentiation scheme possesses the following characteristics: precise at low frequencies, smooth and guaranteed suppression of high frequencies. The order of the differentiator determines the suppression of the high frequencies. In this algorithm, the fifth order of the differentiator is used, which is also the lowest possible one. At the lower frequencies (0–15 Hz), where the majority of the power of the PPG signal is located, the first-difference differentiator and SNRD frequency responses are practically equal.

In practice, biosignals such as PPG, which are related to heart activity, are recurring but not periodic. This means that the harmonic components of the two consecutive recurrences of the PPG signal and its derivatives can be at different frequencies. In this study, the low-pass filter is used with static

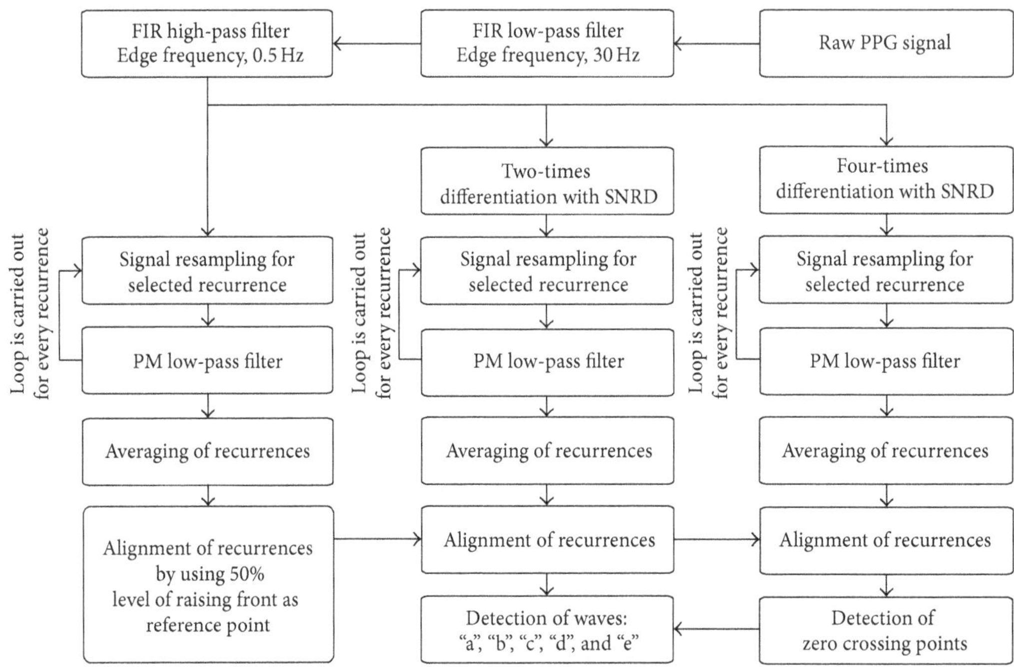

FIGURE 3: Block diagram of the signal processing for the second derivative analysis.

edge frequency. Accordingly, certain numbers of harmonic components are passed and all others are suppressed. The lengths of the PPG signal recurrences are then normalized to ensure that all the harmonic components are processed in the same way (Figure 3).

The PPG signal is resampled in such a way that one of the selected recurrence lengths is 1 s, which corresponds to the pulse frequency of 1 Hz. In this case, the fundamental frequency is situated at 1 Hz. All the other components lay at the frequency multiples of the fundamental frequency. In the next step, the signal is filtered with the 1 Hz wide transition-band PM filter [19]. The maximum allowable errors for the pass and stop bands are set at 0.001. The resampling and filtering are also carried out with the second and fourth derivatives of the PPG signal. After filtering, the copy of selected recurrence is aligned with other normalized and filtered recurrences from this PPG signal. The recurrences of PPG signals can be aligned by using different distinct points from the signal as reference, for example, the recurrence maximal or minimal point of the raising front. The recurrence minimal point can be difficult to determine, because of the wavy ending of the diastole phase. It is also difficult to determine the PPG signal maximal point as it depends on the state of the cardiovascular system [20]. The 50 percent level of the PPG signal raising front is used as the reference point for the alignment of the recurrences. Furthermore, the second and fourth derivatives are moved according to the movement of the PPG signal recurrences.

The resampling, filtering, and aligning processes outlined previously are carried out separately for every recurrence in PPG signals. The averaged waveform for one subject with its 9 recurrences is given in Figure 4.

Subsequently, the peaks of waves "a", "b", "c", "d", and "e" are found from the averaged SDPPG waveform. Firstly, the zero crossings of the averaged fourth derivative waveform are found. The peaks of waves "a", "b", "c", "d", and "e" are between zero crossings of the fourth derivative waveform as it is revealed in Figure 4. Secondly, the minimal and maximal points of the SDPPG waveform are located between the zero crossings of the fourth derivative waveform. There can be waveforms of the SDPPG, where the peaks of the "c" and "d" waves do not appear. In this case, the "c" and "d" waves are determined in the places of the SDPPG waveform, where the fourth derivative is maximal or minimal between zero crossings.

2.2. Optimization of PM Low-Pass Filter Edge Frequency. The recurrences and averaged waveform of the SDPPG are affected by the edge frequency of the PM low-pass filter. The optimal edge frequency of the PM low-pass filter was optimized in order to achieve the lowest standard deviation of the SDPPG wave amplitudes, which ultimately minimizes the standard deviation of *AGI*. In addition, the variation in the placement of waves "a", "b", "c", "d", and "e" on time domain has to be minimal throughout all the periods for one subject. Here, it is assumed that the cardiovascular system does not change over short periods in cases of healthy subjects. The optimization of the PM low-pass filter edge frequency was carried out on signals from a group of healthy subjects.

The width of the PM low-pass filter transition-band was 1 Hz and the edge frequency was changed between 4 and 14 Hz with a step of 1 Hz. The order of the corresponding PM filter was 3255 at sampling frequency 1 kHz. The 3–13 harmonic components in addition to the fundamental harmonic are passed through the filter as the recurrences of the SDPPG were normalized to the frequency of 1 Hz. In this way, the influence of each harmonic component to waves "a", "b", "c", "d", and "e" can be analyzed.

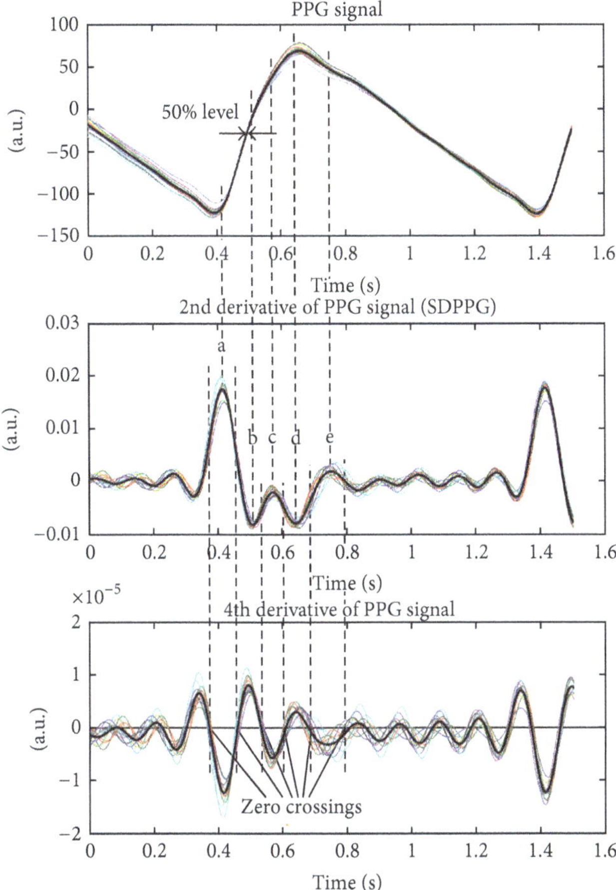

FIGURE 4: The averaged PPG and its second derivative and fourth derivative waveforms (black bold line) with filtered and normalized recurrences (thin lines). The recurrences are aligned according to 50% of the PPG signal raising front and the distinct waves "a", "b", "c", "d", and "e" are detected by using the zero crossings of the fourth derivative.

The standard deviations for the normalized amplitudes, b/a, c/a, d/a, e/a, and AGI were calculated for each edge frequency. For the standard deviation calculation, the normalized amplitudes, x_a, from normalized SDPPG recurrences and from the averaged SDPPG waveform were used. The normalized amplitudes from the averaged SDPPG waveform were taken as average values x_{avg}. The standard deviations were calculated for signals from each healthy subject, k, by using following equation:

$$SD(k) = \sqrt{\frac{\sum_{i=1}^{n}\left(x_a(i) - x_{avg}\right)^2}{n-1}}, \quad (1)$$

where i is the number of period and n is the total number of periods in the processed signal. Similarly, the standard deviation of the detected wave peaks on the time domain was calculated. The standard deviations were averaged over the group of subjects by using following equation:

$$SD_{avg} = \frac{1}{m} \cdot \sum_{k=1}^{m} SD(k), \quad (2)$$

where m is the total number of healthy subjects.

2.3. Pilot Study on Patients. The improved SDPPG signal analysis algorithm was tested on the signals from a group of healthy and diabetes patients. The optimal PM low-pass filter edge frequency was used for the analysis. The SDPPG waves were detected and AGI values were calculated with standard deviations.

2.4. Subjects. The study was performed after approval of the protocol by the Tallinn Ethics Committee on Medical Research at the National Institute for Health Development, Estonia. The PPG signals for the analysis were registered from healthy subjects and diabetes patients.

All the subjects in the healthy group were aged from 21 to 66 years. They were not on permanent medication and they were dealing with various levels of physical activity in their everyday lives. As the waveform of the PPG signal varies with age, the subjects were divided into the following age groups: 20–30, 30–40, 40–50, 50–60, and 60–70. Each age group, except the groups of 50–60 and 60–70, comprised five subjects. Those age groups comprised three subjects, because it was difficult to find healthy subjects to fulfill our criteria. In all, 21 healthy subjects ($m = 21$) participated in the study.

All subjects in the group of diabetes patients had received diagnosis from a medical doctor. In all, 20 diabetes patients participated in this pilot study. The patients were aged between 27 and 66 years. The diabetes patients may have increased arterial stiffness due to the sclerotic processes in the vessels.

2.5. Instrumentation. All signal processing was carried out in MATLAB. The high- and low-pass filter coefficients were calculated by using the "fir1" function. A separate function was written for calculating coefficients of the SNRD [18]. The coefficients of PM filter were calculated using functions "firpmord" and "firpm."

The PPG signals were registered from the index finger by using an experimental measurement complex [21], which included a Nellcor finger clip sensor and lab-built PPG module, among other devices. The PPG signal was digitized with a PCI MIO-16-1 data acquisition card and registered with LabVIEW environment. The sampling frequency was 1 kHz. The 1-minute long signal was recorded, and a 15-period long segment ($n = 15$) was selected for the SDPPG analysis. The signal registration was carried out, while the subject was in a resting position. The subject was in a resting position at least 10 minutes prior to the measurements. The room temperature was around 23 degrees during the experiments.

3. Results

The general change in standard deviation of the normalized amplitudes and AGI in cases of different edge frequencies is illustrated in Figures 5(a)–5(e). For each edge frequency, the given standard deviation is averaged over the group of healthy subjects. The minimal average standard deviation for AGI, b/a, c/a, d/a, and e/a is 0.06, 0.02, 0.04, 0.03, and 0.02 respectively. For all parameters, the minimal standard

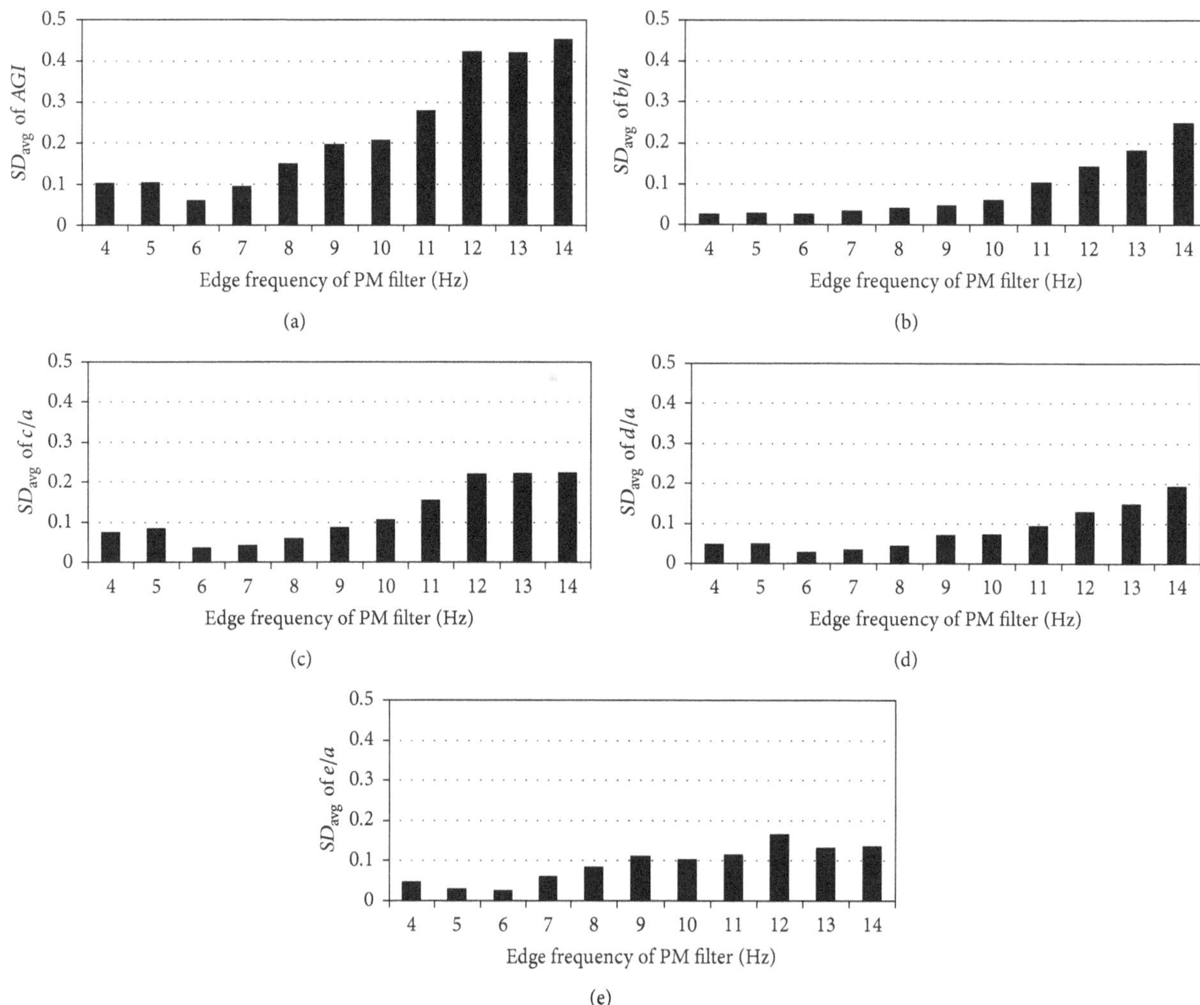

FIGURE 5: The average standard deviations (SD_{avg}) of AGI and normalized amplitudes at different edge frequencies for 21 healthy subjects. (a) AGIs at different edge frequencies; (b) normalized amplitudes b/a at different edge frequencies; (c) normalized amplitudes c/a at different edge frequencies; (d) normalized amplitudes d/a at different edge frequencies; (e) normalized amplitudes e/a at different edge frequencies.

deviation was found where the edge frequency of the PM filter is 6 Hz and transition band is 1 Hz.

Similarly, in Figures 6(a)–6(e), the standard deviations to characterize the dispersion of wave peaks "a", "b", "c", "d", and "e" in the time domain are given. The given standard deviations are averaged over the group of healthy subjects. The minimal average standard deviations for wave peaks "a", "b", "c", "d", and "e" in the time domain are 2.2 ms, 1.9 ms, 4.6 ms, 2.8 ms, and 5.0 ms, respectively. The minimal standard deviations can be found for the edge frequency of 6 Hz for all waves, except for wave "b". In the case of wave "b", the minimal standard deviation was at edge frequency of 4 Hz.

For the purpose of comparison, the same PPG signals were also processed with the algorithm described by Millasseau et al. [17]. The average standard deviation for AGI value is 0.12.

The AGI and age relationship for the healthy subjects and diabetes patients with standard deviation bars are illustrated in Figure 7(a). The PM filter edge frequency and transition-band were 6 Hz and 1 Hz, respectively, which according to the previously presented results seems to be optimal for the SDPPG analysis. In addition, regression analysis was carried out in order to estimate the relationship between AGI and age by using generalized linear model. As a first approach the general linear model was used, which is a case of the generalized linear model with identity link function. A following regression model was proposed: $y_{\text{AGI}} = 0.019x - 1.556$. Despite of relatively simple model high correlation $r = 0.91$ was found between AGI and age for the healthy group, which shows the strong linear relationship between two variables.

In Figure 7(b) it can be seen the Bland-Altman plot for the proposed model y_{AGI}. The standard deviations for the model a $SD_{\text{AGI}} = 0.126$. For diabetes patients the mean AGI value difference from the proposed model y_{AGI} is $mean_{\text{Dia}} = 0.359$. The AGI differences from the proposed model y_{AGI} were compared between healthy and diabetes patients groups. Paired t-test (two-sample assuming unequal variances) was performed in MS Excel with $\alpha = 0.05$. The significance level of paired t-test was $P < 0.0005$, which shows the difference between two groups.

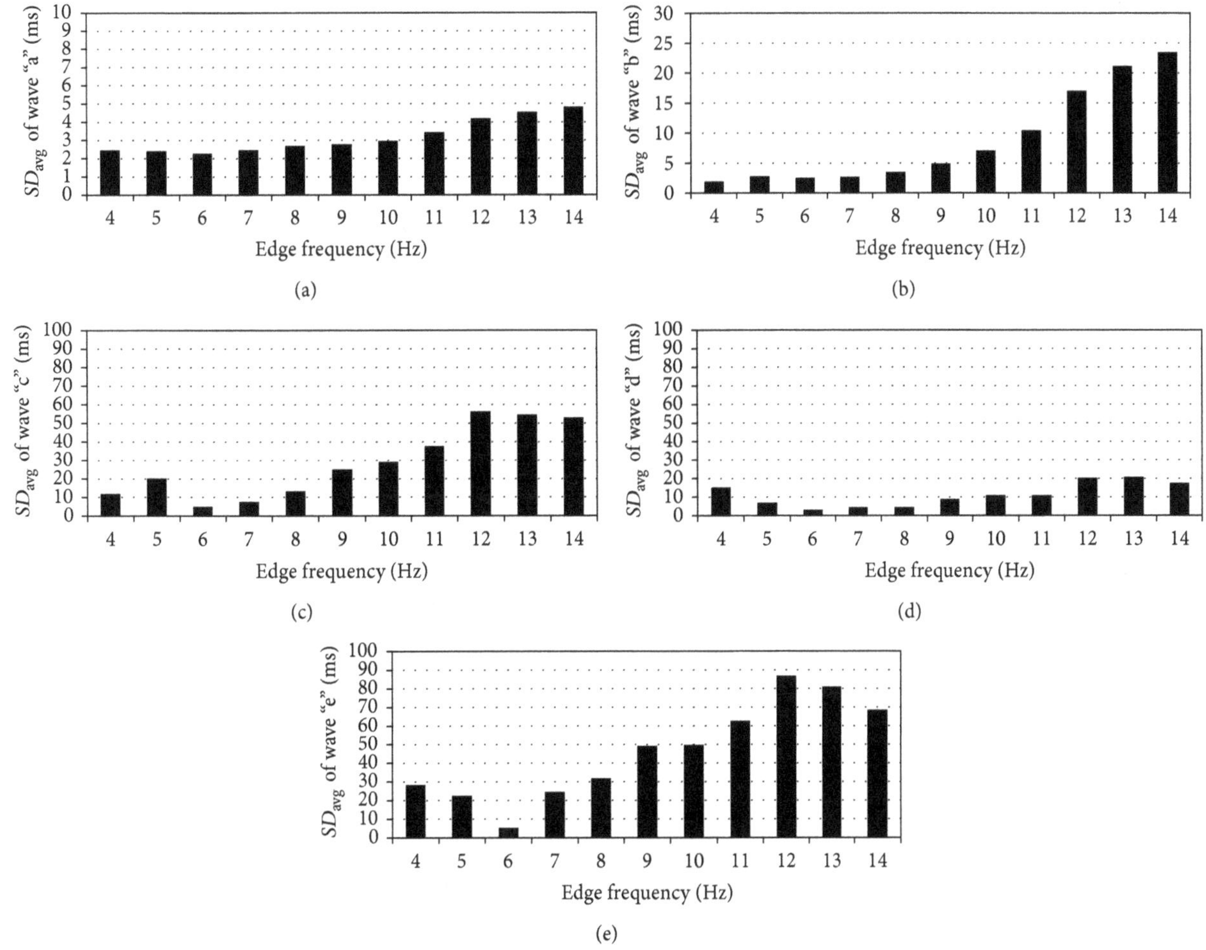

FIGURE 6: The average standard deviations (SD_{avg}) of the waves "a", "b", "c", "d", and "e" on the time domain at different edge frequencies for 21 healthy subjects. (a) standard deviation of wave "a" at different edge frequencies; (b) standard deviation of wave "b" at different edge frequencies; (c) standard deviation of wave "c" at different edge frequencies; (d) standard deviation of wave "d" at different edge frequencies; (e) standard deviation of wave "e" at different edge frequencies.

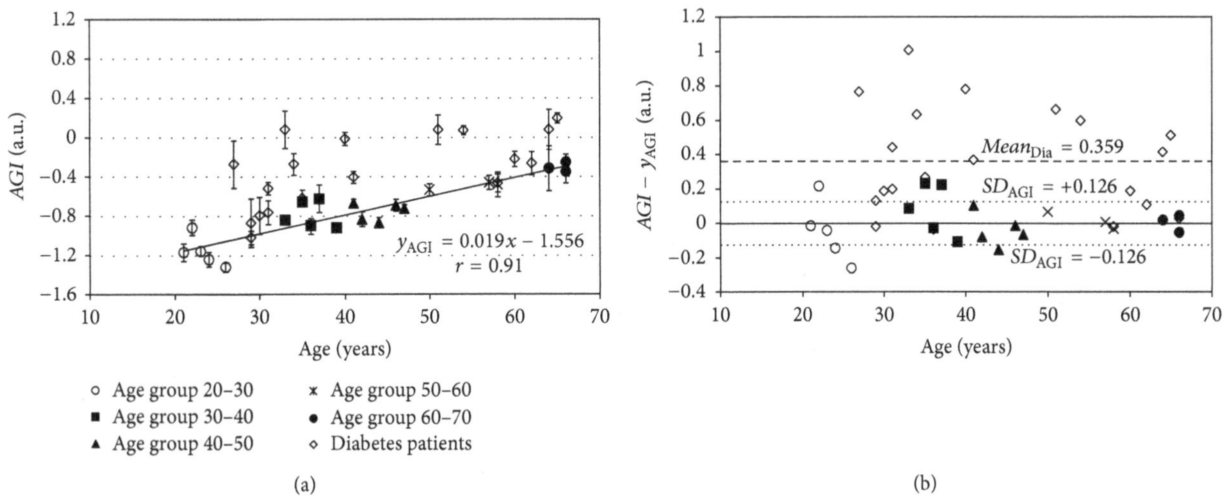

FIGURE 7: (a) The AGI data points with standard deviation (SD) bars for groups of healthy subjects and diabetes patients. The linear model line y_{AGI} is constructed for a group of healthy subjects. (b) Bland-Altman plot for constructed model. With dotted line the standard deviation levels for group of healthy subjects are given. With dashed line is given the mean AGI difference from linear model for diabetes patients.

4. Discussions

With an improved SDPPG analysis algorithm, the average standard deviation for the *AGI* value is 0.06, which constitutes about 5% of the whole scale of *AGI* [12]. Compared to the algorithm of Millasseau et al. [17], the average standard deviation is twice lower. As a result, subjects with increased arterial stiffness can be more easily differentiated from healthy subjects, and the prevention of cardiovascular disease can be improved.

The relatively high correlation relation was found between *AGI* and age by using the algorithm with optimal edge frequency (Figure 7(a)). This is in relation to previously published results by Takazawa et al. [12], in which a good correlation between *AGI* and age among healthy subjects was shown. There are still some deviations from the regression model line, y_{AGI}, which can be caused by the impact of cardiovascular deficiencies and the subject's biological age. In addition model can be more complex and dependent on additional variables, such as blood pressure. However, this should be considered in the scope of future studies.

The noticeably higher *AGI* values, compared to the healthy group of subjects, were found for diabetes patients (Figure 7(a)). The same behavior is also visible in Figure 7(b). Furthermore, the statistically significant difference was found between the healthy subjects and diabetes patients. The higher *AGI* values are caused by the increased arterial stiffness of diabetes patients. Nevertheless, some of the diabetes patients have similar *AGI* values compared to healthy subjects. It can be caused by the early diagnosis of diabetes mellitus, which is followed with efficient therapy, and as a result premature stiffening of the arteries has been stopped.

It can be seen from Figures 5(a)–5(e) that the lowest average standard deviation was achieved when the edge frequency of PM filter is 6 Hz. Close to 6 Hz, the normalized recurrences start to resemble. The larger standard deviations on higher edge frequencies are caused by the noise and unwanted frequency components of the PPG signal, which are situated at higher frequencies and amplified through differentiation. This causes the faulty detection of the waves from the single normalized recurrence and averaged SDPPG waveform. At lower edge frequencies, the harmonic components are suppressed, which form waves "a", "b", "c", "d", and "e", and the peaks of waves, "c" and "d", were missing in single normalized recurrences. As a result, the amplitude of the waves in single normalized recurrences and averaged SDPPG waveform are different, which caused the increase in standard deviation (Figures 5(a)–5(e)). This means that it is necessary to have a fundamental harmonic with 5 higher harmonic components in order to detect waves "a", "b", "c", "d", and "e" from the PPG signal.

The dispersion of wave peaks on time domain decreases, similar to the results seen in Figure 5, when the edge frequency of PM filter approaches 6 Hz (Figures 5(a)–6(e)). As in Figure 5, the detection of the waves from the single normalized recurrences and from the averaged SDPPG waveform can be different at higher frequencies, which increases the standard deviation. At frequencies lower than 6 Hz, the wave peaks can be missing in single normalized recurrence and the detection point is shifted compared with the averaged SDPPG waveform.

5. Conclusions

In conclusion, it can be said that the standard deviation of *AGI* values is minimized by using the improved SDPPG algorithm. Furthermore, the diabetes patients have noticeably higher *AGI* values, which are caused by an increase in the arterial stiffness. As a result, the subjects, with increased arterial stiffness can be more easily differentiated from healthy subjects and the prevention of cardiovascular disease can be improved. As a future study the more complex model should be considered in order to enhance the discrimination of the healthy subjects and patients with increased stiffness by taking into account additional physiological variables. In addition proposed algorithm should be compared with similar arterial stiffness estimation reference methods.

Acknowledgments

This work was supported by the Estonian Science Foundation Grant no. 7506, by the Estonian Targeted Financing Project SF0140027s07, and by the European Union through the European Regional Development Fund.

References

[1] M. W. Rajzer, W. Wojciechowska, M. Klocek, I. Palka, M. Brzozowska-Kiszka, and K. Kawecka-Jaszcz, "Comparison of aortic pulse wave velocity measured by three techniques: complior, sphygmocor and arteriograph," *Journal of Hypertension*, vol. 26, no. 10, pp. 2001–2007, 2008.

[2] S. S. Silvera, H. E. Aidi, J. H. F. Rudd et al., "Multimodality imaging of atherosclerotic plaque activity and composition using FDG-PET/CT and MRI in carotid and femoral arteries," *Atherosclerosis*, vol. 207, no. 1, pp. 139–143, 2009.

[3] M. R. Skilton, L. Boussel, F. Bonnet et al., "Carotid intima-media and adventitial thickening: comparison of new and established ultrasound and magnetic resonance imaging techniques," *Atherosclerosis*, vol. 215, no. 2, pp. 405–410, 2011.

[4] C. F. Clarenbach, A. Stoewhas, A. J. R. Van Gestel et al., "Comparison of photoplethysmographic and arterial tonometry-derived indices of arterial stiffness," *Hypertension Research*, vol. 35, no. 2, pp. 228–233, 2012.

[5] J. Allen, "Photoplethysmography and its application in clinical physiological measurement," *Physiological Measurement*, vol. 28, pp. R1–R39, 2007.

[6] A. A. R. Kamal, J. B. Harness, G. Irving, and A. J. Mearns, "Skin photoplethysmography—a review," *Computer Methods and Programs in Biomedicine*, vol. 28, no. 4, pp. 257–269, 1989.

[7] L. G. Lindberg and P. Å. Öberg, "Optical properties of blood in motion," *Optical Engineering*, vol. 32, pp. 253–257, 1993.

[8] K. Pilt, K. Meigas, R. Ferenets, and J. Kaik, "Photoplethysmographic signal processing using adaptive sum comb filter for pulse delay measurement," *Estonian Journal of Engineering*, vol. 16, no. 1, pp. 78–94, 2010.

[9] M. Huotari, N. Yliaska, V. Lantto, K. Määttä, and J. Kostamovaara, "Aortic and arterial stiffness determination by photoplethysmographic technique," *Procedia Chemistry*, vol. 1, no. 1, pp. 1243–1246, 2009.

[10] S. C. Millasseau, J. M. Ritter, K. Takazawa, and P. J. Chowienczyk, "Contour analysis of the photoplethysmographic pulse measured at the finger," *Journal of Hypertension*, vol. 24, no. 8, pp. 1449–1456, 2006.

[11] U. Rubins, "Finger and ear photoplethysmogram waveform analysis by fitting with Gaussians," *Medical and Biological Engineering and Computing*, vol. 46, no. 12, pp. 1271–1276, 2008.

[12] K. Takazawa, N. Tanaka, M. Fujita et al., "Assessment of vasoactive agents and vascular aging by the second derivative of photoplethysmogram waveform," *Hypertension*, vol. 32, no. 2, pp. 365–370, 1998.

[13] J. Hashimoto, D. Watabe, A. Kimura et al., "Determinants of the second derivative of the finger photoplethysmogram and brachial-ankle pulse-wave velocity: The Ohasama study," *American Journal of Hypertension*, vol. 18, no. 4, pp. 477–485, 2005.

[14] T. Otsuka, T. Kawada, M. Katsumata, and C. Ibuki, "Utility of second derivative of the finger photoplethysmogram for the estimation of the risk of coronary heart disease in the general population," *Circulation Journal*, vol. 70, no. 3, pp. 304–310, 2006.

[15] T. Otsuka, T. Kawada, M. Katsumata, C. Ibuki, and Y. Kusama, "Independent determinants of second derivative of the finger photoplethysmogram among various cardiovascular risk factors in middle-aged men," *Hypertension Research*, vol. 30, no. 12, pp. 1211–1218, 2007.

[16] A. Grabovskis, Z. Marcinkevics, Z. Lukstina et al., "Usability of photoplethysmography method in estimation of conduit artery stiffness," in *Novel Biophotonic Techniques and Applications*, vol. 8090 of *Proceedings of SPIE*, May 2011.

[17] S. C. Millasseau, R. P. Kelly, J. M. Ritter, and P. J. Chowienczyk, "The vascular impact of aging and vasoactive drugs: comparison of two digital volume pulse measurements," *American Journal of Hypertension*, vol. 16, no. 6, pp. 467–472, 2003.

[18] P. Holoborodko, "Smooth noise robust differentiators," http://www.holoborodko.com/pavel/numerical-methods/numerical-derivative/smooth-low-noise-differentiators/, 2008.

[19] T. W. Parks and J. H. McClellan, "Chebyshev approximation for nonrecursive digital filters with linear phase," *IEEE Transactions Circuit Theory*, vol. 19, no. 2, pp. 189–194, 1972.

[20] W. W. Nichols and M. F. O'Rourke, *McDonald's Blood Flow in Arteries: Theoretic, Experimental and Clinical Principles*, Hodder-Arnold, London, UK, 5th edition, 2005.

[21] K. Pilt, K. Meigas, M. Viigimaa, K. Temitski, and J. Kaik, "An experimental measurement complex for probable estimation of arterial stiffness," in *Proceedings of the 32nd International Conference of the IEEE Engineering in Medicine and Biology Society*, pp. 194–197, Buenos Aires, Argentina, 2010.

4

A Novel Image Encryption Algorithm Based on DNA Subsequence Operation

Qiang Zhang, Xianglian Xue, and Xiaopeng Wei

Key Laboratory of Advanced Design and Intelligent Computing, Ministry of Education of Dalian University, Dalian 116622, China

Correspondence should be addressed to Qiang Zhang, zhangq30@yahoo.com

Academic Editors: P. K. Egbert and M. Engin

We present a novel image encryption algorithm based on DNA subsequence operation. Different from the traditional DNA encryption methods, our algorithm does not use complex biological operation but just uses the idea of DNA subsequence operations (such as elongation operation, truncation operation, deletion operation, etc.) combining with the logistic chaotic map to scramble the location and the value of pixel points from the image. The experimental results and security analysis show that the proposed algorithm is easy to be implemented, can get good encryption effect, has a wide secret key's space, strong sensitivity to secret key, and has the abilities of resisting exhaustive attack and statistic attack.

1. Introduction

Nowadays, the computer network has changed the mode of people's communication. People can easily transfer the various multimedia information through the network. However, because of the openness of the network, people have to take more and more attention on security and confidentiality of multimedia information. The digital image is an important information vector of multimedia communication, thus how to protect the image information becomes a universal concern problem for people. The traditional image encryption methods (such as DES, IDEA, and AES) are not suitable for image encryption due to the different storage format of an image. The new research algorithms of image encryption are needed urgently.

As an example, due to chaotic system with like-random, high sensitivity to initial value, and unforeseeable properties, chaos based cryptosystems have become current research hotspot. According to the object of scrambling, the chaos-based algorithms operate in two stages: the shuffling stage and the substitution stage. In the shuffling stage, the position of pixels from original image is changed by chaotic sequences [1] or by some matrix transformation, such as Arnold transform, magic square transform, and so forth. These shuffling algorithms are easy to be realized and have better encryption effect. Due to these shuffling algorithms, just changing the position of pixels but not changing the pixel values leads to the histogram of the encryption image is the same as the original image, and thus its security is threatened by the statistical analysis. In the substitution stage, the pixel values are changed by chaotic sequences. Most of these encryption methods are directly implemented by overlaying a chaotic sequence generated by a single chaotic map and the pixel grey value from the image. Comparing with the method of the shuffling, the method of value substitution may lead to higher security, but from the vision angle the encryption effect is not good. Thereby, in order to improve the security and the encryption effect, the shuffling and the value substitution are combined by some researchers, the readers can refer to [2, 3]. However, only using a single chaotic map to encrypt image may result in lower security and smaller key space. Ren et al. [1] presented a chaotic algorithm of image encryption based on dispersion sampling; their algorithm has better scrambling effect, but has small key space. Zhang et al. [4] use logistic and standard systems to scramble the location and the value of pixel points from the image, and they have got better result; however, no security analysis in their paper was given. Recently, Lian et al. [5–7] used some multidimensional chaotic maps (such as the chaotic standard map, the chaotic neural networks, and spatiotemporal chaos system) to encrypt images and make a detailed analysis for the security of algorithms. Their algorithms have satisfactory security with a low cost. A new image encryption algorithm based on multiple chaos system is proposed by Zuo et al. [8]. Similarly, Liu et al. [9] also used multiple chaotic maps to encrypt image. All of their algorithms have a large key space, high sensitivity to key variation, and unforeseeable and have the ability of resisting traditional attacks. Generally speaking, to improve the security of encryption algorithm, researchers usually try to use more complex chaotic system or combine some new encryption methods with the existing chaotic systems to implement image encryption. However, some chaotic systems have been proven to be insecure [10–12].

With the rapid development of DNA computing, DNA cryptography, as a new field, has come into being. A method for hiding message in DNA microdots was proposed by Clelland et al. [13]. Clelland used DNA microdots to hide message to implement the protection of information. For instance, letter A is expressed as DNA sequence GGT by complex biological operation. Obviously, it is difficult to be implemented and is not suitable for image encryption. Gehani et al. presented an encryption algorithm of the one-time pad cryptography with DNA strands [14]; Gehani's method is effective, but the process of encryption must utilize complex biological operations, which are difficult to be controlled under the experimental environment. So the method is not easy to be realized. In fact, since the high-tech laboratory requirements and computational limitations, combining with the labor intensive extrapolation means, researches of DNA cryptography are still much more theoretical than practical. Recently, Kang presented a pseudo DNA cryptography method [15]. Kang's method not only has the better encryption effect, but also does not require complex biological operation. However, it was only used for encrypting text files.

In this paper, we do not use biological operation to implement image encryption, but adopt the rule of DNA subsequence operation such as truncation operation, deletion operation, transformation operation and so forth, then combine DNA subsequence operation with chaos system to scramble the location and the value of pixel point from the image.

The structure of this paper is as follows. In Section 2, we will introduce the basic theory of the proposed algorithm. The design of the proposed image encryption scheme is discussed in the Section 3. In Section 4, some simulation results and security analysis are given. In Section 5, we compare our algorithm with other encryption algorithms. Section 6 draws the conclusion.

2. Basic Theory of the Proposed Algorithm

2.1. Generation of the Chaotic Sequences. The chaotic system is a deterministic nonlinear system. It possesses a varied characteristics, such as high sensitivity to initial conditions and system parameters, random-like behaviors, and so forth. Chaotic sequences produced by chaotic maps are pseudo-random sequences; their structures are very complex and difficult to be analyzed and predicted. In other words, chaotic systems can improve the security of encryption systems. Thus, it is advisable to encrypt digital image with chaotic systems [16–24]. Here, we introduce the following two chaotic maps, one is logistic map, and the other is 2D logistic map. In the paper, we use 2D logistic map to produce the eight parameters as the initial values and system parameters of four logistic maps.

Logistic map is an example for chaotic map, and it is described as follows:

$$x_{n+1} = \mu x_n (1 - x_n), \quad (1)$$

where $\mu \in [0,4]$, $x_n \in (0,1)$, and $n = 0,1,2,\ldots$. The research result shows that the system is in chaotic state under the condition that $3.56994 < \mu \le 4$.

2D logistic map is described in (2) [25] as follows:

$$\begin{aligned} x_{i+1} &= \mu_1 x_i (1 - x_i) + \gamma_1 y_i^2 \\ y_{i+1} &= \mu_2 y_i (1 - y_i) + \gamma_2 (x_i^2 + x_i y_i). \end{aligned} \quad (2)$$

When $2.75 < \mu_1 \le 3.4$, $2.75 < \mu_2 \le 3.45$, $0.15 < \gamma_1 \le 0.21$ and, $0.13 < \gamma_2 \le 0.15$, the system is in chaotic state and can generate two chaotic sequences in the region (0,1]. Due to the system parameter γ_1 and γ_2 which have smaller value range, we set $\gamma_1 = 0.17$ and $\gamma_2 = 0.14$, other parameters can be seen as secret keys.

2.2. DNA Sequence Encryption

2.2.1. DNA Encoding and Decoding for Image. A single DNA sequence is made up of four nucleic acid bases: A (adenine), C (cytosine), G (guanine), and T (thymine), where A and T are complements, and C and G are complements. Let binary number 0 and 1 be complements, so 00 and 11 are complements, and 01 and 10 are complements. Thus we can use these four bases: A, T, G, and C to encode 01, 10, 00, and 11, respectively. The encoding method still satisfies the Watson-Crick complement rule [25]. Usually, each pixel value of the 8 bit grey image can be expressed to 8 bits binary stream. The binary stream can be encoded to a DNA sequence whose length is 4. For example: if the first pixel value of the original image is 75, convert it into a binary stream [01001011]. By using the above DNA encoding rule to encode the stream, we can get a DNA sequence [$AGTC$], whereas we use A, T, G, and C to express 01, 10, 00, and 11, respectively. We can get a binary sequence [01001011].

2.2.2. DNA Subsequences Operation. In this section we use the idea of [26] to define the DNA subsequence and the corresponding operation. We define that a DNA sequence P_k contains m strands of DNA subsequences according to the order, in the P_k, the number of bases is k $(m \le k)$. The expression is $P_k = P_m P_{m-1} \cdots P_2 P_1$. The number of bases for the corresponding DNA subsequences is $l_m l_{m-1} \cdots l_2 l_1$, respectively. Apparently, $k = l_m + l_{m-1} + \cdots l_2 + l_1$. Based on the above DNA subsequence expression, we described the following five kinds of DNA subsequence operation; they are elongation operation, truncation operation, deletion operation, insertion operation, and transformation operation.

(1) DNA subsequence elongation operation.

Definition 1. We suppose that there is an original DNA sequence P_1, the subsequence P_2, whose length is l_1, is elongated to the tail of P_1. After elongation operation, we can get a new DNA sequence $P' = P_1P_2$. The expression is as follows:

$$P_1 + P_2 \longrightarrow P_1P_2. \quad (3)$$

(2) DNA subsequence truncation operation.

Definition 2. The truncation operation and the elongation operation are contrary. Truncating the end of the subsequence P_2 in the DNA sequence P_1P_2, we will obtain a new DNA sequence $P' = P_1$. The expression is as follows:

$$P_1P_2 - P_2 \longrightarrow P_1. \quad (4)$$

(3) DNA subsequence deletion operation.

Definition 3. We suppose that there is an original DNA sequence $P = P_3P_2P_1$. Deleting the subsequence P_2, then we will obtain a new DNA sequence $P' = P_1P_3$. The expression is as follows:

$$P_3P_2P_1 - P_2 \longrightarrow P_3P_1. \quad (5)$$

(4) DNA subsequence insertion operation.

Definition 4. The deletion operation and the insertion operation are contrary. We suppose that there is an original DNA sequence $P = P_3P_1$, inserting a subsequence P_2, whose length is l_2, into P. The expression is as follows:

$$P_3P_1 + P_2 \longrightarrow P_3P_2P_1. \quad (6)$$

(5) DNA subsequence transformation operation.

Definition 5. In brief, the locations of two subsequences are transformed. If the original DNA sequence is $P = P_5P_4P_3P_2P_1$. Transforming the locations of P_4 and P_2, we will get a new DNA sequence $P' = P_5P_2P_3P_4P_1$. The expression is as follows:

$$P_5P_4P_3P_2P_1 \longrightarrow P_5P_2P_3P_4P_1. \quad (7)$$

We introduced five kinds of DNA subsequence operations, where the inverse operation of elongation operation is truncation operation and the inverse operation of deletion operation is insertion operation. In our algorithm, we use elongation operation, truncation operation, deletion operation, and transformation operation and combined with the use of the Logistic chaotic map we will realize the image encryption algorithm. However, the insertion operation is just used in the decryption process.

3. Algorithm Description

3.1. Generation of Chaotic Sequences. Input initial state (x_0, μ_1, y_0, μ_2), by using 2D Logistic to produce eight parameters $(x_1, x_2, x_3, x_4, x_5, x_6, x_7, x_8)$ after iterating 1000 times. We Use the following formulas to produce four groups of parameters:

$$\begin{aligned} x_1 &= x_1, & u_1 &= 3.9 + 0.1 \times x_2, \\ y_1 &= x_3, & u_2 &= 3.9 + 0.1 \times x_4, \\ z_1 &= x_5, & u_3 &= 3.9 + 0.1 \times x_6, \\ q_1 &= x_7, & u_4 &= 3.9 + 0.1 \times x_8. \end{aligned} \quad (8)$$

Then, by using logistic chaotic map to generate four chaotic sequences under the condition that the four groups of initial values are (x_1, u_1), (y_1, u_2), (z_1, u_3), and (q_1, u_4), their length are $m \times n$, respectively.

3.2. Generation of DNA Subsequences

Step 1. Input an 8 bit grey image $A(m, n)$, as the original image, where m and n is rows and columns of the image.

Step 2. Convert image A into binary matrix A' whose size is $(m, n \times 8)$ and divide A' into eight bit-planes. Here, the first bitplanes and the eighth bitplanes, the second bitplanes and the seventh bitplanes, the third bitplanes and the sixth bitplanes, and the forth bit-planes and the fifth bit-planes are composed, respectively. Then we obtain four bit-planes.

Step 3. Carry out DNA encoding operation according to Section 2.2.1 for the four bitplanes, then we get four coding matrices P_1, P_2, P_3, P_4, all of their sizes are (m, n).

Step 4. Convert P_1, P_2, P_3, P_4 into P'_1, P'_2, P'_3, P'_4 whose sizes are $(1, (m \times n))$, then divide P'_1, P'_2, P'_3, P'_4 into DNA subsequence; the average length of subsequences are $l_1 = 128$, $l_2 = 64$, $l_3 = 32$, and $l_4 = 8$, respectively. So, there are the following conclusions:

$$\begin{aligned} P'_1 &= p_{11}p_{12}\cdots p_{1(mn/l_1)}, \\ P'_2 &= p_{21}p_{22}\cdots p_{2(mn/l_2)}, \\ P'_3 &= p_{31}p_{32}\cdots p_{3(mn/l_3)}, \\ P'_4 &= p_{41}p_{42}\cdots p_{4(mn/l_4)}, \end{aligned} \quad (9)$$

where p_{ij} are DNA subsequences, l_i are lengths of these subsequences, $i \in [1, 4]$, and $j \in [1, mn/l_i]$.

3.3. Deletion Operation

Step 1. We suppose that there is a chaotic sequence $X = \{x_1, x_2 \cdots x_{mn/l_i}\}$.

Step 2. If $x_i < 0.5$, delete the ith subsequence according to Section 2.2.2, otherwise save the subsequence.

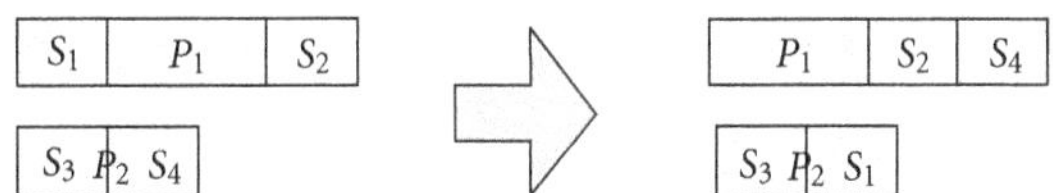

FIGURE 1: Elongation and truncation operation of DNA subsequences.

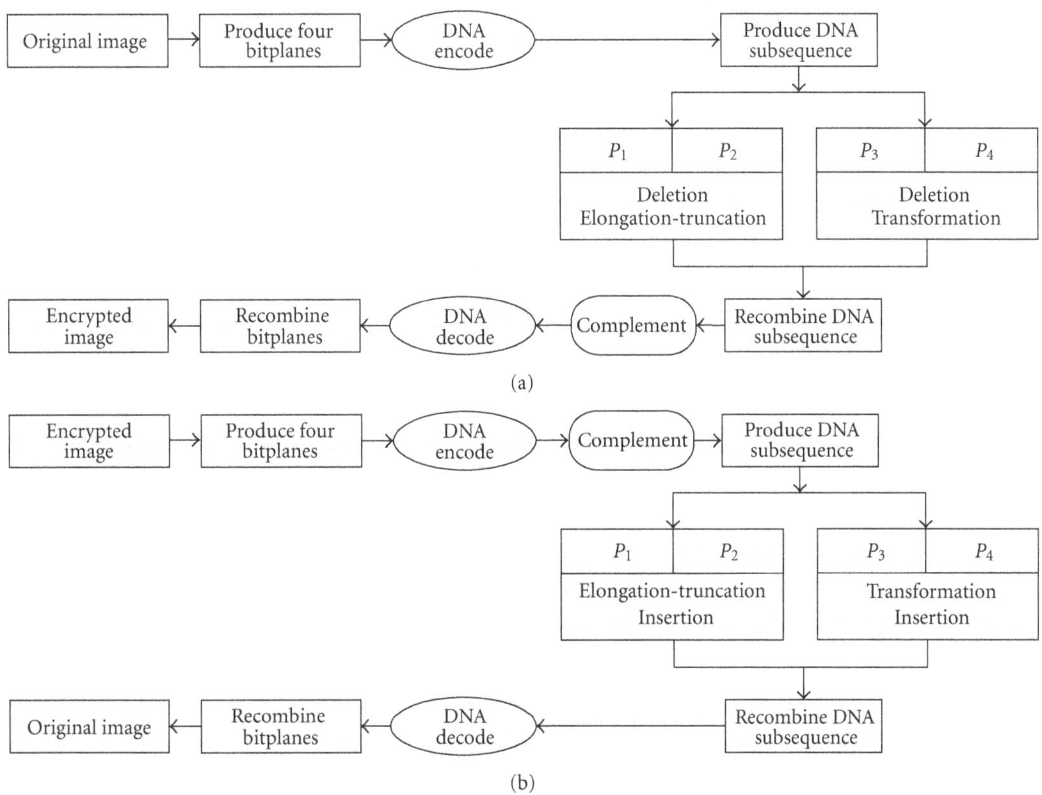

FIGURE 2: The block diagram of the proposed algorithm. (a) The block diagram of the encryption algorithm. (b) The block diagram of the decryption algorithm.

Step 3. Those deleted subsequences are moved to the end of the saved subsequences.

3.4. Transformation Operation

Step 1. We suppose that there is a chaotic sequence $X = \{x_1, x_2 \cdots x_{mn/l_i}\}$.

Step 2. To sort X by ascending, we get a new sequence $X' = \{x'_1, x'_2 \cdots x'_{mn/l_i}\}$.

Step 3. If $x_i < 0.5$, the ith subsequence and the i'th subsequence from the location of X' are transformed according to Section 2.2.2.

3.5. Elongation and Truncation Operation. As shown in Figure 1, P_1 and P_2 are two DNA subsequences from any of two bit-planes, we suppose that the length of P_1 is 128, the length of P_2 is 64, S_1 and S_2, S_3, and S_4 are DNA subsequences of P_1 and P_2, respectively. First, we truncate S_1 and S_4, then elongate S_1 to the tail of P_2, elongate S_4 to the tail of P_1.

3.6. Complement Operation. Complement operation is carried out for every one dimension bit-plane whose size is $(1, m \times n)$, we suppose there is a chaotic sequence $X = \{x_1, x_2 \cdots x_{mn/l_i}\}$. If $x_i < 0.5$, the nucleic acid base of the ith location is complemented, otherwise, it is unchanged.

3.7. The Procedure of Image Encryption and Decryption. The proposed encryption algorithm includes three steps: first, by using the method proposed in the Section 3.1 to produce four groups of DNA sequences P_1, P_2, P_3, and P_4, where P_i $(i = 1, 2, 3, 4)$ is made up of many DNA subsequences. Then, to disturb the position and the value of pixel points from image by combining the logistic map, generate chaotic sequences and DNA subsequence operations (such as elongation operation, truncation operation, deletion operation, transformation, etc.). At last, the encrypted image is obtained by DNA decoding and recombining bit-planes. The block diagram of the proposed algorithm is shown in Figure 2, the block diagram of the encryption algorithm is shown in Figure 2(a), and Figure 2(b) shows the block diagram of the decryption algorithm. We can see that the procedure of

Input: An 8-bit image A and chaotic parameters $(x_0, \mu_1, y_1, y_0, , \mu_2, y_2)$
Output: The encrypted image B
(1) $[S_1, S_2, S_3, S_4]$:= four DNA sequences obtained from image A;
(2) $[P_1, P_2, P_3, P_4]$:= four groups of DNA subsequences obtained from image $[S_1, S_2, S_3, S_4]$;
(3) $[x_1, u_1, y_1, u_2, z_1, u_3, q_1, u_4]$:= eight chaotic parameters obtained by 2D Logistic map under chaotic initial parameters $(x_0, \mu_1, y_1, y_0, , \mu_2, y_2)$;
(4) $[X, Y, Z, Q]$:= four chaotic sequences obtained by Logistic map under the chaotic parameters $(x_1, u_1, y_1, u_2, z_1, u_3, q_1, u_4)$;
(5) $A_1 = Deletion(P_1, X)$;
(6) $A_2 = Deletion(P_2, Y)$;
(7) $[E_1, E_2] = Elongation - truncation(A_1, A_2)$;
(8) $A_3 = Deletion(P_3, Z)$;
(9) $A_3^{'} = Transformation\ (A_3, Z)$;
(10) $A_4 = Deletion(P_4, Q)$;
(11) $A_4^{'} = Transformation\ (A_4, Q)$;
(12) $[B_1, B_2, B_3, B_4] = Recombine - subseqence(E_1, E_2, A_3^{'}, A_4^{'})$;
(13) $[B_1^{'}, B_2^{'}, B_3^{'}, B_4^{'}] = Complement(B_1, B_2, B_3, B_4)$;
(14) B := carry out DNA decoding and recombining binary bitplanes for $B_1^{'}, B_2^{'}, B_3^{'}, B_4^{'}$;

Algorithm 1: An image encryption algorithm based on DNA subsequence operation.

Input: The decrypted image B and chaotic parameters $(x_0, \mu_1, y_1, y_0, \mu_2, y_2)$
Output: The encrypted image A
(1) $[B_1, B_2, B_3, B_4]$:= four DNA sequences obtained from image B;
(2) $[B_1^{'}, B_2^{'}, B_3^{'}, B_4^{'}] = Complement(B_1, B_2, B_3, B_4)$;
(3) $[P_1, P_2, P_3, P_4]$:= four groups of DNA subsequences obtained from image $[B_1^{'}, B_2^{'}, B_3^{'}, B_4^{'}]$;
(4) $[x_1, u_1, y_1, u_2, z_1, u_3, q_1, u_4]$:= eight chaotic parameters obtained by 2D Logistic map under chaotic initial parameters $(x_0, \mu_1, y_1, y_0, \mu_2, y_2)$;
(5) $[X, Y, Z, Q]$:= four chaotic sequences obtained by Logistic map under the chaotic parameters$(x_1, u_1, y_1, u_2, z_1, u_3, q_1, u_4)$;
(6) $[E_1, E_2] = Elongation - truncation(P_1, P_2)$;
(7) $M_1 = Insertion(E_1, X)$;
(8) $M_2 = Insertion(E_2, Y)$;
(9) $E_3 = Transformation\ (P_3, Z)$;
(10) $M_3 = Insertion(E_3, Z)$;
(11) $E_4 = Transformation\ (E_4, Q)$;
(12) $M_4 = insertion(E_4, Q)$;
(13) $[A_1, A_2, A_3, A_4] = Recombine - subseqence(M_1, M_2, M_3, M_4)$;
(14) A := carry out DNA decoding and recombining binary bit-planes for A_1, A_2, A_3, A_4;

Algorithm 2: An image decryption algorithm based on DNA subsequence operation.

image decryption is inverse procedure of image encryption from Figure 2. The detailed procedure of our encryption and decryption algorithms are explained in the following pseudo-codes (Algorithms 1 and 2).

The functions and parameters in Algorithms 1 and 2 are the same as Sections 3.1–3.6, where DNA decoding and recombining are the inverse process of Steps 3 and 2 in Section 3.2. The procedure of acquiring the original image from the encryption image is an inverse operation according to Algorithm 2, where deletion operation is replaced by insertion operation.

4. Simulation Result and Security Analysis

4.1. Simulation Result. In this paper, for standard 256×256 gray image Lena, we use Matlab 7.1 to simulate experiment. In our experiment, we set $x_0 = 0.95$, $\mu_1 = 3.2$, $y_1 = 0.17$, $y_0 = 0.25$, $\mu_1 = 3.3$, $y_2 = 0.14$. The original image is shown

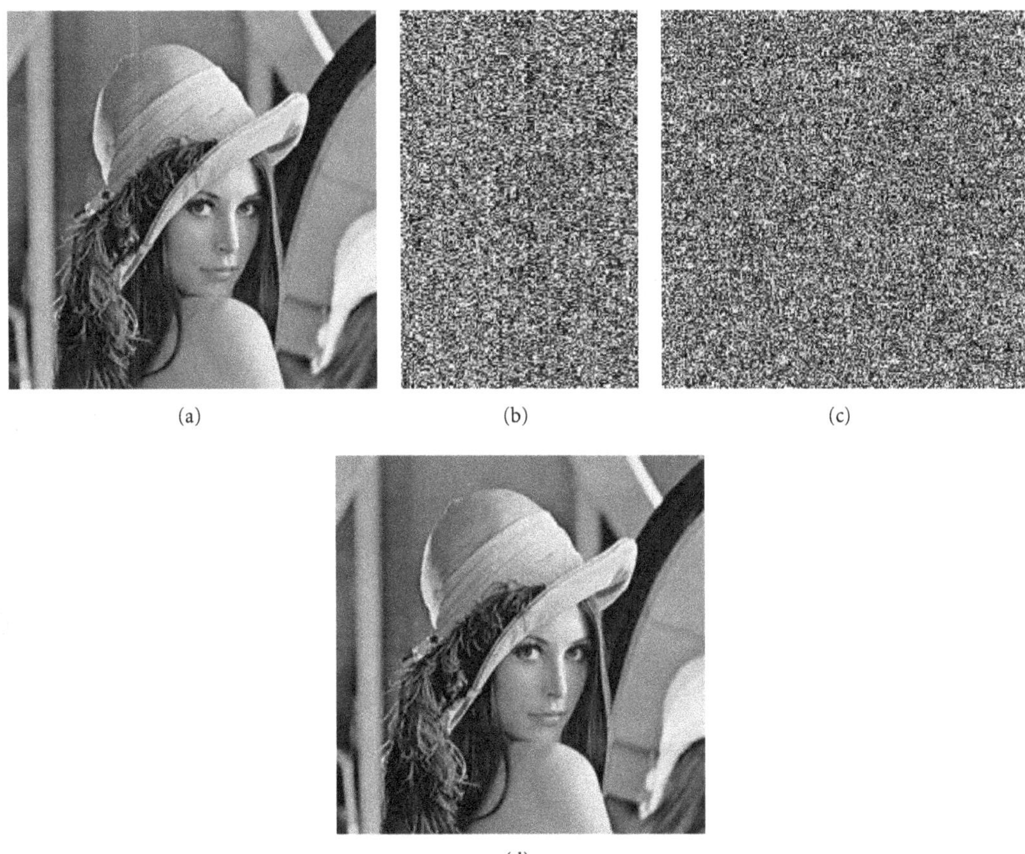

FIGURE 3: Encrypted image and decrypted image. (a) The original image. (b) The encrypted image. (c) The decrypted image under the wrong secret keys. (d) The decrypted image under the correct secret keys.

in Figure 3(a), Figure 3(b) shows encrypted image, and Figure 3(b) points out that it is difficult to recognize the original image. Figures 3(c) and 3(d) show the decrypted image under the wrong secret keys and the right secret keys, respectively. From Figure 3(c), we know that it has not any connection with the original image, but Figure 3(d) is as same as the original image.

4.2. Secret Key's Space Analysis. In the proposed algorithm, the initial value and the parameter of the system of 2D logistic are identified as secret keys of this algorithm. Therefore, our algorithm has six secret keys $x_0, \mu_1, \gamma_1, y_0, \mu_2, \gamma_2$. If the precision is 10^{-14}, the secret key's space is $10^{14} \times 10^{14} \times 10^{14} \times 10^{14} \times 10^{14} \times 10^{14} = 10^{84} \approx 2^{279}$. It is shown that the secret key's space is large enough to resist exhaustive attack.

4.3. Secret Key's Sensitivity Analysis. The chaotic map is very sensitive to the initial value in chaotic state, in other words, it also ensured the sensibility of this encryption algorithm to the secret key. In this paper, if the initial values from three chaotic maps are changed a little, the recovering image is not allowed to be read, but we can get the original image from the encrypted image by using the correct secret keys. The experiment results are shown in Figure 4, where Figure 4(a) shows the decrypted image under the secret keys $(0.95000000000001, 3.2, 0.17, 0.25, 3.3, 0.14)$. The corresponding histogram is shown in Figure 4(b), and we can see that the histogram of the decrypted image is very uniform. The sensitivity of other parameters is similar. From Figure 4, we can see that only when all secret keys (the chaotic initial value and system parameter) are correct, the original image can be obtained. Otherwise the decrypted image will have no connection with the image. Based on the above argument, our algorithm has strong sensitivity to secret key and we can say again that our algorithm can resist exhaustive attack.

4.4. Statistical Analysis

4.4.1. The Grey Histogram Analysis. We compare the grey histogram of the image before and after encryption to analyze the statistical performance. Figure 5(a) shows the grey histogram of the original image and Figure 5(b) shows the grey histogram of the encrypted image. From the two figures, we can see that the original pixel grey values are concentrated on some value, but the pixel grey values after the encryption are scattering in the entire pixel value space, namely, two images have lower similarity. Clearly, it is difficult to use the statistical performance of the pixel grey value to recover the original image. Thereby, our algorithm has strong ability of resisting statistical attack.

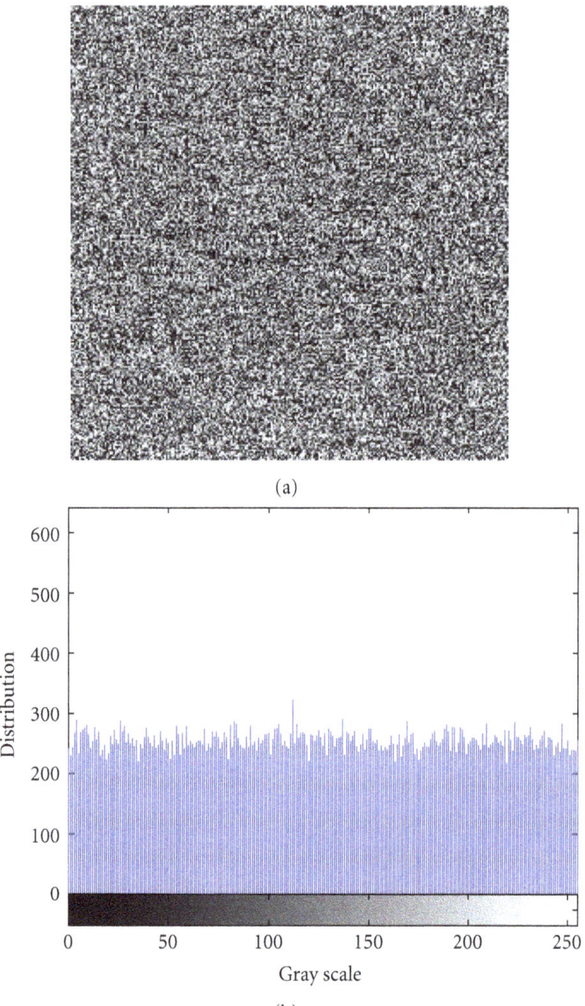

FIGURE 4: The sensitivity of secret key x_0. (a) The decrypted image with secret key $(0.95000000000001, 3.2, 0.17, 0.25, 3.3, 0.14)$. (b) The corresponding histogram.

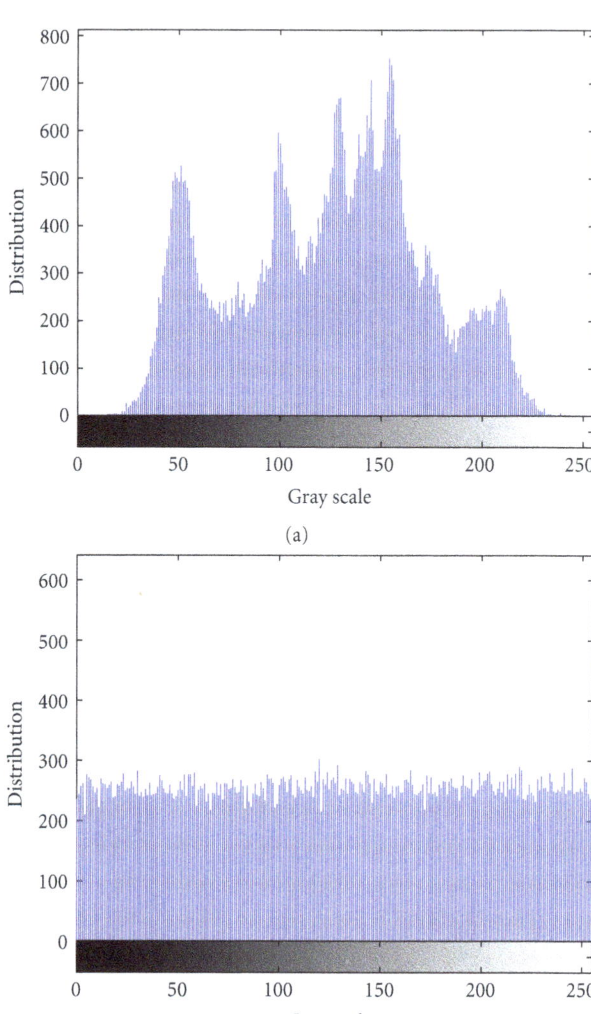

FIGURE 5: The grey histogram of the original image and the encrypted image. (a) The grey histogram of the original image. (b) The grey histogram of the encrypted image.

4.4.2. Correlation Coefficient Analysis. The correlation of the adjacent pixels in original image is very high, an effective encryption algorithm can reduce the correlation of between adjacent pixels. Here, we randomly select 3000 pairs (horizontal, vertical and diagonal) of adjacent pixels from the original image and the encrypted image, then by using the following formulas to calculate the correlation coefficient:

$$\begin{aligned} E(x) &= \frac{1}{N}\sum_{i=1}^{N} x_i, \\ D(x) &= \frac{1}{N}\sum_{i=1}^{N} (x_i - E(x))^2, \\ \operatorname{cov}(x, y) &= \frac{1}{N}\sum_{i=1}^{N} (x_i - E(x))(y_i - E(y)), \\ r_{xy} &= \frac{\operatorname{cov}(x, y)}{\sqrt{D(x)} \times \sqrt{D(y)}}, \end{aligned} \tag{10}$$

where x and y are grey value of two adjacent pixels in the image.

Figures 6(a) and 6(b) show the correlation of two horizontally adjacent pixels in the original image and that in the encrypted image, where the correlation coefficients are 0.9432 and 0.1366, respectively. Other results are shown in Table 1. From Figure 6(b) and Table 1, we can see that the correlation coefficient of the adjacent pixels in encrypted image is low, which is close to 0. It follows from Figure 6(b) and Table 1 that the proposed image encryption algorithm has strong ability of resisting statistical attack.

4.4.3. Information Entropy. It is well known that information entropy can measure the distribution of grey value in the image. We can make sure that the bigger information entropy the more uniform for the distribution of grey value. The definition of information entropy is as follows:

$$H(m) = -\sum_{i=0}^{L} P(m_i)\log_2 P(m_i), \tag{11}$$

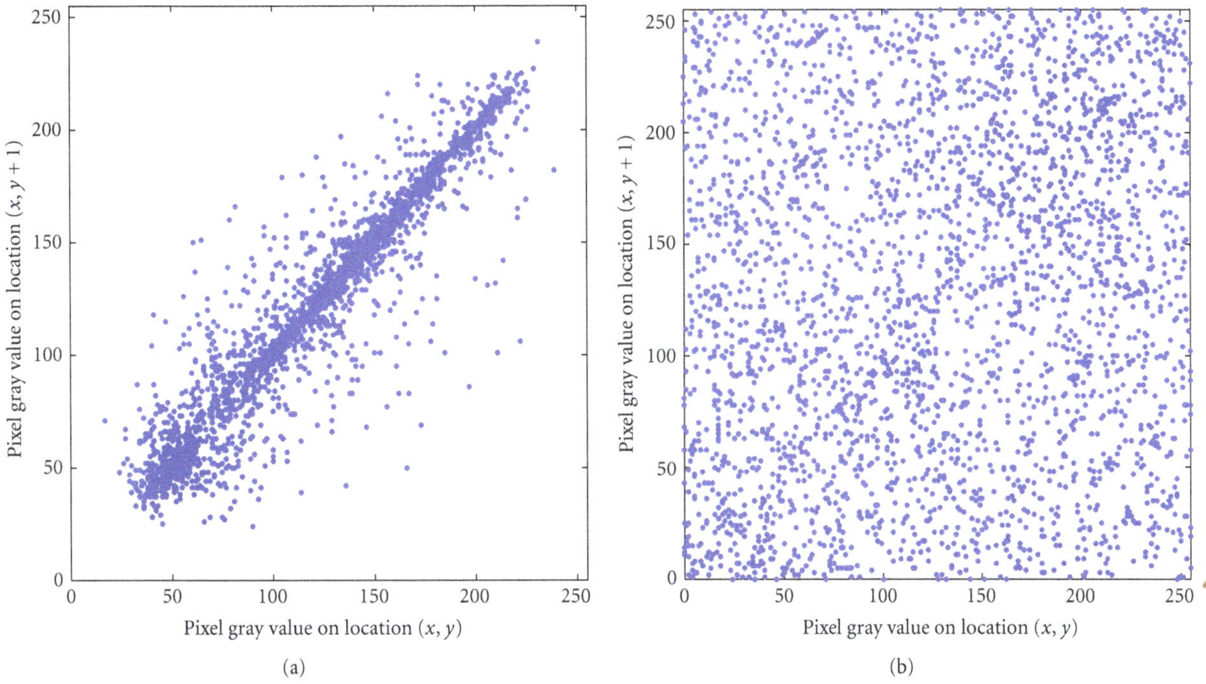

FIGURE 6: Correlation of two horizontally adjacent pixels in the original image and in the encrypted image.

TABLE 1: Correlation coefficients of two adjacent pixels in two image.

Model	The original image	The encrypted image
Horizontal	0.9432	0.1366
Vertical	0.9688	0.0166
Diagonal	0.9148	0.0021

TABLE 2: Comparison with other chaos-based encryption algorithms.

Considered items for 256 × 256 lena image	Proposed	Reference [16]	Reference [17]	Reference [18]	Reference [1]
Key space	2^{279}	2^{233}	N/A	2^{260}	2^{99}
Key sensitivity	Yes	Yes	N/A	Yes	Yes
information entropy	7.9975	N/A	7.99732	7.9968	N/A
Chaotic system used	DNA operation and logistic system	Chen's chaotic system	Chen's chaotic system	Coupled chaotic system	Logistic chaotic system

TABLE 3: Comparison with other DNA-based encryption algorithms.

Considered items	Proposed	Clelland et al. [13]	Gehani et al. [14]	Kang [15]
Image	Yes	No	Yes	No
Text	Yes	Yes	Yes	Yes
Security analysis	Yes	No	No	Yes
Biology operation	No	Yes	Yes	No
Method	DNA subsequence operation chaotic maps	Messages encoded as DNA stands	Use micro-array technology	A pseudo DNA cryptography

where m_i is the ith grey value for L level grey image and $P(m_i)$ is the emergence probability of m_i. The information entropy of an idea random image is 8. For the proposed algorithm, the information entropy is 7.9975. It is very close to 8.

5. Comparison with Other Encryption Algorithms

In this section, we will compare our proposed algorithm with existing chaos-based and DNA-based encryption algorithms. We focus on the security consideration in the comparative aspects of chaos-based and focus on the encryption objects and environment in the comparative aspects of DNA-based. The comparison results are shown in Tables 2 and 3. From Table 2, we can see that the key space and the information entropy of our proposed algorithm are larger than others. However the methods in [17, 18] can resist differential attack, the proposed method in this paper cannot resist differential attack. From Table 3, we easily found that only our algorithm and [14] can implement image encryption. But the algorithm proposed in [14] is difficult to be implemented owing to the complex biologic operation. Kang's encryption effect [15] is better than others. However, his algorithm can only encrypt the text. After comparing with other encryption algorithms proposed in Tables 2 and 3, the proposed algorithm is better than other DNA-based encryption algorithm and has larger key space and high key sensitivity, but the disadvantage is that the algorithm cannot resist differential attack. This is our next study work.

6. Conclusion

A novel image encryption algorithm based on DNA subsequence operation is proposed in this paper. The simulation experimental results and security analysis show that the encryption algorithm is effective, easy to be realized, has larger key space, and is sensitive to the secret key. Our algorithm can also resist statistical analysis and exhaustive attacks. Furthermore, it avoids complex biological experiment in traditional DNA cryptography. But because DNA subsequence operation is based on horizontal, or the length of the subsequences selected is longer, it may lead to the horizontal correlation of the adjacent pixels in original image a bit high. We can improve the horizontal correlation through changing the lengths of DNA subsequences from each bit-planes. In addition to that, the weak ability of resisting differential attack is also a defect of this algorithm. They are our next research works.

Acknowledgments

This work is supported by the National Natural Science Foundation of China (nos. 31170797, 30870573), Program for Changjiang Scholars and Innovative Research Team in University (no. IRT1109), the Program for Liaoning Innovative Research Team in University (no. LT2011018), the Program for Liaoning Excellent Talents in University (no. LR201003), the Program for Liaoning Innovative Research Team in University (no. LT2011018) and the Program for Liaoning Science and Technology Research in University (no. LS2010179).

References

[1] H. Ren, Z. Shang, Y. Wang, and J. Zhang, "A chaotic algorithm of image encryption based on dispersion sampling," in *Proceedings of the 8th International Conference on Electronic Measurement and Instruments (ICEMI '07)*, vol. 2, pp. 836–839, August 2007.

[2] C. Fu and Z. Zhu, "A chaotic image encryption scheme based on circular bit shift method," in *Proceedings of the 9th International Conference for Young Computer Scientists (ICYCS '08)*, vol. 522, pp. 3057–3061, November 2008.

[3] H. E. Ren, J. Zhang, X. J. Wang, and Z. W. Shang, "Block sampling algorithm of image encryption based on chaotic scrambling," in *Proceedings of the International Conference on Computational Intelligence and Security Workshops (CIS '07)*, vol. 109, pp. 773–776, December 2007.

[4] Y. H. Zhang, B. S. Kang, and X. F. Zhang, "Image encryption algorithm based on chaotic sequence," in *Proceedings of the 16th International Conference on Artificial Reality and Telexistence—Workshops (ICAT '06)*, vol. 29, pp. 221–223, December 2006.

[5] S. Lian, "Efficient image or video encryption based on spatiotemporal chaos system," *Chaos, Solitons and Fractals*, vol. 40, no. 5, pp. 2509–2519, 2009.

[6] S. Lian, J. Sun, and Z. Wang, "A block cipher based on a suitable use of the chaotic standard map," *Chaos, Solitons and Fractals*, vol. 26, no. 1, pp. 117–129, 2005.

[7] S. Lian, "A block cipher based on chaotic neural networks," *Neurocomputing*, vol. 72, no. 4–6, pp. 1296–1301, 2009.

[8] Y. Z. F. Zuo, Z. Zhai, and C. Xiaobin, "A new image encryption algorithm based on multiple chaos system," in *Proceedings of the International Symposium on Electronic Commerce and Security (ISECS '08)*, vol. 142, pp. 347–350, August 2008.

[9] J. M. Liu, S. S. Qiu, F. Xiang, and H. J. Xiao, "A cryptosystem based on multiple chaotic maps," in *Proceedings of the International Symposium on Information Processing (ISIP '08) and International Pacific Workshop on Web Mining and Web-Based Application (WMWA '08)*, vol. 99, pp. 740–743, May 2008.

[10] G. Jakimoski and L. Kocarev, "Analysis of some recently proposed chaos-based encryption algorithms," *Physics Letters A*, vol. 291, no. 6, pp. 381–384, 2001.

[11] G. Álvarez, F. Montoya, M. Romera, and G. Pastor, "Cryptanalysis of an ergodic chaotic cipher," *Physics Letters A*, vol. 311, no. 2-3, pp. 172–179, 2003.

[12] I. S. Mohamed and R. S. Alaa-eldin, "Methods of attacking chaotic encryption and countermeasures," in *Proceedings of the IEEE Interntional Conference on Acoustics, Speech, and Signal Processing*, pp. 1001–1004, Salt Lake, Utah, USA, May 2001.

[13] C. T. Clelland, V. Risca, and C. Bancroft, "Hiding messages in DNA microdots," *Nature*, vol. 399, no. 6736, pp. 533–534, 1999.

[14] A. Gehani, T. LaBean, and J. Reif, "DNA-based cryptography," in *Proceedings of the DIMACS Workshop on DNA Based Computers*, June 1999.

[15] N. Kang, "A pseudo DNA cryptography method," http://arxiv.org/abs/0903.2693.

[16] T. Gao and Z. Chen, "A new image encryption algorithm based on hyper-chaos," *Physics Letters A*, vol. 372, no. 4, pp. 394–400, 2008.

[17] S. J. Xu, J. Z. Wang, and S. X. Yang, "An improved image encryption algorithm based on chaotic maps," *Chinese Physics B*, vol. 17, no. 11, pp. 4027–4032, 2008.

[18] S. Behnia, A. Akhshani, H. Mahmodi, and A. Akhavan, "A novel algorithm for image encryption based on mixture of chaotic maps," *Chaos, Solitons and Fractals*, vol. 35, no. 2, pp. 408–419, 2008.

[19] Y. Xiao and L. Xia, "A new hyper-chaotic algorithm for image encryption," in *Proceedings of the 9th International Conference for Young Computer Scientists (ICYCS '08)*, pp. 2814–2818, November 2008.

[20] Y. Zhai, S. Lin, and Q. Zhang, "Improving image encryption using multi-chaotic map," in *Proceedings of the Workshop on Power Electronics and Intelligent Transportation System (PEITS '08)*, vol. 10, pp. 143–148, August 2008.

[21] X. Zhang and W. Chen, "A new chaotic algorithm for image encryption," in *Proceedings of the International Conference on Audio, Language and Image Processing (ICALIP '08)*, pp. 889–892, July 2008.

[22] L. Wang, Q. Ye, Y. Xiao, Y. Zou, and B. Zhang, "An image encryption scheme based on cross chaotic map," in *Proceedings of the 1st International Congress on Image and Signal Processing (CISP '08)*, vol. 3, pp. 22–26, May 2008.

[23] J. Peng, D. Zhang, and X. Liao, "A digital image encryption algorithm based on hyper-chaotic cellular neural network," *Fundamenta Informaticae*, vol. 90, no. 3, pp. 269–282, 2009.

[24] C. Çokal and E. Solak, "Cryptanalysis of a chaos-based image encryption algorithm," *Physics Letters A*, vol. 373, no. 15, pp. 1357–1360, 2009.

[25] J. D. Watson and F. H. C. Crick, "Molecular structure of nucleic acids: a structure for deoxyribose nucleic acid," *Nature*, vol. 171, no. 4356, pp. 737–738, 1953.

[26] W. C. Chen, Z. Y. Chen, Z. H. Chen et al., "Operational rules of the digital coding of DNA sequences in high dimension space," *Acta Biophysica Sinica*, vol. 17, no. 3, pp. 542–549, 2001.

5

A Comparative Analysis of Burned Area Datasets in Canadian Boreal Forest in 2000

Laia Núñez-Casillas,[1] José Rafael García Lázaro,[2] José Andrés Moreno-Ruiz,[2] and Manuel Arbelo[1]

[1] *Grupo de Observación de la Tierra y la Atmósfera (GOTA), Universidad de la Laguna, 38200 San Cristóbal de la Laguna, Spain*
[2] *Departamento de Informática, Universidad de Almería, 04120 Almería, Spain*

Correspondence should be addressed to Laia Núñez-Casillas; lnunez@ull.es

Academic Editors: A. E. Cetin, I. Korpeoglu, and S. Verstockt

The turn of the new millennium was accompanied by a particularly diverse group of burned area datasets from different sensors in the Canadian boreal forests, brought together in a year of low global fire activity. This paper provides an assessment of spatial and temporal accuracy, by means of a fire-by-fire comparison of the following: two burned area datasets obtained from SPOT-VEGETATION (VGT) imagery, a MODIS Collection 5 burned area dataset, and three different datasets obtained from NOAA-AVHRR. Results showed that burned area data from MODIS provided accurate dates of burn but great omission error, partially caused by calibration problems. One of the VGT-derived datasets (L3JRC) represented the largest number of fire sites in spite of its great overall underestimation, whereas the GBA2000 dataset achieved the best burned area quantification, both showing delayed and very variable fire timing. Spatial accuracy was comparable between the 5 km and the 1 km AVHRR-derived datasets but was remarkably lower in the 8 km dataset leading, us to conclude that at higher spatial resolutions, temporal accuracy was lower. The probable methodological and contextual causes of these differences were analyzed in detail.

1. Introduction

Different research centers and institutions have worked on burned area (BA) mapping on regional and global scales. However, there are still technical limitations linked to the remote sensing imagery, classification methods, and classification challenges based on differing environments. The spectral, spatial, and temporal resolutions of the remote sensing data; the trade-off between accuracy and adaptability of BA detection algorithms; and different vegetation dynamics and postfire recovery patterns are some of the factors that may result in varying levels of efficiency of a given methodology. Some periods of time were particularly unfavorable to BA detection for different reasons, affecting the continuity of BA time series datasets and highlighting the strengths and weaknesses of BA detection methods at the same time. For instance, the year 2000 was a year of low global fire activity [1], and fires in Canada did not produce extremely large burn scars to the large boreal forest belt compared to other years. Instead, fires tended to affect several areas of moderate expanse. Furthermore, during the year 2000 NASA's Moderate-Resolution Imaging spectroradiometer (MODIS) and the Advanced Very High Resolution (AVHRR) instruments produced some unreliable or insufficient information. On the other hand, the year 2000 was considered a turning point for coarse to moderate-resolution datasets in BA mapping, with the advent of 500 m MODIS data after the first two years of 1 km imagery from the VEGETATION (VGT) sensor onboard SPOT-4 satellite. This combination of facts makes it interesting to assess the accuracy of the BA data available for this particular year in the North-American boreal forest, whereas a comparative study would help in analyzing different methodologies during a period of transition to finer spatial resolution imagery.

NASA's projects of the Pathfinder 8 km Land dataset (PAL) and the subsequent Land Long Term Data Record (LTDR) were two significant attempts at achieving a consistent coarse-resolution imagery time series on a global scale from 4 km AVHRR data. The PAL time series from 1984 onwards has been used in several works related to BA estimation [2, 3]. The recently released 5 km BA dataset for the Canadian boreal forests from 1984 to 1999 from LTDR data [4] showed a spatial accuracy comparable to the 1 km BA product from AVHRR-Local Area Coverage (LAC) data published by Chuvieco et al. in 2008 [5] for the years 1984–2006. With the latest version of LTDR (which is still not available for the years after 2000) it will be possible to expand the time series until the present day.

Several global initiatives have coordinated the work done in different countries in order to provide world BA maps from other sensors. In 2001, the GLOBSCAR project launched by the European Spatial Agency (ESA) produced monthly 1 km BA maps from the year 2000 onwards by analyzing data from the moderate-resolution sensor Along Track Scanning Radiometer-2 (ATSR-2) [10]. The following year, the Joint Research Centre (JRC) and other research centers from different countries launched the Global Burned Area 2000 (GBA2000) initiative, a BA map of global coverage from SPOT-VGT data [11]. In reality, however, both projects were in fact limited to the year 2000 and remarkable discrepancies were found between them [1]. The GBA2000 dataset progressively evolved into a new product named L3JRC which would provide a wider temporal coverage until 2007 [8]. Soon after, ESA's GLOBCARBON project integrated efforts from different workstations and the acquired knowledge from GLOBSCAR to provide BA datasets from 2000 to 2007, generated from the same ATSR-2, AATSR, and MERIS sensors on the Envisat platform in addition to the VGT instrument [12]. However, the project ended in 2006 and BA data is no longer available. Finally, the MODIS Collection 5 BA product (MCD45A1) was to become the first global BA data provided from satellite imagery at 500 m spatial resolution and an approximate date of burn in Julian days from the year 2000 [13]. Nevertheless, the MODIS sensor is known to have presented calibration problems during its first year, suggesting that an accuracy assessment would be useful as well.

An actual comparative analysis of spatial and temporal accuracy from the diverse BA products available for the year 2000 in the Canadian boreal forest has not yet been conducted. Pros and cons of each dataset will be strongly related to inherent features of the corresponding remote sensing imagery, as well as to the particular approaches chosen to solve BA detection challenges and to the regional differences of forest ecology and fire dynamics. From these observations the questions of where, when, and how efficiently each dataset detects BA in the boreal forest of Canada are discussed. Firstly, some ecological patterns of the Canadian boreal forest that might influence BA detection are described, as well as the different methodologies that have been used to detect BA in each analyzed dataset. Secondly, a spatial accuracy assessment in a global, fire-by-fire, and regional approach is presented. Finally, the burn date estimates are considered by means of a spatiotemporal analysis and an overall accuracy assessment. The spatiotemporal accuracy patterns found in each dataset, as well as their probable causes, were summarized in five sections: the relationship between fire size and imagery spatial resolution, the different combinations of remote sensing data depending on each BA detection method, use of fixed and dynamic radiometric thresholds, geographic patterns that might be linked to ecological factors; and considerations regarding the temporal shifts and variability.

2. Canadian Boreal Forest

Coniferous species are dominant in the North-American Boreal forests and they vary with latitude and from east to west. Moisture and temperature are general constraints that determine the stand density and species composition. In the highest latitudes, there is a transition zone between the boreal forest and tundra biome. Predominant tree species are black spruce (*Picea mariana* (Mill.) BSP), paper birch (*Betula papyrifera* Marsh.), trembling aspen (*Populus tremuloides* Michx.), balsam fir (*Abies balsamea* (L.)), jack pine (*Pinus banksiana* Lamb.), white spruce (*Picea glauca* (Moench) Voss.), and tamarack (*Larix laricina* (Du Roi) K. Koch). Black spruce, white birch, and aspen are found across the entire boreal forest. Northward, black spruce forms open forests [14], balsam fir is present to the east, lodgepole pine (*Pinus contorta* Dougl. ex. Loud.) is widely distributed in the west, and jack pine is found on both sides [15]. Fire plays an important role in the landscape and has different effects depending on the region by favoring the development of different species after disturbance [16]. Highly energetic crown fires and species with regeneration capacity are the most frequent combination [17]. However, surface fires of medium intensity may occur where trees are sparse or crowns are high above the ground [18].

3. Burned Area Data

Active fire detection is based on changes in the thermal bands, whereas BA mapping usually depends on the availability of radiometric data in the green (0.5-0.6 μm) and red (0.6-0.7 μm) bands plus the NIR (0.75–0.9 μm) and/or SWIR (1.5–2.0 μm) bands. BA mapping algorithms take into account the differences in these spectral regions between prefire and postfire situations, either through abrupt changes within a sliding temporal window, between discrete units of time, between dates before and after the end of the fire season, or the same dates in different years. The datasets assessed were as follows: the Riaño et al. [2] algorithm applied to PAL data; the Moreno-Ruiz et al. [4] algorithm applied to LTDR v.3 data, the Chuvieco et al. [5] BA dataset from LAC data, the GBA2000 and L3JRC datasets from SPOT-VGT, and the MCD45A1 dataset from MODIS (Table 1). The first three will hereafter be referred to as PAL.ba, LTDR.ba, and LAC.ba, respectively.

3.1. PAL.ba from the Riaño et al. Algorithm (8 km). Riaño et al. [2] used 7-day composites from the daily PAL dataset [19, 20]. The BA algorithm (BAA) used three different types of threshold in order to detect changes along the time series: a temporal threshold checked a simultaneous decrease of vegetation indices VI3T [21] and GEMI [22] indices and

TABLE 1: Original datasets features grouped by sensor. Spectral bands in bold are involved in the corresponding burn scar detection algorithms.

Dataset	Spatial		Temporal		Spectral		
	Coverage	Resolution	Coverage (BA time series)	Resolution (Date of burn)	Spectral bands used (2000)*	Compositing criteria	BA detection
AVHRR							
BA-PAL	Global	0.1° (~8 km)	1984–2000	Monthly	**ρ1 (~0.63 μm)** **ρ2 (~0.83 μm)** **T3 (~3.75 μm)** T4 (~10.8 μm) T5 (~12.0 μm)	7-day composites; minimum albedo	Multitemporal and multithreshold BA Algorithm (BAA) [2]
BA-LTDR	Global	0.05° (~5 km)	1984–2000	10 days	**ρ1 (~0.63 μm)** **ρ2 (~0.83 μm)** ρ3 (~1.61 μm) **T3 (~3.75 μm)** T4 (~10.8 μm) T5 (~12.0 μm)	10-day composites; maximum T3	Bayesian network classification [4]**
BA-LAC	Canada, northern USA, Alaska, and Greenland	0.01° (~1 km)	1984–2006	Daily	**ρ1 (~0.63 μm)** **ρ2 (~0.83 μm)** T3 (~3.75 μm) T4 (~10.8 μm) T5 (~12.0 μm)	10-day composites; SPARC algorithm [6]	Multitemporal and multithreshold contextual algorithm [5]**
VEGETATION							
GBA2000	Global	0.01° (~1 km)	2000	Monthly	B1 (~0.45 μm) **B2 (~0.66 μm)** **B3 (~0.83 μm)** **B4 (~1.65 μm)**	10-day composites; maximum NDVI	Multiple logistic regression model [7]**
L3JRC	Global	0.01° (~1 km)	2000–2007	Daily	B1 (~0.45 μm) **B3 (~0.83 μm)** B4 (~1.65 μm)	None	Comparison of daily values and intermediate composites in NIR [8]
MODIS							
MCD45A1	Global	(~500 m)	2000–present	Daily	1 (~0.65 μm) **2 (~0.86 μm)** 3 (~0.47 μm) 4 (~0.56 μm) **5 (~1.24 μm)** **6 (~1.64 μm)** **7 (~2.13 μm)**	None	Bidirectional Reflectance Distribution Function (BRDF) [9]

*Bands in bold were used in change detection.
**BA detection was optimized for the Canadian boreal region.

an increase in AVHRR channel 3 brightness temperature; a fixed threshold was adjusted based on the knowledge of the land cover, separability tests, and postfire conditions; and a third type of threshold was automatically set by statistical calculations from a vegetation index distribution. Monthly burn scar maps with a spatial resolution of 8 km were then created.

3.2. LTDR.ba from the Moreno-Ruiz et al. Algorithm (5 km). The Moreno-Ruiz et al. [4] algorithm was applied to 10-day composites from LTDR version 3 daily data [23, http://ltdr.nascom.nasa.gov/ltdr/ltdr.html]. The algorithm combined radiometric thresholds, a Bayesian network classifier [24–26], and a postclassification analysis of neighborhood conditions. The algorithm identified the date of fire as the first step in calculating a set of prefire and postfire statistics from several vegetation indices and spectral bands and then estimated the probability of a pixel to have been burned. The last step accounted for neighborhood conditions in every burned pixel in order to improve spatial coherence and to avoid false detection. A 5 km BA dataset was generated with a temporal resolution of 10 days.

3.3. LAC.ba from the Chuvieco et al. Algorithm (1.1 km). The Chuvieco et al. [5] algorithm was applied to daily AVHRR-LAC data from Latifovic et al. [27]. Spectral bands and derived indices from 10-day composites [6] were fed into the algorithm, a multivariable thresholding approach adapted to Canadian territory. Different thresholds accounted for considerable prefire activity of vegetation; sharp and persistent changes before/after burn; higher temperatures and lower reflectance (bands 1 and 2) after fire; separability with respect to water, dark soils, and so forth. Postprocessing steps eliminated small polygons and applied more relaxed thresholds to pixels surrounding burned detections.

3.4. GBA2000 (1.1 km). This regional algorithm used daily SPOT-VGT-S1 data in Canada [28, 29]. Training areas were obtained by applying the Li et al. [18] boreal active fire detection algorithm to daily AVHRR and a postfire season SPOT-VGT image. BA detection was based on a multiple logistic regression analysis that used a set of variables to explain a dichotomous independent variable with a value of 1 (burned) or 0 (unburned). The predictor variables were chosen from 10-day and 30-day changes in SPOT-VGT red, NIR, and SWIR bands; NDVI; and a SWIR-based vegetation index (SWVI) [7]. Each change was previously normalized with respect to similar background vegetation (in order to distinguish changes due to seasonality of phenological cycle) and was differentiated according to ecozones.

3.5. L3JRC (1 km). Modifications from the GBA2000 method were developed by Ershov and colleagues [29, 30]. A daily pixel contamination product was created accounting for snow, clouds, shadows, satellite view angle, and fire smoke. Abrupt change detection was analyzed by iterative comparisons of daily NIR (0.83 μm) reflectance values to an aggregation of historical NIR values, according to a temporal index [8]. Tests were applied to the eligible pixels based on statistics of the temporal index within a 200 × 200 spatial window of valid pixels and on fixed thresholds at 0.83 μm and 1.66 μm. False detections for that year were amended by comparison with the Global Land Cover 2000 database [31]. The resulting product was a 1 km daily BA dataset.

3.6. MCD45A1 (500 m). Imagery from the year 2000 shows striping in bands 5, 6, and 2, which constitute some of the key data in burned area discrimination (http://modis-fire.umd.edu/BA_methodology.html; http://mcst.gsfc.nasa.gov/index.php?section=15). The BRDF (Bi-directional Reflectance Distribution Function) [9] is the basis of change detection here, whereas Z-score calculations account for the persistence of change. Burned cell classification is based on temporal changes in daily surface reflectance of NIR and MIR bands. An iterative process predicts reflectance values for each day based on a number of previous days within a temporal window, and the error-weighted difference between predicted and observed values acts as a sign of change. A temporal window of ±8 days establishes the detection precision of burn date estimates, in order to compensate for the effects of missing data [13].

3.7. Canadian Forest Service National Fire Database (CFSNFD). All of this data was compared to the CFSNFD [32]. Only fires larger than 5,000 ha occurring in the year 2000 were studied. Burn scars in this database are mapped by different Canadian fire management agencies using a variety of means including remote sensing, helicopter GPS, ground-based GPS, aerial photography, and aerial sketching. However, the mapping detail in the CFSNFD should be comparable to the detail that would be expected from coarse to moderate-resolution remote sensing data sources, as performed in this study, and has the advantage of providing burn dates that do not depend exclusively on the availability of satellite data on the date the fire started. CFSNFD polygons typically map the outer fire perimeters and do not remove unburned islands and lakes. In this regard, fires selected from the CFSNFD in this study were checked against postfire Landsat imagery in order to visually confirm that no inner lakes or green islands of significant size were included. We used updated BA information from the CFSNFD point version (last released in August 2010), where initial BA estimates were improved with respect to initial fire perimeter delineation.

4. Accuracy Assessment Methods

Selected BA fires from the CFSNFD were identified in every dataset assessed. The test sites were numbered eastwards from west to east (1 to 20). Sites named as #1.1, #2.1, #2.2, and #13.2 did not appear in the polygon version of the database and were added from the point version. Fire-by-fire comparisons of BA estimates allowed the detection of four regional patterns, which were analyzed as well: West coast, North-West, mid-West, and East coast (Figure 1).

4.1. Spatial Analysis. First spatial accuracy assessment was performed on a grid of 40 km squared cells using the Lambert Conformal Conic (LCC) projection, in order to minimize distortions and to keep the shape of the squares equal [5]. BA was calculated for every 40 km cell in the reference map and was contrasted with the same calculations in each of the other BA maps by means of a simple regression model [33].

The remaining procedures were implemented in a fire-by-fire analysis approach and based on relative analyses rather than absolute BA detections. The different spatial coverage of the LAC.ba determined the repetition of all the analyses for both the entire Canada and the central spatial subset corresponding to the LAC data frame. First, the relative error was calculated for every fire by comparison against the same fire scene in the reference map, as follows:

$$\text{RelativeError}_i = \frac{(\text{ClassifiedBurnedArea}_i - \text{BurnedArea}_i)}{\text{BurnedArea}_i}. \quad (1)$$

Then, for the relative error distribution, we calculated the weighted means and standard deviations, where each scene's weight was considered as a proportion of the BA of the scene with respect to the sum of all scenes:

$$\mu = \sum \frac{\text{RelativeError}_i * \text{BurnedArea}_i}{\text{TotalBurnedArea}},$$
$$\sigma = \sqrt{\sum \frac{(\text{RelativeError}_i - \mu)^2 * \text{BurnedArea}_i}{\text{TotalBurnedArea}}}. \quad (2)$$

The error matrix (P_{ij}) [34] and its derived indices (commission and omission errors) were computed at scene level, so as to prevent georeference-derived errors. Considering a dichotomous classification of classes, C_1 (burned) and C_2 (unburned), each element of the error matrix was obtained from the expression

$$P_{ij} = \frac{\sum_{k=1}^{N} (p_{ij})_k}{N}, \quad (3)$$

with $(p_{ij})_k$ representing the proportion of the BA in the zone k assigned to the class C_i which actually belongs to class C_j. To compute the error matrix elements we compared the classified percentage of BA (c_1) and the reference percentage (r_1) and assigned the least value as the percentage of the correctly classified (p_{11}) class [35]. The absolute value of the difference is distributed, either in the percentage of the burned class classified incorrectly (p_{21}) or in the percentage of the nonburned class classified as burned (p_{12}), by means of a simple algorithm:

IF ($c_1 \geq r_1$)

THEN

$p_{11} = r_1$

$p_{12} = c_1 - r_1$

$p_{21} = 0$

$p_{22} = c_2$

ELSE

$p_{11} = c_1$

$p_{12} = 0$

$p_{21} = r_1 - c_1$

$p_{22} = r_2$

END_IF.

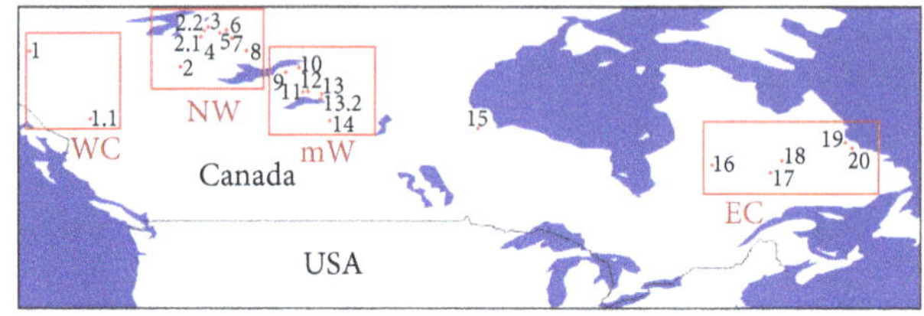

FIGURE 1: Study area: test sites location and distribution of regional patterns in the entire Canada. The LAC region (upper left corner: 62°16′N, −122°20′E; bottom right corner: 53°N, −96°30′E) contains fires #2 and #8 to #14. Note: fire #11 includes two adjacent fires which were registered separately in the database. Sites were grouped into four geographic regions: West Coast (WC), North-West (NW), mid-West (mW), and East Coast (EC).

Thus, the derived indices of interest from the error matrix [36] were computed:

$$\begin{aligned}
&\text{user's accuracy of class } C_1 = \frac{P_{11}}{P_{1+}}, \text{ where} \\
&P_{1+} = P_{11} + P_{12}, \\
&\text{producer's accuracy of class } C_1 = \frac{P_{11}}{P_{+1}}, \text{ where} \\
&P_{+1} = P_{11} + P_{21}, \\
&\text{commission error of class } C_1 \\
&\quad = 1 - \text{user's accuracy of class } C_1, \\
&\text{omission error of class } C_1 \\
&\quad = 1 - \text{producer's accuracy of class } C_1.
\end{aligned} \quad (4)$$

4.2. Timing Analysis. The temporal distribution of estimated BA of each dataset was represented in a timing histogram obtained from all of the burned cells in each dataset. As temporal resolution varies between different datasets, both monthly and 10-day-composite distributions were built getting the best possible level of detail in each case for the temporal analysis. Burn dates were missing in some of the eastern sites in the reference data, so the 5th percentile from all other products was used for the time distribution. For the largest sites we assumed that the entire area could not have burned on the starting date of the fire, so a proportional amount of the BA estimation was shifted to the following time period.

5. Results

5.1. Spatial Accuracy Assessment. In this study the area burned in the boreal forests of Canada was underestimated by all the remote sensing products when compared to the CFS-NFD for the same year, 2000. The coefficient of determination from the regression model was under 0.6 in all cases, as shown in the scatter plots depicted in Figure 2. The best spatial correlation was found in GBA2000; LAC.ba and LTDR.ba showed very similar behavior in the common areas; and the remaining datasets presented lower spatial correlation, which

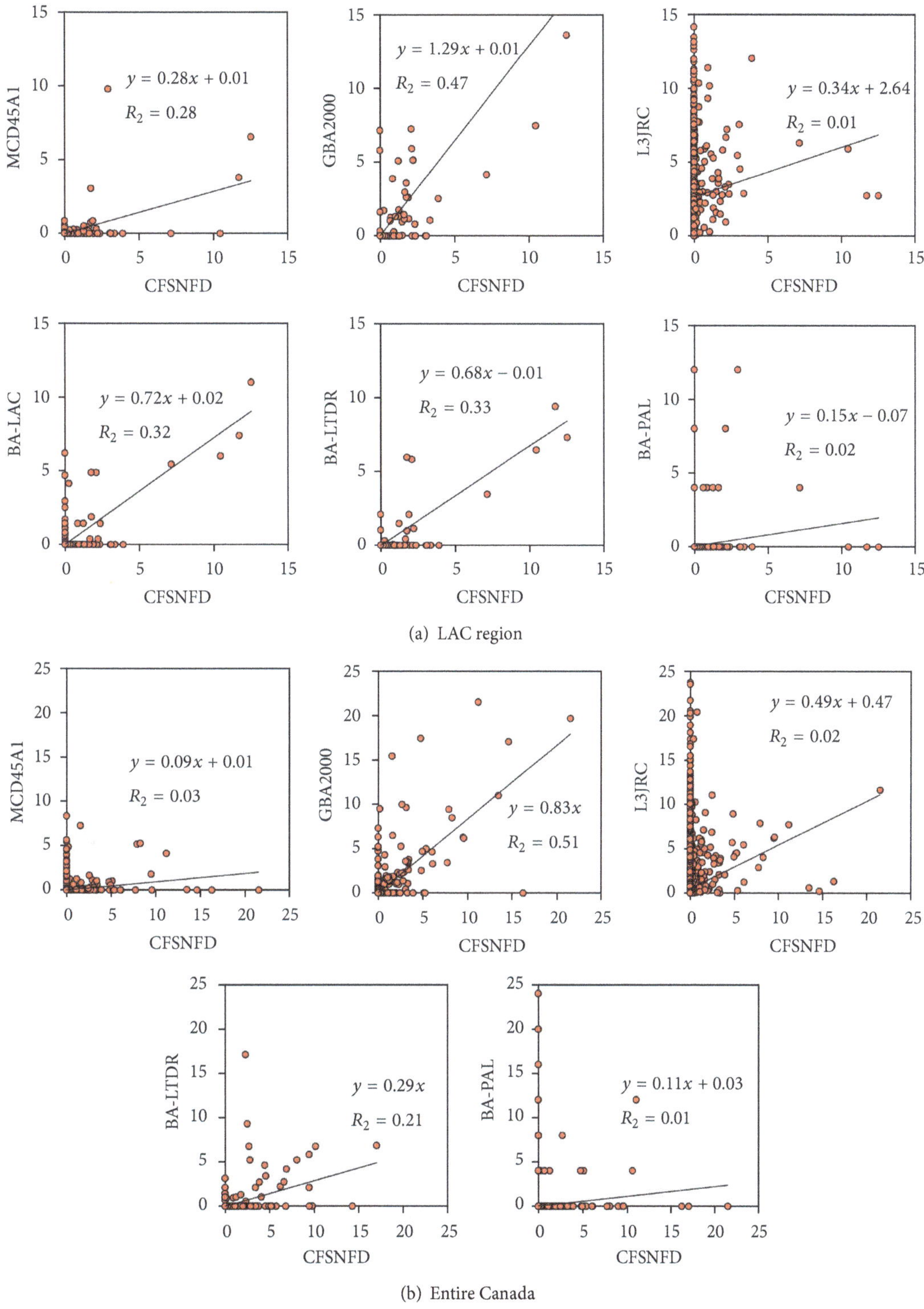

FIGURE 2: Linear regression of BA proportion (%) in 40 km^2 from MCD45A1, GBA2000, BA-LAC, BA-LTDR and BA-PAL against BA proportion (%) in CFSNFD for the year 2000: (a) in the LAC region; (b) in Canada.

normally improved in the central spatial subset with respect to the whole territory, except for GBA2000 by overestimation and L3JRC by underestimation.

Relative error distributions of BA estimations in the fire-by-fire analysis are specified in Table 2, by their corresponding weighted mean (μ) and standard deviation (σ) values. The smallest μ was found in GBA2000 (−16%, the negative sign meaning underrating) and the greatest was found in PAL.ba and MCD45A1 (<−75%), the two extremes in spatial resolution. In general, the lower the μ, the higher the σ, which in GBA2000 was remarkably higher than the others (around 40%).

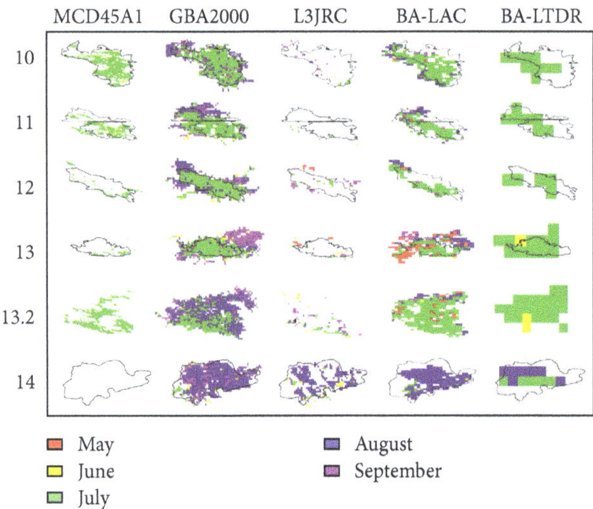

FIGURE 3: Spatiotemporal discrepancies among BA datasets (except for BA-PAL) in burn scar mapping in the mid-West region: monthly timing depiction. Pixel sizes keep the original proportion among different spatial resolutions, and color coding reflects monthly timing. Fire contours from CFSNFD polygon version are added only as means of guidance except for #13.2 where it did not exist.

Table 3 shows BA estimations for every fire, and the regional patterns are shown in Table 4. In general, fires below 8,000 ha were hardly detected by any AVHRR-derived datasets. However, the latter together with GBA2000 usually gave the best estimates of large burned areas. On the contrary, L3JRC represented small fires proportionally better than large fires. Some moderate size fires were captured only by one of the algorithms, depending on the region. Overestimation in GBA2000 seemed to be more important in small central fires; however, the two VGT-derived datasets provided the best overall representation in terms of the number of fires detected.

Burn detections from the mid-West region in all datasets except for PAL.ba are depicted scene by scene on a monthly basis in Figure 3, where some spatiotemporal patterns may now be observed. Burn date variability within the same fire tended to be higher in VGT-derived data and LAC.ba, and it was mostly associated to burn date delay or temporal range extension. Moreover, L3JRC presented simultaneously high temporal and spatial variability, with a considerable proportion of isolated cells or groups of cells. On the contrary, MCD45A1 and LTDR.ba maintained a rather homogeneous and consistent spatiotemporal distribution for most scenes.

Some other spatiotemporal patterns were found in the rest of the regions: (a) northern fires were well represented in MCD45A1 and GBA2000 but were delayed by up to one month in the latter; (b) West coast fires—only estimated by the VGT-derived datasets—showed highly variable burn date estimates compared to other regions; and (c) heterogeneity of timing estimates in a site varied considerably in LTDR.ba between different regions (higher heterogeneity in mid-West and eastern sites).

TABLE 2: Mean and standard deviation of relative errors in BA estimations in all BA estimates for the year 2000, for Canada and the LAC region (fires larger than 5,000 ha).

	Entire Canada		Local Coverage subset	
	μ	σ	μ	σ
MCD45A1	−82.66	21.51	−75.73	22.4
GBA2000	−16.3	41.57	3.92	38.49
L3JRC	−79.41	20.43	−78.47	18.00
BA-LAC	—	—	−34.04	28.14
BA-LTDR	−58.94	34.2	−35.48	28.54
BA-PAL	−86.89	17.67	−85.6	15.41

5.2. Temporal Accuracy Assessment. Monthly distributions of BA percentages were depicted in Figures 4(a) and 4(b). GBA2000 presented a monthly pattern almost matching that of the CFSNFD, except for a +1 month temporal shift. Both of the VGT-derived datasets showed a delayed maximum in August, although for the rest the maximum BA percentage was registered in July of that year. However, the most conspicuous temporal shifts were found in the LAC.ba dataset, with +1 month in 30% of the BA at the end of the fire season and −1 month in 10% during May.

With a finer temporal resolution (Figure 4(c)), BA was mainly distributed along composites 18 (June), 19, and 21 (July) according to the CFSNFD, with a major peak at the end of June and a secondary peak at the end of July. Temporal shifts range now from +1 composite—the two maxima in LTDR.ba and only the first one in MCD45A1—and up to +3 composites in LAC.ba. The time range was moderately extended in LTDR.ba and much more in LAC.ba, where nearly 40% of the area burned in the central subset was located between the 24th and 25th composites; in contrast, the MCD45A1 time range narrowed with respect to the reference data.

6. Discussion

Comparative fire-by-fire BA estimations shown in Table 3 provide an idea of accuracy in terms of BA quantification rather than actual spatial accuracy, for which validation with 30 m Landsat TM/ETM+ imagery or other moderate-resolution data might be preferable rather than the CFSNFD. Although both concepts (sensitivity and spatial accuracy) should correlate, the first one allowed us to preclude errors caused by geospatial processing, such as georeferencing issues or border effects linked to large pixels.

6.1. Patch Size versus Pixel Size. According to the results shown in Tables 3 and 4, PAL.ba should never be used for BA monitoring when fires are smaller than 16,000 ha. In larger areas, underestimation ranges from −56% to −100%, and sensitivity decreases in large fires in the mid-West, indicating that there are other factors involved apart from patch size. The use of three datasets obtained from the same sensor with different pixel sizes allows us to weight the influence of the spatial resolution in BA detection. Starting with

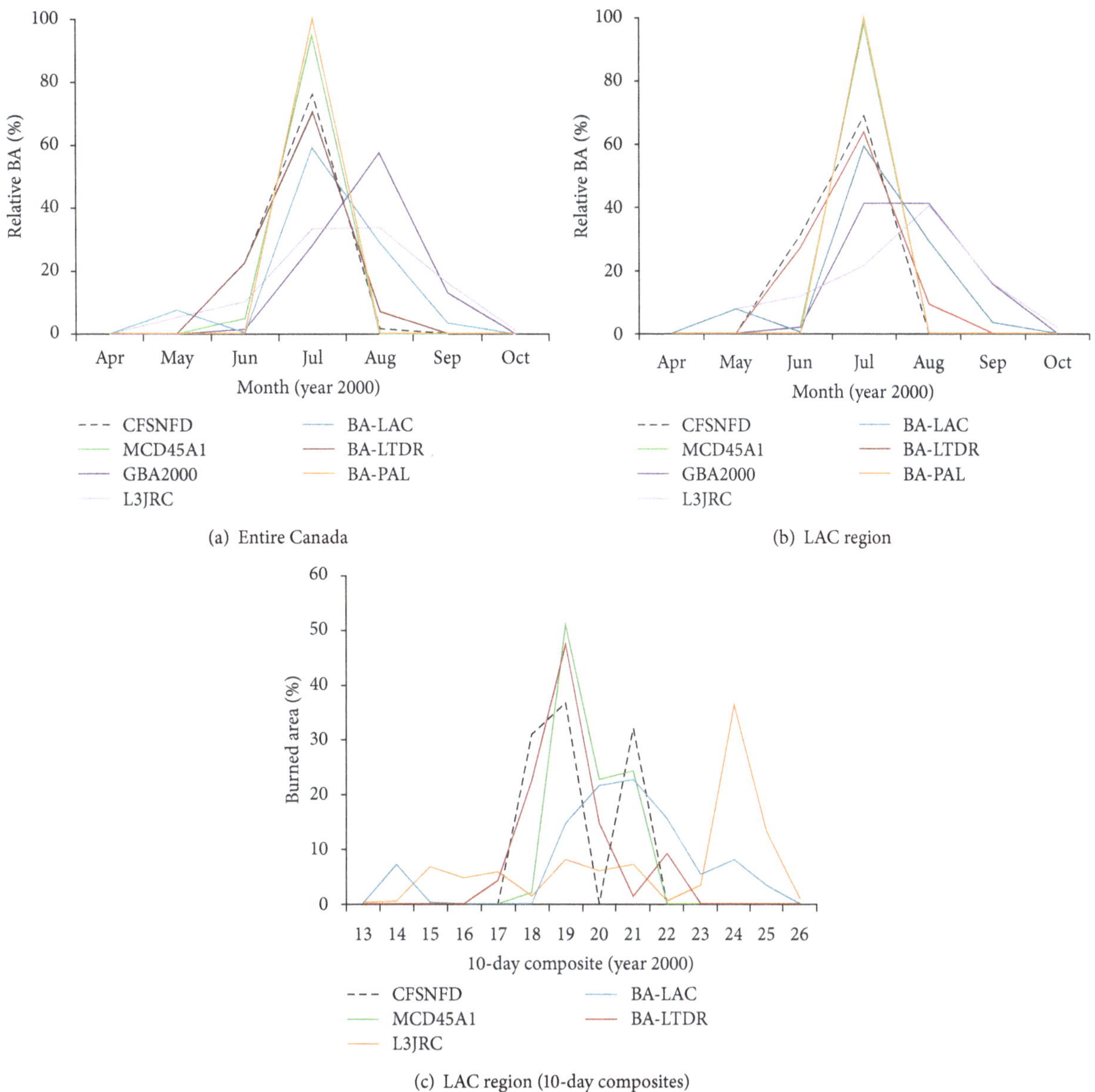

FIGURE 4: Temporal evolution of BA of Canadian fires greater than 5,000 ha during the year 2000, relative to total BA estimated by each dataset: (a) monthly, on global scale; (b) monthly, in the LAC; (c) every 10 days in the LAC for BA datasets when the temporal resolution allows it.

the lowest spatial resolution, the 8 km PAL.ba dataset showed an underestimation greater than −85%, in accordance with Carmona-Moreno et al. [3], who pointed out that the BA omissions for boreal forest in higher latitudes were deemed to be greater than 80%. Next, the Bayesian network algorithm used in the 5 km LTDR.ba dataset increased the total amount of BA, with an underestimation of below −60%. Three important factors contributed to these improvements are (a) improved spatial resolution, (b) improved data quality of LTDR data (calibration, geolocation, and BRDF corrections), and (c) a new algorithm used. However, the BA overall quantification does not improve much from this dataset to the 1 km LAC.ba data, in the central spatial subset, although the spatial resolution is five times more detailed in the latter.

On the other hand, MCD45A1 has the finest spatial resolution (500 m) and yet showed substantial underestimation (more than 80%) with no detections in half of the fires larger than 5,000 ha, leading us to conclude that the (fire size) : (pixel size) ratio is not necessarily the most limiting factor in BA detection. Moreover, during the same year in Southern Africa the MCD45A1 dataset only omitted 25% of the area burned [33], which is considerably better result than that in North-American boreal forest, pointing to a variable efficiency of the product in different environments.

6.2. Multitemporal, Multisensor, and Multispectral Approaches. Calibration problems and different operational events occurred during this year which caused the loss of a certain

TABLE 3: Comparative BA estimates in fires larger than 5,000 ha that occurred in the year 2000, for all datasets in a Lon/Lat WGS84 reference system.

Scene ID	BA estimations (ha)						
	CFSNFD	MCD45A1	GBA2000	L3JRC	BA-LAC	BA-LTDR	BA-PAL
LAC region							
2	5 450	0	1 903	1,485	0	0	0
8	6 500	0	4 523	4,744	0	0	0
9	7 693	0	0	802	0	0	0
10	20 040	11 116	24 761	1,391	16 928	13 107	0
11	16 452	6 432	22 340	496	11 353	10 904	6 000
12	9 249	1 737	18 394	2,225	6 651	7 765	0
13	19 154	1 436	24 112	1,494	18 764	18 776	6 026
13.2	43 701	20 081	45 023	6,044	36 391	37 943	12 181
14	39 850	0	33 628	17,502	20 790	19 963	0
Subtotal	168 089	40 802	174 684	36,184	110 877	108 458	24 207
1	5 668	0	5 751	0		0	0
1.1	9 000	0	2 501	266		0	0
2.1	6 800	636	0	2,287		0	0
2.2	5 800	272	0	0		0	0
3	6 500	1 762	0	2,342		2 578	0
4	15 600	2 805	10 633	6,973		0	0
5	7 500	3 588	5 959	839		0	0
6	8 000	1 044	6 381	4,930		0	0
7	15 500	8 898	15 590	8,155		5 333	0
15	5 182	0	0	0		0	0
16	13 288	0	13 598	59		9 194	0
17	7 483	0	0	298		0	0
18	48 079	70	39 977	872		9 118	21 070
19	10 125	0	5 197	1,285		0	0
20	12 636	0	8 704	6,612		7 093	0
Total	339 582	59 877	288 975	71,102	110 877	141 774	45 277

TABLE 4: Proportion of BA estimates with respect to BA registered in CFSNFD, presenting different detectability patterns among BA products related to the location of fires within the Canadian territory.

	Regional BA estimates with respect to CFSNFD (%)				
	West coast (WC)	North-West (NW)	Mid-West (mW)	Central Canada (cC)	East coast (EC)
MCD45A1	0	24	26	0	<1
GBA2000	56	58	108*	0	74
L3JRC	1	20	41	0	10
BA-LAC	—	—	71	—	—
BA-LTDR	0	10	69	0	28
BA-PAL	0	0	16	0	23

*GBA2000 overestimation of total BA in mW.

quantity of data available from this sensor (http://mcst.gsfc.nasa.gov/index.php?section=15) and may explain to some extent the overall BA underestimation of the MCD45A1 product. Nevertheless, several radiometric constraints are combined with temporal constraints (the temporal window of ±8 days used in change detection) and with the circumstantial lack of MODIS data that year. Pixel quality requirements might be too restrictive in this case, causing great omission of burned cells. Otherwise, the multitemporal 500 m approach (where no data compositing is needed and every valid daily data file is used in order to compute presence and persistence of sharp radiometric changes) might have been more accurate, spatially and temporally. The temporal window is longer in other datasets: GBA2000 carries out 10- to 30-day change observations, and LTDR.ba checks up to ±60-day and even pre- and postfire years observations.

As well as the differences in the use of multitemporal data for change detection, all BA algorithms studied in this work except for MCD45A1, use spectral indices in addition to band reflectances, as they are known to reduce noise and sharpen both vegetation activity and BA signals. Most relevant indices used in LTDR.ba were GEMI, NDVI, and Vi3T as vegetation indices using the NIR and red regions, whereas the BBFI (Burned Boreal Forest Index) added the MIR region for better detection of bare soil caused by fire. However, several fires of differing sizes were not represented by any of the AVHRR-derived datasets. This probably reflects some advantage of the spectral data used in the VGT- and MODIS-derived datasets over the others. One aspect that these three products do have in common is the use of a SWIR band centered at ~1.65 μm, from VGT channel 4 and MODIS channel 6, respectively. This region of the spectrum would approximately match the AVHRR channel 3A reflectance, which was not included in either of the BA algorithms working with this sensor. Moreover, the SWIR spectral regions were used in MCD45A1 span to MODIS bands 5 (1.24 μm) and 7 (2.13 μm), eventually increasing the probability of discrimination between different BA patterns and vegetation covers (e.g., in presence of forest discontinuities or less foliage density). In contrast, the AVHRR-derived datasets used the Red, the NIR, and—except for LAC.ba—the MWIR spectral bands; the latter is centered at 3.75 μm, corresponding to the thermal AVHRR channel 3B emissivity.

On the other hand, GBA2000 somehow included both SWIR and MWIR spectral information, as it used the AVHRR active fire product (based on emissivity rather than reflectance) in addition to the VGT postfire season data. It consists of a different approach to obtain more spectral information, by means of two remote sensing sources and combining the advantages of both sensors in the detection of fire-related processes. However, it embraces pros and cons of both sources, which could be a reason why GBA2000 overestimated BA and biased the date of burn of several fires.

6.3. Thresholding Concept. As shown previously, PAL.ba presented the least variable total error and never detected burn scars smaller than 16,000 ha, making it the most size restricted and predictable dataset at the same time. Exceptionally, two large fires were not detected by PAL.ba (#10 and #14), which might be more related to the risk associated with the thresholding approach, rather than to its coarse spatial resolution. Fixed radiometric thresholds are usually quite tolerant in order to exclude land covers that are very different to the target feature; as such, dynamic thresholds are able to refine the selections (e.g., decrease of GEMI after fire). Although fixed and even dynamic thresholds may be adapted to the study area (e.g., North-American boreal forest), once established they may not actually adapt to the peculiarities of different sites. In several cases, undetected fires were geographically close to others that were detected, meaning that the same threshold criteria did not operate in the same way within the same area. Thus, in absence of other discriminating mechanisms (such as the Bayesian network in LTDR.ba or neighborhood tests in LTDR.ba and L3JRC), the algorithm may fail by underestimation when thresholds are too restrictive, or by overestimation in the opposite case. For example, LTDR.ba with a relatively similar spatial resolution showed more adaptability than the previous PAL.ba, by detecting seven additional regions (including sites #10 and #14) and by reducing underestimation in practically all of them. In this case, the Bayesian network could be improved by using training areas taken from data of the year analyzed, instead of using the same model for the time series as described by Moreno-Ruiz et al. [4]. However the more we increase the adaptability to a particular year, the lower its capacity to adapt to different years in the same area of study may be.

6.4. Geographic Variations Linked to Ecological Factors. Moving on now to analyze the ecological factors causing the geographic variability of BA detection, there seems to be a latitudinal variation of BA detection efficiency: mid-West fires are mostly detected by all the algorithms whereas North-West fires are hardly detected by a handful of them. This pattern coincides with the ecological changes from boreal forest to tundra in the transition belt between both ecosystems. Mid-West sites correspond with medium to tall, closed stands of white spruce, black spruce, and balsam fir in late successional stages. In such environment, once fire has reached the tree cover it will spread through the crowns very rapidly, releasing great energy; strong radiometric signals emitted during the fire will be easily detected in thermal bands, and the BA signal will persist longer after the fire than in an open forest, where shrub and herbaceous species are predominant and postfire recovery is faster. Therefore, BA mapping based on active fire detection may benefit from this regional feature.

The case of fire #15, undetected by all BA algorithms, is coherent with this interpretation, since the forest structure is very similar to the transition zone of northern sites: open stunted stands of black spruce with significant understory composed of ericaceous shrubs and other species such as dwarf birch and Labrador tea. Furthermore, southern site #14, which caused detection problems in PAL.ba and MCD45A1, follows the same structural pattern, more similar to northern locations and actually placed in a different ecozone to the other mid-West sites: jack pine is the dominant tree, the understory is mostly formed by ericaceous shrubs and lichen, and earlier successional states are predominant as fires occur very often. Something similar is observed in West coast sites except for the fact that ecological particularities are caused by an altitudinal variation, as they are placed on the Cordilleran System at more than 1,500 m in height; there are open forests of white spruce, black spruce, and an understory of deciduous shrubs in addition to ericaceous, dwarf willows, and birch. Only VGT-derived datasets were able to detect them.

Finally, eastern fires were intermittently detected by different BA algorithms. The smallest eastern fire was exclusively represented in L3JRC; the largest (#18) was only reasonably well captured by GBA2000 (−17%), partially detected by PAL.ba (−56%), barely by LTDR.ba (−81%) and L3JRC (−98%), and only a trace in MCD45A1. Here, very open forests of black spruce are found, where 50% of vegetation cover is composed of nonericaceous shrubs; a transition zone to tundra and mixed evergreen and deciduous shrubs appears in the last two test sites, influenced by the Atlantic Ocean. Thus, LTDR.ba may have been more efficient in areas where the understory was not dominated by ericaceous type of shrub, whereas MCD45A1 worked in the opposite way, being more efficient in the northwestern open forests than in the eastern ones. The overall efficiency of GBA2000 can be partially caused by taking its ecozone features into account.

6.5. Temporal Shifts and Variability. Several factors contributed to producing temporal shifts of BA estimates with respect to the reference data temporal distribution, for instance, the unequal spatial representation of different fires. Figures 4(a) and 4(b) show how temporal shifts shorten as spatial variability lowers by subsampling (making up the central subset) allowing LAC.ba data to be added to the comparison as well. On the other hand, the fact that most fires in the central spatial subset occurred at the end of the month according to the CFSNFD may have increased the temporal shift observed at a monthly resolution (Figures 4(b) and 4(c)), especially in the case of MCD45A1 and LTDR.ba, which were the two datasets that best reproduced the reference temporal pattern. MCD45A1 was particularly influenced by the temporal resolution used, as the two major peaks fell within the same month. At site level, the spatiotemporal depiction of central fires showed that GBA2000 presented the most clearly defined and consistent shapes. However, timing differences between close pixels were up to four months, an incoherence also found in LAC.ba. In these cases, earliest and latest burn dates seemed to be linked to some peripheral distribution and spatial discrepancies with the other products. However, these two datasets agreed with the predominant date in the site core of LTDR.ba and MCD45A1, which thus showed the most consistent temporal distribution in terms of realistic burning duration of a fire and burn date variability among adjacent cells.

7. Conclusions

The fire-by-fire approach in the accuracy assessment of BA datasets revealed more details about differences in BA detection than the global statistics based on a regular grid. It also avoided errors linked to georeferencing and upscaling issues.

The ratio between fire size and pixel size is one of the factors influencing BA detection, but once over a minimum size threshold of the fire, other factors related to the BA detection method appear to be more influential. Among them are the width of the temporal window chosen for the change detection, the wavelengths used (especially in the MIR spectral region), the use of vegetation indices or BA-correlated indices, and the integration of different remote sensing sources.

Other factors are contextual and linked to particular ecosystems within the boreal forest biome, as well as to the transition zone between boreal forest and tundra. Thus, different vertical structures of the forest, species composition, fire regimes, and postfire recovery will affect the capacity for mapping the area burned. For example, highly energetic crown fires were the most common, benefiting the BA mapping methods that are based on the emissivity from active fires; however, in more open forests, less energetic surface and smoldering fires take place and different approaches are needed.

In terms of overall BA quantification, the estimated ranking of the datasets assessed in this work for the year 2000 was GBA2000 > LTDR.ba > MCD45A1 > L3JRC > PAL.ba. The first one also showed, however, the highest error variability and occasional overestimation, whereas the other VGT-derived dataset L3JRC, in spite of its rather poor BA estimation in the study area, was able to capture areas that passed unnoticed to others. Thus, regarding the overall representativeness of BA sites in the entire territory, our ranking would change with L3JRC in the first place. The MODIS-derived dataset was not the best BA estimator in year 2000, mostly caused by a circumstantial lack of data, but possibly also due to a lower adaptability of its change detection method to the boreal forests of Canada compared to other datasets. Regarding the AVHRR-derived datasets, BA estimations generally improved from coarser to finer spatial resolution, particularly from 8 km to 5 km; however, there was no substantial improvement from 5 km to 1 km, indicating that differences in the classification method between LTDR.ba and LAC.ba were more important than pixel size.

On the other hand, from a temporal accuracy perspective, burn date results in increasing order of spatiotemporal coherence and temporal accuracy was MCD45A1 > LTDR.ba~ PAL.ba > LAC.ba > GBA2000 > L3JRC. The LTDR.ba dataset struck a balance that year between MCD45A1 (spatial and temporal accuracy but great omission) and GBA2000 (best BA quantification but local overestimation and burn date retardation). For the rest of the datasets assessed, the higher the spatial accuracy is, the lower the temporal accuracy, and *vice versa*.

Acknowledgments

This work was supported by the Spanish Ministry of Science and Innovation under Grant CGL2010-22189-C02/CLI. They also thanks go to Emilio Chuvieco for providing the burned area record generated from AVHRR-LAC data and to Mike Flannigan and John Little for providing the fire perimeter dataset for Canada. They thank the agencies and services for the processing and distribution of satellite data of NASA and NOAA, from whom we have obtained the greatest part of the images used in this work (MCD45A1, AVHRR-LAC, and LTDR). SPOT 4-VEGETATION images were made available in the framework of the GBA2000 project.

References

[1] L. Boschetti, H. D. Eva, P. A. Brivio, and J. M. Grégoire, "Lessons to be learned from the comparison of three satellite-derived biomass burning products," *Geophysical Research Letters*, vol. 31, no. 21, 2004.

[2] D. Riaño, J. A. Moreno Ruiz, D. Isidoro, and S. L. Ustin, "Global spatial patterns and temporal trends of burned area between 1981 and 2000 using NOAA-NASA pathfinder," *Global Change Biology*, vol. 13, no. 1, pp. 40–50, 2007.

[3] C. Carmona-Moreno, A. Belward, J. P. Malingreau et al., "Characterizing interannual variations in global fire calendar using data from Earth observing satellites," *Global Change Biology*, vol. 11, no. 9, pp. 1537–1555, 2005.

[4] J. A. Moreno-Ruiz, D. Riaño, M. Arbelo, N. H. F. French, S. L. Ustin, and M. L. Whiting, "Burned area mapping time series in Canada (1984-1999) from NOAA-AVHRR LTDR: a comparison with other remote sensing products and fire perimeters," *Remote Sensing of Environment*, vol. 117, pp. 407–414, 2012.

[5] E. Chuvieco, P. Englefield, A. P. Trishchenko, and Y. Luo, "Generation of long time series of burn area maps of the boreal forest from NOAA-AVHRR composite data," *Remote Sensing of Environment*, vol. 112, no. 5, pp. 2381-2396, 2008.

[6] K. V. Khlopenkov and A. P. Trishchenko, "SPARC: new cloud, snow, and cloud shadow detection scheme for historical 1-km AVHHR data over Canada," *Journal of Atmospheric and Oceanic Technology*, vol. 24, no. 3, pp. 322–343, 2007.

[7] R. H. Fraser, Z. Li, and R. Landry, "SPOT VEGETATION for characterizing Boreal forest fires," *International Journal of Remote Sensing*, vol. 21, no. 18, pp. 3525-3532, 2000.

[8] K. Tansey, J. M. Grégoire, P. Defourny et al., "A new, global, multi-annual (2000–2007) burnt area product at 1 km resolution," *Geophysical Research Letters*, vol. 35, no. 1, Article ID L01401, 2008.

[9] D. P. Roy, Y. Jin, P. E. Lewis, and C. O. Justice, "Prototyping a global algorithm for systematic fire-affected area mapping using MODIS time series data," *Remote Sensing of Environment*, vol. 97, no. 2, pp. 137–162, 2005.

[10] M. Simon, S. Plummer, F. Fierens, J. J. Hoelzemann, and O. Arino, "Burnt area detection at global scale using ATSR-2: the GLOBSCAR products and their qualification," *Journal of Geophysical Research D*, vol. 109, no. 14, 2004.

[11] J. M. Grégoire, K. Tansey, and J. M. N. Silva, "The GBA2000 initiative: developing a global burnt area database from SPOT-VEGETATION imagery," *International Journal of Remote Sensing*, vol. 24, no. 6, pp. 1369–1376, 2003.

[12] S. Plummer, O. Arino, F. Ranera et al., "The GLOBCARBON initiative: global biophysical products for terrestrial carbon studies," in *Proceedings of the IEEE International Geoscience & Remote Sensing Symposium (IGARSS '07)*, pp. 2408–2411, 2008.

[13] D. P. Roy, L. Boschetti, C. O. Justice, and J. Ju, "The collection 5 MODIS burned area product—global evaluation by comparison with the MODIS active fire product," *Remote Sensing of Environment*, vol. 112, no. 9, pp. 3690–3707, 2008.

[14] J. S. Rowe and G. W. Scotter, "Fire in the boreal forest," *Quaternary Research*, vol. 3, no. 3, pp. 444–464, 1973.

[15] J. L. Farrar, *Trees in Canada*, Fitzhenry & Whiteside, Canadian Forest Service, Natural Resources Canada, Toronto, Canada, 1995.

[16] E. S. Kasischke, D. Williams, and D. Barry, "Analysis of the patterns of large fires in the boreal forest region of Alaska," *International Journal of Wildland Fire*, vol. 11, no. 2, pp. 131–144, 2002.

[17] C. E. Van Wagner, "Simulating the effect of forest fire on long-term annual timber supply," *Canadian Journal of Forest Research*, vol. 13, no. 3, pp. 451–457, 1983.

[18] Z. Li, Y. J. Kaufman, C. Ichoku et al., "A review of AVHRR-based active fire detection algorithms: principles, limitations, and recommendations," in *Global and Regional Vegetation Fire Monitoring From Space: Planning and Coordinated International Effort*, F. J. Ahern, J. G. Goldammer, and C. O. Justice, Eds., pp. 199–225, SPB Acadamic, The Hague, The Netherlands, 2001.

[19] P. A. Agbu and M. E. James, *The NOAA/NASA Pathfinder AVHRR Land Data Set User's Manual*, Goddard Distributed Active Archive Center, NASA, Goddard Space Flight Center, Greenbelt, MD, USA, 1994.

[20] M. E. James and S. N. V. Kalluri, "The pathfinder AVHRR land data set: an improved coarse resolution data set for terrestrial monitoring," *International Journal of Remote Sensing*, vol. 15, no. 17, pp. 3347–3363, 1994.

[21] P. M. Barbosa, J. M. Grégoire, and J. M. C. Pereira, "An algorithm for extracting burned areas from time series of AVHRR GAC data applied at a continental scale," *Remote Sensing of Environment*, vol. 69, no. 3, pp. 253–263, 1999.

[22] B. Pinty and M. M. Verstraete, "GEMI: a non-linear index to monitor global vegetation from satellites," *Vegetatio*, vol. 101, no. 1, pp. 15–20, 1992.

[23] J. Pedelty, S. Devadiga, E. Masuoka et al., "Generating a long-term land data record from the AVHRR and MODIS instruments," in *Proceedings of the IEEE International Geoscience & Remote Sensing Symposium (IGARSS '07)*, pp. 1021–1025, 2007.

[24] T. Bayes, "An essay towards solving a problem in the doctrine of chances," *Philosophical Transactions of the Royal Society of London*, vol. 53, pp. 370–418, 1763.

[25] H. Langseth and T. D. Nielsen, "Classification using Hierarchical Naïve Bayes models," *Machine Learning*, vol. 63, no. 2, pp. 135–159, 2006.

[26] L. Giglio, T. Loboda, D. P. Roy, B. Quayle, and C. O. Justice, "An active-fire based burned area mapping algorithm for the MODIS sensor," *Remote Sensing of Environment*, vol. 113, no. 2, pp. 408–420, 2009.

[27] R. Latifovic, A. P. Trishchenko, J. Chen et al., "Generating historical AVHRR 1 km baseline satellite data records over Canada suitable for climate change studies," *Canadian Journal of Remote Sensing*, vol. 31, pp. 324–346, 2005.

[28] R. H. Fraser, R. Fernandes, and R. Latifovic, "Multi-temporal mapping of burned forest over Canada using satellite-based change metrics," *Geocarto International*, vol. 18, pp. 37–48, 2003.

[29] K. Tansey, J. M. Grégoire, D. Stroppiana et al., "Vegetation burning in the year 2000: global burned area estimates from SPOT VEGETATION data," *Journal of Geophysical Research D*, vol. 109, no. 14, 2004.

[30] D. V. Ershov and V. P. Novik, "Mapping burned areas in Russia with SPOT4 VEGETATION (S1 product) imagery," type, Joint Research Centre of the European Commission, Brussels, Belgium, 2001, Contract 18176-2001-07-F1EI ISP RU.

[31] R. Latifovic, Z. L. Zhu, J. Cihlar, C. Giri, and I. Olthof, "Land cover mapping of North and Central America—global land cover 2000," *Remote Sensing of Environment*, vol. 89, no. 1, pp. 116–127, 2004.

[32] J. Little, Wildland Fire Information Systems Group. Natural Resources Canada, and Canadian Forest Service, Northern Forestry Centre, "Personal communication: electronic mail, National Fire Database—Agency fire polygon data," Edmonton, Alberta, Canada, September 2009.

[33] D. P. Roy and L. Boschetti, "Southern Africa validation of the MODIS, L3JRC, and GlobCarbon burned-area products," *IEEE Transactions on Geoscience and Remote Sensing*, vol. 47, no. 4, pp. 1032–1044, 2009.

[34] R. G. Congalton, "A review of assessing the accuracy of classifications of remotely sensed data," *Remote Sensing of Environment*, vol. 37, no. 1, pp. 35–46, 1991.

[35] R. G. Pontius and M. L. Cheuk, "A generalized cross-tabulation matrix to compare soft-classified maps at multiple resolutions," *International Journal of Geographical Information Science*, vol. 20, no. 1, pp. 1–30, 2006.

[36] S. V. Stehman, “Selecting and interpreting measures of thematic classification accuracy,” *Remote Sensing of Environment*, vol. 62, no. 1, pp. 77–89, 1997.

6

Multisensor Network System for Wildfire Detection Using Infrared Image Processing

I. Bosch, A. Serrano, and L. Vergara

Signal Processing Group, Institute of Telecommunications and Multimedia Applications (iTEAM), Universitat Politècnica de València, Camino de Vera, S/N, 46022 Valencia, Spain

Correspondence should be addressed to I. Bosch; igbosroi@dcom.upv.es

Academic Editors: A. E. Cetin, I. Korpeoglu, B. U. Toreyin, and S. Verstockt

This paper presents the next step in the evolution of multi-sensor wireless network systems in the early automatic detection of forest fires. This network allows remote monitoring of each of the locations as well as communication between each of the sensors and with the control stations. The result is an increased coverage area, with quicker and safer responses. To determine the presence of a forest wildfire, the system employs decision fusion in thermal imaging, which can exploit various expected characteristics of a real fire, including short-term persistence and long-term increases over time. Results from testing in the laboratory and in a real environment are presented to authenticate and verify the accuracy of the operation of the proposed system. The system performance is gauged by the number of alarms and the time to the first alarm (corresponding to a real fire), for different probability of false alarm (PFA). The necessity of including decision fusion is thereby demonstrated.

1. Introduction

Conserving unique natural areas should be a priority for advanced societies in our time. One of the biggest threats faced by these natural areas is wildfire devastation. The unfortunate reality is that most of these areas are unprotected, or at most, only monitored during certain months of the year and then, only during certain times of day, leaving the nighttime periods more vulnerable without proper monitoring. The entire system suffers from teams of workers woefully ill equipped in terms of manpower and technology.

In response to these limitations, we have developed different ways to help these teams in their complex yet tedious task of forest monitoring. The literature has focused extensively on technical aspects of the problem with the aim of discovering solutions.

Various authors have focused on solutions derived from specialized satellite infrastructure available today [1, 2]. Due to the nature of nongeosynchronous satellites [3], these proposals present four principal technical difficulties: the limited availability to cover the desired area, the effective resolution cell (taking into account the distances at which the sensors are positioned), (especially), the effective detection times and the times between satellite positioning.

Another option includes ground implementation, which entails designing specialized systems for the desired coverage area [4]. These designs employ different processing techniques that are typically divided into two major families (based on the type of information processed): the first is limited to collecting data with infrared sensors [5, 6]; the second encompasses working with visible images (such as [7, 8]), looking for specific types of fire in these images (as in [9] or [10]) and improving computer vision [11–13].

As part of ground implementation, additional consideration must often be given to expanding the inherently limited coverage area of these systems [14], thereby creating opportunities for wireless sensor networks as in [15, 16] with cameras or other specialized sensors [17, 18].

Another broad field includes the efforts of researchers to detect smoke [19] in visible images, [20, 21], to distinguish between the flame of the fire focus and smoke [22], and to use video to detect fires at night [23].

To address these issues, this paper presents the next step in the evolution of multisensor wireless network systems employed in terrestrial forest fire detection. This system has been under development for the last ten years as part of multiple research projects within the Signal Processing Group (GTS), part of the Institute of Telecommunication and Multimedia Applications (iTEAM) at the Universitat Politècnica de València (UPV). Our system exploits different expected characteristics of a real fire, including persistence and increases over time [24], in infrared images, while concurrently detecting smoke in visible images.

Research in the area of fire detection began with an initial processing scheme, as presented in [25]. It employed infrared radar as part of a linear scanning surveillance designed to detect wide-area, uncontrolled fires. The proposed scheme includes a linear predictor, and a subspace model with a prewhitening filter for the signal to be detected and introduces a simple procedure for improving linear prediction, as described in [26]. This scheme was applied to real infrared data collected by a passive infrared radar, located in a mountainous area in Southeast Spain (Alcoy, Alicante). Electronic range scanning and an azimuth mechanical system were likewise used.

In [27], we presented a general scheme for the automatic detection of events in surveillance systems; it consisted of the initial basic scheme but extended to include nonlinear prediction and an increase detector. As part of the same project [27], exhaustive research was conducted on the design of the predictor, with the first theoretical considerations on the matched subspace detector and the increase detector being subsequently introduced. The need for decision fusion for the two detectors to make a final decision was likewise presented for the first time. Real-data experiments validated the interest of the proposed scheme. Results in a real operating system were shown, specifically those from several tests with real fires and from day-to-day operations in the Albufera Natural Park (Valencia).

Once the proposed schemes were installed in several real scenarios, we realized that the processing times of the various detectors needed to be considered. Consequently, detection algorithms were the focus of [28], with a special emphasis on the fusion of different decisions in order to exploit both the short-term persistence and the long-term increases found in uncontrolled fires. In [29, 30], we added a linear predictor to use a reference image for prediction, rather than previous images (used in earlier systems). System delays in alarm detection of controlled fire were also evaluated. Temporary evolution of false and true alarms is presented in [31, 32], part of a long-term performance evaluation carried out in the Font Roja Natural Park in Alcoy (Alicante, Spain).

In this paper, we focus on verifying the improvements made in the processing scheme for real fire signals. Section 2 presents a description of the system, with real-data results presented in Section 3. Finally, conclusions about the improvements of the proposed scheme are then offered.

2. System Detector Scheme

The proposed system consists of a wireless sensor network with a central monitoring station. This sensor network is strategically positioned to significantly expand the effective coverage of the system, with several areas of overlap between the different coverages to verify alarms, especially when the distances increase considerably (tens of kilometers).

Each sensor is comprised of two cameras (thermal and visible); a motor with different presets to sweep a larger area of coverage; and an integrated system of capture, processing and communication (see Figure 1). This sensor scheme allows autonomous monitoring of portions of the coverage area as well as in situ processing, generation, and transmission of alarms to the other elements in the wireless sensor network and to the central station. The said station can monitor the proper operation of the system and locate the position of each sensor with a geographic information system (GIS).

It is important to note that the system requirements are minimal: it is not necessary to use high-resolution cameras or show temperatures values, it does not require fast processing times, since it is better that the times between capture are seconds apart to have a margin for growth.

As mentioned above, the original system has been implemented and tested in real scenarios [27, 31] and shown to operate properly [32]. Thus, this paper proposes upgrading the original processing scheme with some improvements described in detail below and further verification with controlled fire experiments.

The new processing scheme is shown in Figure 2, where each infrared image is converted into a matrix of pixels. Each pixel is associated with a resolution cell corresponding to certain coordinates of rank and azimuth; pixel-by-pixel processing is performed to then generate vectors describing the time history of each resolution cell.

The sensor motor is initially placed in one of the presets and, assuming no fire, the pattern $\mathbf{w}$ is calculated by acquiring a predefined number of images, which in turn, are used to generate the vector $\mathbf{w}_\mathrm{D}$ (ideally composed only of noise). This vector $\mathbf{w}_\mathrm{D}$ is sorted from low to high, the least representative of the extremes are removed, and the average of the remaining values is calculated, thereby yielding the searching pattern $\mathbf{w}$. The system is also calibrated with the same images used in the pattern to generate the variables required in the subsequent processing stages.

This pattern $\mathbf{w}$ is now introduced in the linear prediction stage, represented by matrix $\mathbf{H}$, to obtain the estimated noise signal $x_\mathrm{p} = \mathbf{H} \cdot \mathbf{w}$.

Now, in normal operation, the infrared images are captured and generate the vector $\mathbf{x} = \mathbf{s} + \mathbf{w}_\mathrm{d}$, composed of both signal $\mathbf{s}$ and noise $\mathbf{w}_\mathrm{d}$. These values are then used to form vector $\mathbf{e} = \mathbf{x} - \mathbf{x}_\mathrm{p}$, subtracting the previously estimated vector $\mathbf{x}_\mathrm{p}$ from this vector $\mathbf{x}$, which will ideally contain only signal $\mathbf{s}$ if they are predicted correctly.

This vector $\mathbf{e}$ has a Gaussian probability distribution [25], therefore a prewhitening stage must be performed using the matrix, $\mathbf{Rzz}$, to optimize the calculation of the threshold for a given PFA. Thus, we obtain the vector $\mathbf{u} = \mathbf{Rzz} \cdot \mathbf{e}$, which is used as input for the subsequent detection stages.

We established four levels for risk of fire detection (ranging from low to high), with each corresponding to the following four alarms:

(i) Type 1: signal level alarm,

(ii) Type 2: persistence in the signal level alarm (green in the figures of results),

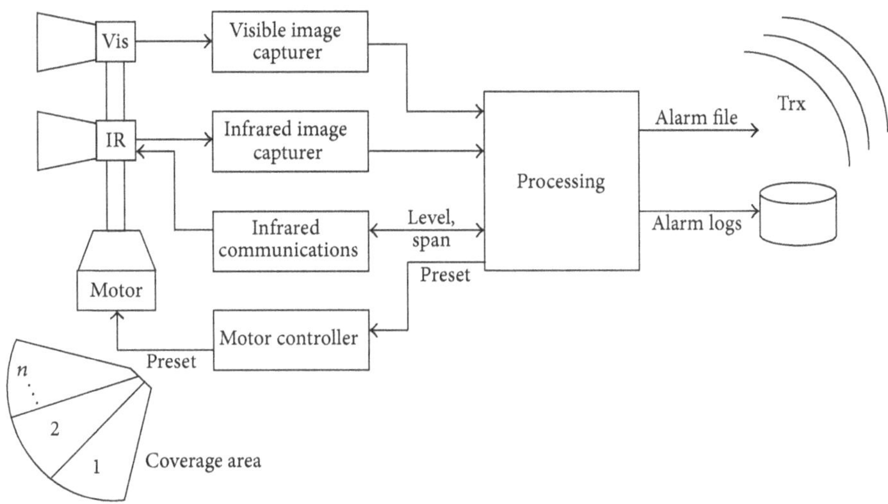

FIGURE 1: Sensor scheme.

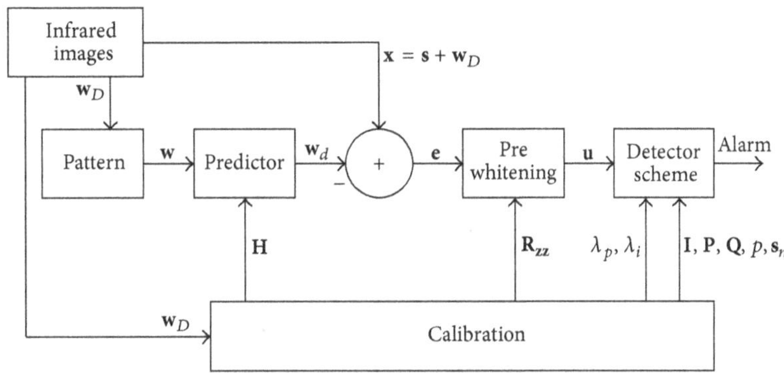

FIGURE 2: Processing Scheme.

(iii) Type 3: increasing alarm (orange),

(iv) Type 4: thermal saturation alarm (red).

The first type of alarm is designed to detect any change in the signal level. It is calculated from the vector **u** with a matched subspace filter (1), which uses an identity matrix **I** as a signal estimator. It thus becomes a simple signal-level detector to compare against a threshold, λ_c, optimally calculated for a given PFA_c [25]:

$$c_i = \mathbf{u}_i^T \mathbf{I} \mathbf{u}_i \gtrless \lambda_c. \quad (1)$$

The second type of alarm is designed to observe the permanence of the change in the signal level, thus avoiding false alarms triggered by random changes or low persistence elements (e.g., a hot element moving into the infrared coverage area). It is calculated again from the vector **u** using a matched subspace filter, but now a projection matrix, **P**, is employed as a signal estimator, as designed in [25]. Assuming that the fire signature is inside a "low pass" subspace, the resulting estimator, r (2), with a χ_p^2 *distribution* (chi-square probability density function (pdf) with p degrees of freedom, where p is the subspace dimension), is compared with a second threshold, λ_p, optimally calculated for a given PFA_p [25]:

$$r_i = \mathbf{u}_i^T \mathbf{P} \mathbf{u}_i \gtrless \lambda_p. \quad (2)$$

The third type of alarm is designed to detect the presence of increasing trends over a longer term. Thus, the observation period of time is increased from the D images used in the previous alarms to the L overlapped groups of D images, as seen in Figure 3. To accomplish this, we first generate an estimator vector $z = [r_1 \cdots r_L]^T$, from the L previous *persistence detectors* results, $\mathbf{r_i}$, leaving only a margin of nu images without persistence detection to avoid sporadic decreases, according to the fusion rule implemented in [28].

Then, an increase estimator is generated from the decision fusion of the L *persistence detectors*, **z**, each compared with the threshold λ_i. As in the previous cases, this threshold is optimally obtained from [27] for a required PFA_i, according to the following expression (3):

$$\frac{\mathbf{z}^T \mathbf{Q}^{(n)T} \left(\mathbf{Q}^{(n)} \mathbf{Q}^{(n)T} \right)^{-1} \mathbf{s_n}}{\sqrt{2 p \mathbf{s}_n^T \left(\mathbf{Q}^{(n)} \mathbf{Q}^{(n)T} \right)^{-1} \mathbf{s_n}}} \gtrless \lambda_i, \quad (3)$$

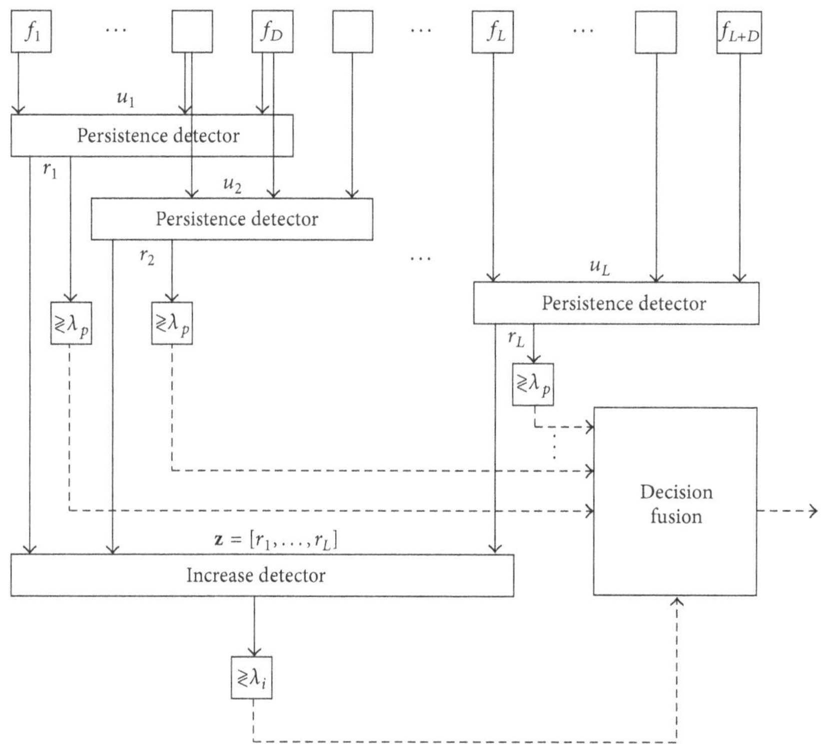

FIGURE 3: New detector scheme.

where $s_n = \underbrace{[1 \cdots 1]}_{L-n}^T$, and the difference matrix is defined by $\mathbf{Q}^{(n)} = \mathbf{Q}_{L-n+1} \cdots \mathbf{Q}_{L-1} \cdot \mathbf{Q}_L$. Where matrix $\mathbf{Q}_L$ is defined by (4)

$$\mathbf{Q}_L = \underbrace{\left.\begin{bmatrix} -1 & 1 & 0 & \cdots & \cdots & 0 \\ 0 & -1 & 1 & \cdots & \cdots & 0 \\ & & \vdots & & & \\ 0 & 0 & 0 & \cdots & -1 & 1 \end{bmatrix}\right\} L-1}_{L} \quad (4)$$

Finally, the fourth type of alarm is the *thermal saturation alarm*, which is activated if the saturation level of the IR camera is surpassed.

These four types of detectors generate the corresponding four alarm types, but in practice it has been observed that they may be fused together (see detection scheme in Figure 3) as follows.

(i) Type 1, signal level alarm, is used as a requirement for all other alarms, thus preventing any observation of a low-level signal.

(ii) Type 2, persistence level signal alarm, is used as a condition for checking the increase in the persistence in the type 3 alarm. With the *nu* parameter, it allows a number of controlled images without increased detection, and thereby avoids being too restrictive. A more comprehensive study on this condition can be found in [31].

(iii) Type 3, increasing alarm, is activated when Type 1 (signal level alarm) and Type 2 (persistence alarm) have each been previously activated. This is a good indication of a possible source of fire.

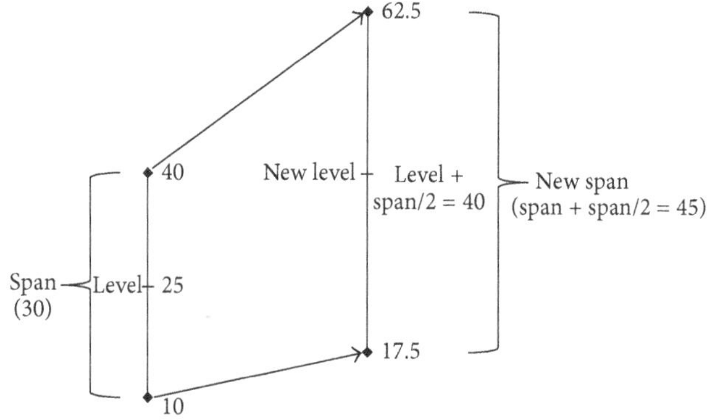

FIGURE 4: Change of level and span to leave sufficient margin of fire growth.

(iv) Type 4, thermal saturation alarm, is also used as an indication of fire if Types 1 and 2 have been previously activated. This is because, if the system is calibrated correctly, the level and span parameters of the IR camera must be artificially increased (e.g., 50% of the span, as shown in Figure 4). The fire level will then have a margin for growth. In Type 4, the thermal saturation level is only achievable if the signal level has been growing. In this case, Type 3 likewise fails because it does not possess any margin for growth.

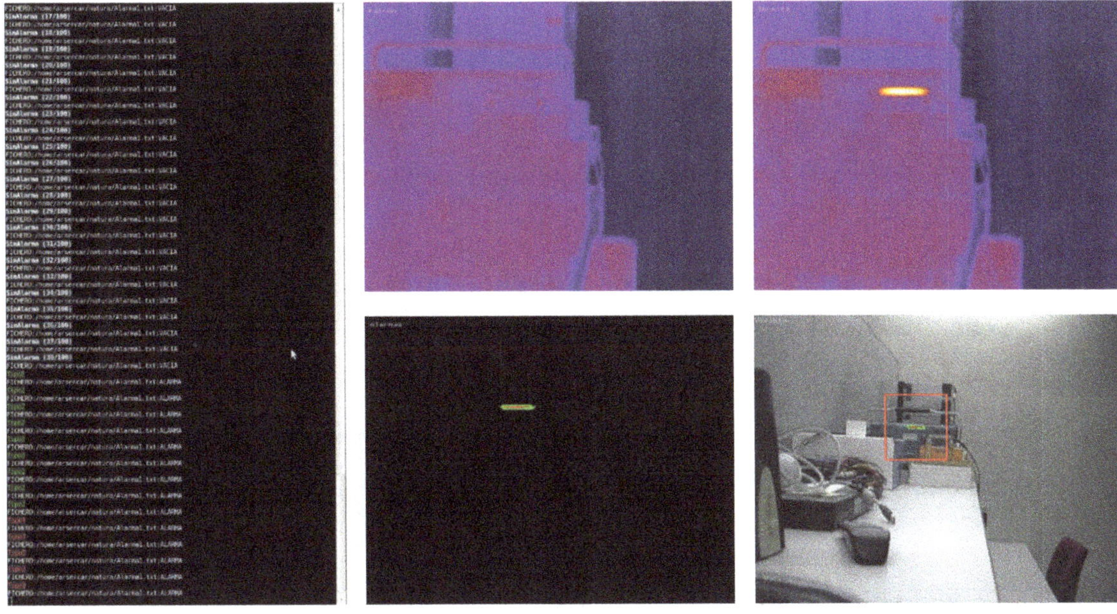

FIGURE 5: Capture of processing example in laboratory simulation.

3. Experiments

The system improvements were tested both in the laboratory, with known conditions, and in a controlled fire test. Additionally, results from a live-burn test in a real environment are shown to demonstrate the effectiveness of the proposed system.

In the laboratory experiments, the system was tested using a high power resistor supplied by a DC power source. Initially, with the power off, the system was calibrated and the pattern was generated. Subsequently, we activated the power supply and increased the voltage applied to the resistor; the increased radiated temperature simulated a possible source of heat.

In this case, we had to adjust the system parameters to simulate a fire at close range. Also, we had to add Gaussian noise onto the images with which the pattern **w** was calculated, because the span values were too small, and this caused numerical errors in the calculation of the calibration matrices. The added noise had a zero mean, and the variance was adjusted by taking into account the range of values of the camera signal (span).

Figure 5 shows captures of processing examples in the laboratory simulation. The left section displays a log with each iteration of the normal program operation. The right contains four images as follows: in the top left, the pattern used to detect the alarms is displayed; the top right shows the image captured by the thermal camera, as it is being processed; the bottom left shows the detected alarms; and the bottom right shows these detected alarms over the visible image. We can see how only the simulated fire is detected when generating the different types of alarms. Furthermore, on the bottom right, detected alarms are overlapped in the visible image, in order to locate them more easily. This process was performed using a projective transformation of the thermal image coordinates to the corresponding coordinates of the visible image. This transformation was calculated beforehand by manually defining equivalent locations in both images. In this case, alarms are not perfectly located on the visible image, because the distances with respect to the cameras are small, and the transformation error is large. Finally, a frame is also drawn to delimit the area in which all the alarms occur.

A controlled fire test was then performed in a real environment. The simulated fire was generated in a small container at a distance of 100 m (Figure 6).

In Figure 6, the captured thermal image (FLIR ThermoVision A20-V, Focal Plane Array (FPA), uncooled microbolometer with spectral range: 7,5 to 13 μm) can be seen on the left, and the visible image with overlapped alarms, on the right. Also, a closer view of the camera mounting and the source of the fire is shown in both images, respectively. The figure shows how the system properly detected the simulated fire and, in this case, correctly located it on the visible image. Several tests were performed with different levels of system sensitivity; in all cases, early detection of the fire was attained.

Finally, we tested the system in two real environments, with real controlled fires under firefighter supervision. In this case, the infrared data was recorded, and these recordings were subsequently processed in the laboratory to simulate a real-time operating environment, thereby allowing a comprehensive examination of the system performance to be conducted, based on different parameters.

The first test was held in the Font Roja Natural Park in Alcoy (Alicante). It was a fire at a distance of 800 meters (Figure 7), and the fire can be seen in the image to start approximately 50 seconds into the recording. The second one was held in the Valencian town of Ayora. It was a fire at a greater distance (about 1500 meters), and the fire can be

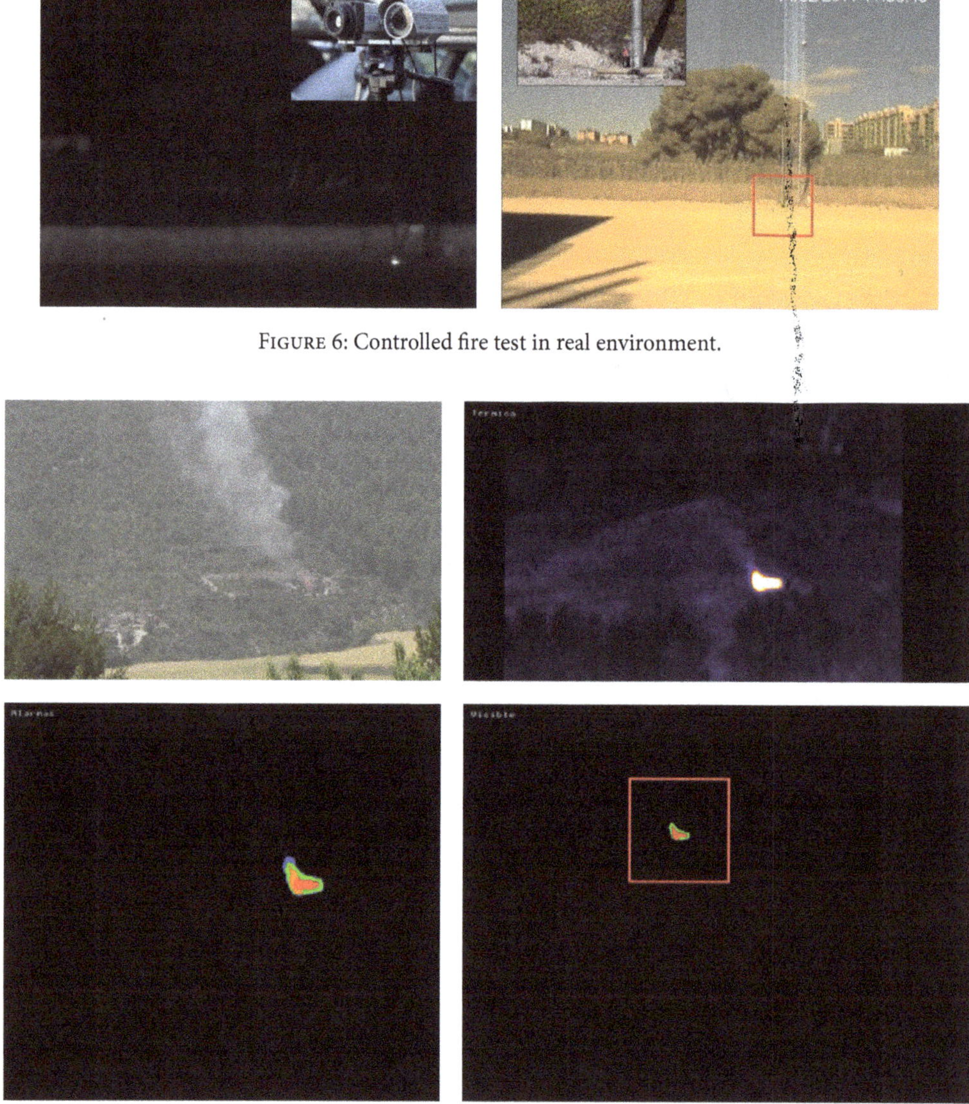

FIGURE 6: Controlled fire test in real environment.

FIGURE 7: Example of controlled fire detection in real environment.

seen in the image to start approximately 30 seconds into the recording.

A comprehensive analysis of the alarm evolution was then carried out for a fixed PFA using independent detectors and the fusion rules implemented in the system (developed and discussed in Section 2). The results of this analysis of the experiments in Font Roja and Ayora are shown in Figures 8 and 9, respectively, as a comparison between independent detectors (a) and fused detectors (b). At the top of each subfigure, the state of the detected alarms for a given time is seen. The middle graph displays the evolution over time of the total number of alarms of different types for the whole image. The bottom subfigure graphs the evolution of alarms for a given pixel over time.

Looking specifically at the alarm images located at the top of these figures, the spatial distribution of the fire at a given instant (indicated by the vertical black line) can be deduced. From these, it can be verified that these are two fires moving spatially, because the *increasing alarms* are located at one end of the fire. Namely, in Figure 8, a fire is moving toward the left, while in Figure 9, displacement is to the right.

The bottom of each figure displays the time evolution for the different alarms of the same fire for a particular pixel, which has been chosen in order to observe all types of alarms. In both cases, it can be seen how, at this point, the fire temperature increased until reaching saturation. During this time, various types of alarms were generated, depending on whether the fusion of detectors was being used or not.

In the nonfusion case, it can be observed how the *increasing alarms* may be activated before *the persistence alarm*, which could lead to an advantage in detection time. However, this may generate a greater number of false detections, as seen in the alarms that appear on the bottom-right portion of the alarms image in Figure 8(a) and, in the early detections

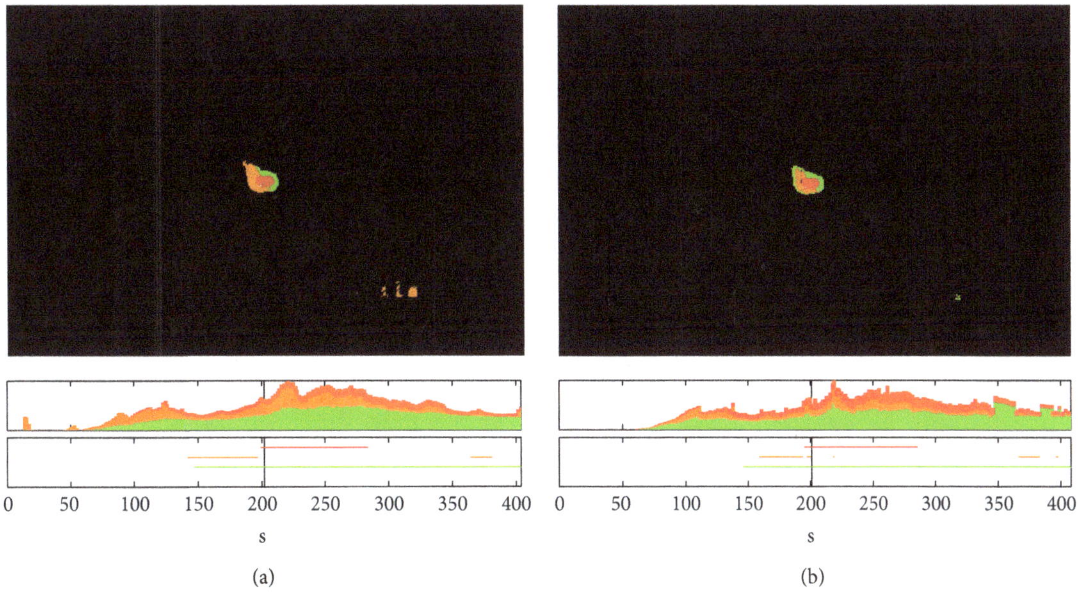

FIGURE 8: Fire alarm evolution for a fixed PFA with independent detectors (a) and fused detectors (b) in Font Roja.

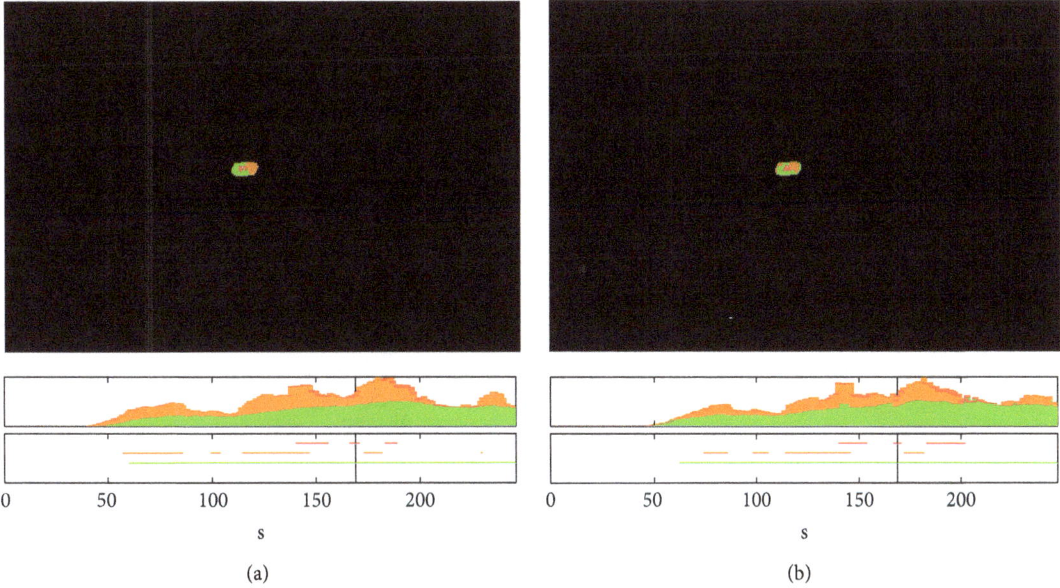

FIGURE 9: Fire alarms evolution for a fixed PFA with independent detectors (a) and fused detectors (b) in Ayora.

appear in the evolution of the number of alarms in the same figure. These are not observed in the case of the fused scheme (see Figure 8(b)).

We can conclude that, for a fixed PFA, the probability of fire detection of the system can be greatly enhanced if these *increasing alarms* are preceded by *persistence alarms*, although a detection delay is introduced. That is, the necessity of using the implemented fusion rules and how this use introduces a significant improvement in a real environment are verified.

While working with the fused detector scheme, an analysis was likewise performed on the evolution of the total number of alarms (Figure 10) and the time to the first alarm (Figure 11), depending on the required system PFA and the type of alarm.

The results shown in Figure 10 verify that, in practice, the number of *persistence* and *increasing alarms* rises as the PFA increases, independently of the *saturation* alarms. This is logical: as the PFA increases, more true or false alarms are present in the system. Thus, once again, the control exerted over the PFA is verified. We can also observe that the number of alarms is considerably greater in the case of Font Roja, since it was a fire at a smaller distance than in Ayora.

From Figure 11 we can verify in practice how, as the PFA increases, the time to the first alarm decreases. This is evident in the two graphs but particularly noticeable in

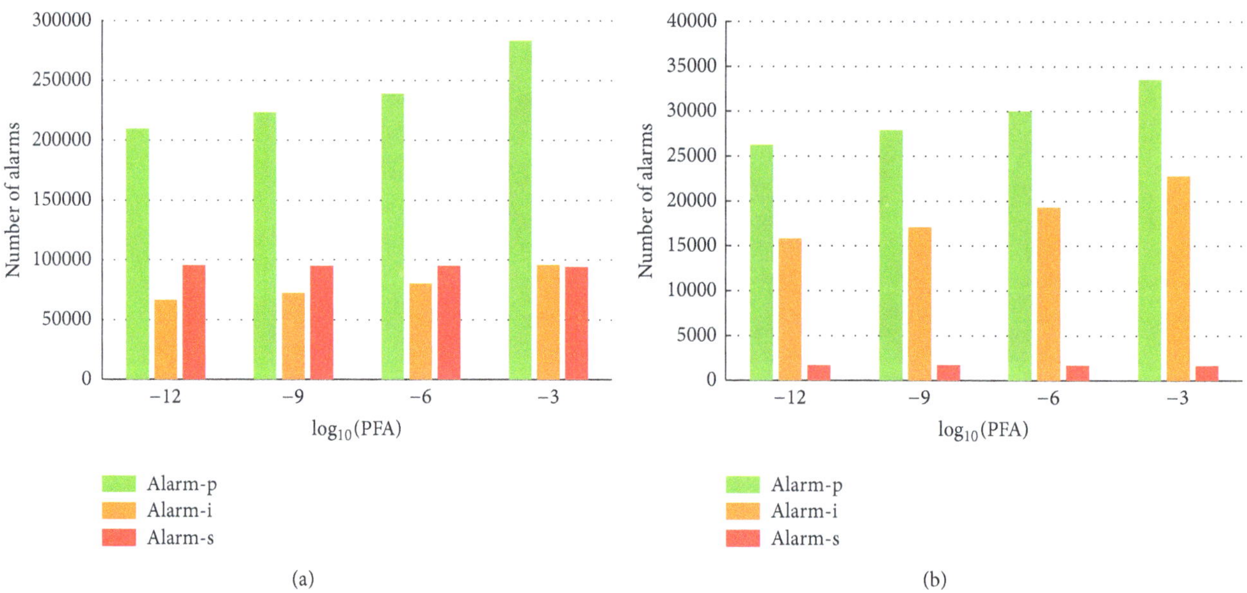

FIGURE 10: Number of alarms versus PFA, for different types of alarms, in Font Roja (a) and Ayora (b).

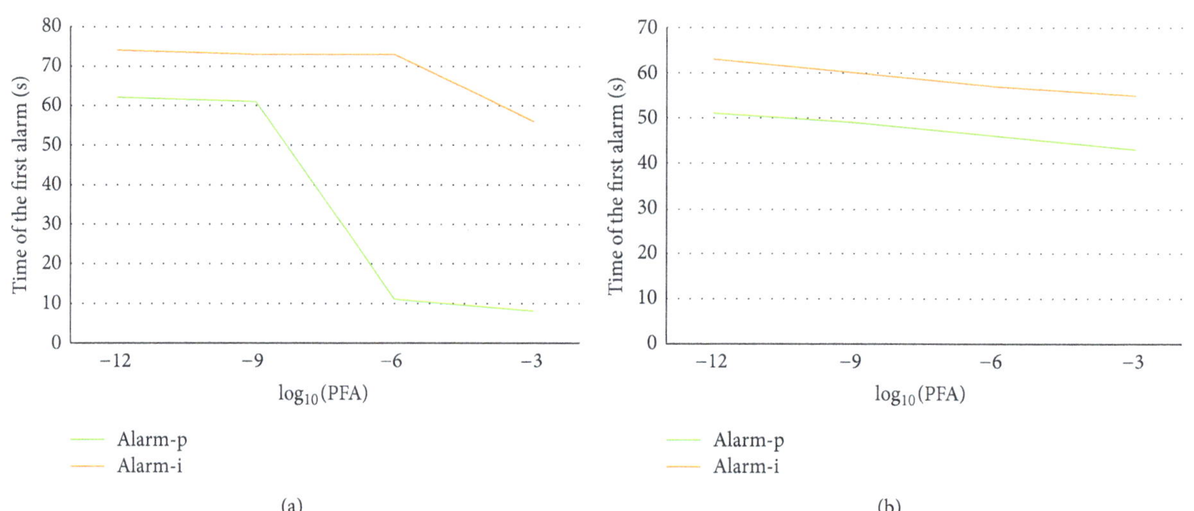

FIGURE 11: Time of the first alarm versus PFA for *persistence* (Alarm-p) and *increasing* (Alarm-i) alarms, in Font Roja (a) and Ayora (b).

the green line (*persistence alarms*) in Font Roja (Figure 11(a)), where the first alarms in the highest PFA were false, likewise observed in Figure 8(a). In Figure 11(b), this aspect is more clearly observed. The delay in the appearance of *increasing alarms* with respect to *persistence* ones is nearly constant and independent of the PFA.

4. Conclusions

We present the next step in the evolution of the multisensor wireless network system, based on infrared and advanced image sensors for automatic wildfire detection. This paper focuses on the description of the sensor and the processing scheme, highlighting the improvements in both.

The different types of detectors are described and special emphasis is given to the decision fusion rules for the *persistence* and *increase detectors*, which can exploit short- and long-term characteristics expected in a real fire.

The functionality of the system is verified in diverse, controlled real-environment tests in order to authenticate the accuracy of the proposed system. Spatial and temporary evolutions of the alarms are likewise shown as part of an evaluation of the system in a real environment. Through a comprehensive analysis of different processing schemes, the necessity of including decision fusion is demonstrated. The performance of the system is also evaluated by measuring the number of alarms and the time to the first alarm corresponding to a real fire, for different PFA.

The results obtained reveal a high potential for this system in aiding human surveillance. Future research will include detecting smoke generated by a fire in the visible image.

Acknowledgments

This work has been supported by Generalitat Valenciana under Grant PROMETEO 2010-040 and Spanish Administration and European Union FEDER Programme under Grant TEC2011-23403 01/01/2012.

References

[1] Y. Rauste, E. Herland, H. Frelander, K. Soini, T. Kuoremäki, and A. Ruokari, "Satellite-based forest fire detection for fire control in boreal forests," *International Journal of Remote Sensing*, vol. 18, no. 12, pp. 2641–2656, 1997.

[2] L. Giglio, J. Descloitres, C. O. Justice, and Y. J. Kaufman, "An enhanced contextual fire detection algorithm for MODIS," *Remote Sensing of Environment*, vol. 87, no. 2-3, pp. 273–282, 2003.

[3] M. J. Carlotto, "Detection and analysis of change in remotely sensed imagery with application to wide area surveillance," *IEEE Transactions on Image Processing*, vol. 6, no. 1, pp. 189–202, 1997.

[4] B. C. Arrue, A. Ollero, and J. R. Martinez De Dios, "An intelligent system for false alarm reduction in infrared forest-fire detection," *IEEE Intelligent Systems and Their Applications*, vol. 15, no. 3, pp. 64–73, 2000.

[5] J. Vicente and P. Guillemant, "An image processing technique for automatically detecting forest fire," *International Journal of Thermal Sciences*, vol. 41, no. 12, pp. 1113–1120, 2002.

[6] S. Briz, A. J. De Castro, J. M. Aranda, J. Meléndez, and F. López, "Reduction of false alarm rate in automatic forest fire infrared surveillance systems," *Remote Sensing of Environment*, vol. 86, no. 1, pp. 19–29, 2003.

[7] E. Kuhrt, J. Knollenberg, and V. Mertens, "An automatic early warning system for forest fires," *Annals of Burns and Fire Disasters*, vol. 14, no. 3, pp. 151–154, 2001.

[8] T.-H. Chen, P.-H. Wu, and Y.-C. Chiou, "An early fire-detection method based on image processing," in *Proceedings of the International Conference on Image Processing (ICIP '04)*, vol. 3, pp. 1707–1710, October 2004.

[9] J. Li, Q. Qi, X. Zou, H. Peng, L. Jiang, and Y. Liang, "Technique for automatic forest fire surveillance using visible light image," in *Proceedings of the IEEE International Geoscience and Remote Sensing Symposium (IGARSS '05)*, vol. 5, pp. 3135–3138, July 2005.

[10] Y. Dedeoglu, B. U. Toreyin, U. Güdükbay, and A. E. Cetin, "Real-time fire and flame detection in video," in *Proceedings of the IEEE 30th International Conference on Acustics, Speech and Signal Processing (ICASSP '05)*, pp. 669–672, 2005.

[11] J. R. Martinez-de Dios, B. C. Arrue, A. Ollero, L. Merino, and F. Gómez-Rodríguez, "Computer vision techniques for forest fire perception," *Image and Vision Computing*, vol. 26, no. 4, pp. 550–562, 2008.

[12] B. U. Töreyin, R. G. Cinbiş, Y. Dedeoğlu, and A. E. Çetin, "Fire detection in infrared video using wavelet analysis," *Optical Engineering*, vol. 46, no. 6, Article ID 067204, 2007.

[13] B. Ugur Toreyin and A. Enis Cetin, "Wildfire detection using LMS based active learning," in *Proceedings of the IEEE International Conference on Acoustics, Speech, and Signal Processing (ICASSP '09)*, pp. 1461–1464, April 2009.

[14] Z. Chaczko and F. Ahmad, "Wireless sensor network based system for fire endangered areas," in *Proceedings of the 3rd International Conference on Information Technology and Applications (ICITA '05)*, vol. 2, pp. 203–207, July 2005.

[15] Y. Li, Z. Wang, and Y. Song, "Wireless sensor network design for wildfire monitoring," in *Proceedings of the 6th World Congress on Intelligent Control and Automation (WCICA '06)*, vol. 1, pp. 109–113, June 2006.

[16] M. Hefeeda and M. Bagheri, "Wireless sensor networks for early detection of forest fires," in *Proceedings of the IEEE Internatonal Conference on Mobile Adhoc and Sensor Systems (MASS '07)*, pp. 1–6, October 2007.

[17] J. Lloret, M. Garcia, D. Bri, and S. Sendra, "A wireless sensor network deployment for rural and forest fire detection and verification," *Sensors*, vol. 9, no. 11, pp. 8722–8747, 2009.

[18] J. Lloret, I. Bosch, S. Sendra, and A. Serrano, "A wireless sensor network for vineyard monitoring that uses image processing," *Sensors*, vol. 11, no. 6, pp. 6165–6196, 2011.

[19] D. Kim and Y.-F. Wang, "Smoke detection in video," in *Proceedings of the WRI World Congress on Computer Science and Information Engineering (CSIE '09)*, vol. 5, pp. 759–763, April 2009.

[20] D. Krstinić, D. Stipaničev, and T. Jakovčević, "Histogram-based smoke segmentation in forest fire detection system," *Information Technology and Control*, vol. 38, no. 3, pp. 237–244, 2009.

[21] T. Jakovčević, L. Šerić, D. Stipaničev, and D. Krstinić, "Wildfire smoke-detection algorithms evaluation," in *Proceedings of the 6 International Conference on Forest Fire Research*, pp. 1–12, 2010.

[22] C.-C. Ho, "Machine vision-based real-time early flame and smoke detection," *Measurement Science and Technology*, vol. 20, no. 4, Article ID 045502, 2009.

[23] O. Günay, K. Taşdemir, B. Uğur Töreyin, and A. Enis Çetin, "Video based wildfire detection at night," *Fire Safety Journal*, vol. 44, no. 6, pp. 860–868, 2009.

[24] E. Pastor, L. Zárate, E. Planas, and J. Arnaldos, "Mathematical models and calculation systems for the study of wildland fire behaviour," *Progress in Energy and Combustion Science*, vol. 29, no. 2, pp. 139–153, 2003.

[25] L. Vergara and P. Bernabeu, "Automatic signal detection applied to fire control by infrared digital signal processing," *Signal Processing*, vol. 80, no. 4, pp. 659–669, 2000.

[26] L. Vergara and P. Bernabeu, "Simple approach to nonlinear prediction," *Electronics Letters*, vol. 37, no. 14, pp. 926–928, 2001.

[27] P. Bernabeu, L. Vergara, I. Bosh, and J. Igual, "A prediction/detection scheme for automatic forest fire surveillance," *Digital Signal Processing*, vol. 14, no. 5, pp. 481–507, 2004.

[28] I. Bosch and L. Vergara, "Forest fire detection by infrared data processing," in *Data Fusion For Situation Monitoring, Incident Detection and Response Management, 198 NATO Science Series: Computer & Systems Sciences*, E. Shahbazian, G. Rovina, and P. Valin, Eds., vol. 6, pp. 931–944, IOS Press, 2006.

[29] I. Bosch, S. Gómez, and L. Vergara, "Automatic forest surveillance based on infrared sensors," in *Proceedings of the International Conference on Sensor Technologies and Applications (SENSORCOMM '07)*, pp. 572–577, October 2007.

[30] I. Bosch, S. Gómez, L. Vergara, and J. Moragues, "Infrared image processing and its application to forest fire surveillance," in *Proceedings of the IEEE Conference on Advanced Video and Signal Based Surveillance (AVSS '07)*, pp. 283–288, September 2007.

[31] I. Bosch and L. Vergara, "Infrared wireless network sensors for imminent forest fire detection," *Internat International Journal on Advances in Networks and Services*, vol. 3, no. 1, pp. 40–49, 2010.

[32] I. Bosch, S. Gómez, and L. Vergara, "A ground system for early forest fire detection based on infrared signal processing," *International Journal of Remote Sensing*, vol. 32, no. 17, pp. 4857–4870, 2011.

7

Robustness of Auditory Teager Energy Cepstrum Coefficients for Classification of Pathological and Normal Voices in Noisy Environments

Lotfi Salhi and Adnane Cherif

Signal Processing Laboratory, Physics Department, Sciences Faculty of Tunis, University of Tunis ElManar, 1060 Tunis, Tunisia

Correspondence should be addressed to Lotfi Salhi; lotfi.salhi@laposte.net

Academic Editors: E. P. Ong and L. Silva

This paper focuses on a robust feature extraction algorithm for automatic classification of pathological and normal voices in noisy environments. The proposed algorithm is based on human auditory processing and the nonlinear Teager-Kaiser energy operator. The robust features which labeled Teager Energy Cepstrum Coefficients (TECCs) are computed in three steps. Firstly, each speech signal frame is passed through a Gammatone or Mel scale triangular filter bank. Then, the absolute value of the Teager energy operator of the short-time spectrum is calculated. Finally, the discrete cosine transform of the log-filtered Teager Energy spectrum is applied. This feature is proposed to identify the pathological voices using a developed neural system of multilayer perceptron (MLP). We evaluate the developed method using mixed voice database composed of recorded voice samples from normophonic or dysphonic speakers. In order to show the robustness of the proposed feature in detection of pathological voices at different White Gaussian noise levels, we compare its performance with results for clean environments. The experimental results show that TECCs computed from Gammatone filter bank are more robust in noisy environments than other extracted features, while their performance is practically similar to clean environments.

1. Introduction

In the objective support of the analysis and the selection of vocal and voice diseases, the automatic evaluation of voice quality based on acoustic analysis stays an efficient tool. In the speech pathology field, on which this work focuses, pathological voices can be evaluated using two approaches that are perceptual analysis and objective analysis. The analysis of pathological voice is a hot topic that has received large attention. There are several medical diseases that harmfully affect our human voice like laryngitis, laryngeal atypia and early/advanced cancer, Reinke's edema, RRP-papillomatosis, spasmodic dysphonia, vocal fold granuloma, vocal fold paresis/paralysis, voice dysfunction in neurological disorders (stroke, Parkinson's disease, benign essential tremor, amyotrophic lateral sclerosis, myasthenia gravis, multiple sclerosis), and pediatric voice disorders. The analysis of the voice disorder stays essentially clinic [1, 2]. The clinician can use the available apparatus for identification of pathological voice. It is usually made by laryngoscopical exams, which are considered invasive, and requires an expert analysis of numerous human speech signal parameters. Automatic analysis of pathological voices has its advantages, such as having its quantitative and noninvasive nature. Furthermore, it allows the detection and supervising of vocal system diseases and reducing analysis charge and time. Based on the voice of a patient, the goal of pathological voice classification is to make a decision whether it is normal or pathological. Successful pathological voice classification will enable an automatic noninvasive device to diagnose and analyze the voice of the patient.

In the current literature, the majority of approved researches in this area have been oriented to the study of acoustic parameters perturbation measurements and noise.

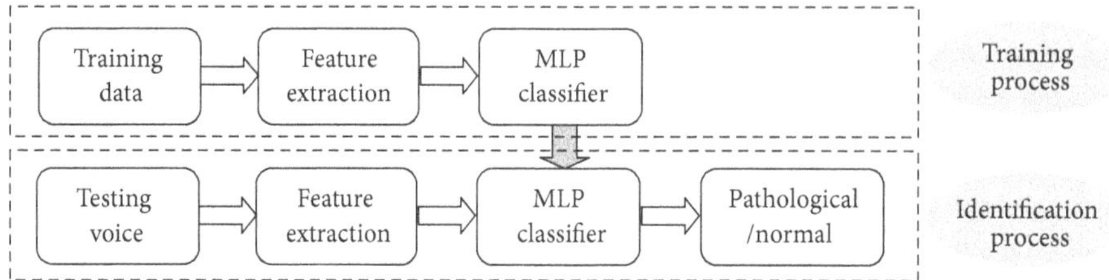

FIGURE 1: Block diagram of the proposed system.

The features used are often extracted from the audio data for voice pathology analysis including the fundamental frequency (F_0), jitter, shimmer, Mel-Frequency Cepstral Coefficients (MFCC), signal-to-noise ratios (SNR), Harmonic-to-Noise Ratios (HNR), and High Order Statistics parameters (HOS) [3]. However, the research for a more detailed and representative acoustic analysis of pathological voice signals is still a promising area. Also, the techniques based on the description of the spectral components to detect the disorder glottal activity have been shown to be consistent in the detection of pathological voices [4].

Regardless of recent advances in the state of the art of automatic classification of pathological and normal voices, identification of pathological voices in noisy conditions remains an open research problem [5]. In general robust identification of pathological voices is an important research area. Performance of all speech parameters deployed in the field often degrades due to adverse and unexpected environmental conditions. Most approaches that have been used in the literature for improving the voice disorders identification in noisy environments mainly fall into three categories: acoustic model adaptation algorithms, speech enhancement algorithms, and robust feature extraction algorithms [6]. In this study, we focus on the topic of robust feature extraction. So, we propose to use the auditory Teager energy feature set for parameterization of voice signal. This proposition is motivated by speech perception consideration that is based on the human perception models and the nonlinear Teager-Kaiser operator that provide a good estimation of the "real" energy of the source of a resonance signal [7, 8]. It is for this reason that this parameter was used recently for identification of pathological voices.

In this study, a parametric analysis based on auditory Teager energy is employed to discriminate pathological voices of speakers affected by different vocal pathologies in noisy environments. In addition to the TECCs performances, the robustness of the proposed system is motivated by the use of human perception models which is a filter bank of one of the three auditory systems: Gammatone or Mel scale triangular. The filter bandwidths are proportional to the auditory Equivalent Rectangular Bandwidth (ERB) function as described in [9–11]. However, the Multilayer Neural Network has been generally used because there is no need to think about the details of the mathematical models of the data and it is reasonably easy to train and has produced a good pathological recognition performance [12, 13]. We admit that a comparison with other feature classification results is needed to evaluate the performance of the proposed feature. In this paper, the MLP method was used to classify the mixed voiced dataset. The proposed features labeled auditory Teager Energy Cepstrum Coefficients (TECCs) are evaluated on tasks of classification of pathological and normal voices in noisy and clean condition. Then, a comparison of the performance of different used features was performed in order to show that it is the most robust in noisy environment. Note that the robustness is shown in terms of correct classification rate (CCR) accuracy.

2. Materials and Methods

2.1. Identification of Pathological Voices System. The proposed approach for the task of automatic classification of pathological and normal voices in noisy environments essentially consists of two parts: feature extraction and classification. Figure 1 illustrates the block diagram of the proposed system.

In the training process, we train the MLP classifier with feature model using the training voice data. We use the supervised training algorithm. So, we give each speech sample with the corresponding nature class label. Then the MLP classifier will be saved with all his specific parameters. In the identification part, the input of this system is .wav files, which come from the testing database or from the real-time speech.

2.2. Corpus. One corpus comprises four sentences produced by 62 normophonic speakers (35 male and 27 female) and 50 dysphonic speakers (28 male and 22 female). The mean age of the selected patients was 53 years (range, 32 to 75 years). Also, the mean age of the volunteers' normophonic speakers working in or around the laboratory was 47 years (range, 28 to 82 years). The patients had been diagnosed on the base of a clinical examination at the ENT Department of the Rabta Hospital in Tunis, Tunisia [14, 15]. Continuous speech samples from patients with a wide variety of organic, neuralgic, traumatic, and psychogenic voice disorders, as well as 62 normal subjects are included. The pathologies had been determined as follows: vocal cord palsy, vocal edema, vocal polyps, vocal nodules, vocal cysts, chronic laryngitis, glottic cancer, Parkinson's, and Alzheimer's. Subjects were instructed to produce four standardized Arabic sentences at a comfortable pitch and volume as naturally as possible without overacting.

A second corpus comprises sustained vowels "a," including onsets and offsets, and four French sentences produced

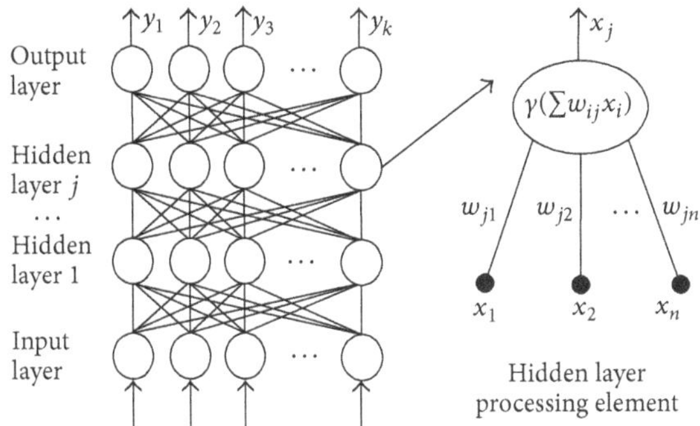

FIGURE 2: General schematic of a neural network (MLP).

by 22 normophonic or dysphonic speakers (10 male and 12 female speakers) [16, 17]. The corpus includes 20 adults (from 20 to 79 years), one boy aged 14 years and one girl aged 10 years. Five speakers are normophonic, the others are dysphonic. The dysphonic speakers were patients of the laryngology department of a university hospital in Brussels, Belgium. The disordered voices range from mildly deviant to very deviant. The pathologies were diagnosed as follows: dysfunctional dysphonia, bilateral nodule, polyp on the left vocal fold, edema of the vocal folds, mutational disorder, dysphonia plica ventricularis, and unilateral vocal fold paralysis. The sentences are referred to as S_1, S_2, S_3, and S_4, respectively. They have the same grammatical structure, the same number of syllables, and roughly the same number of resonants and plosives. Sentences S_1 and S_2 are voiced by default, whereas S_3 and S_4 include voiced and unvoiced segments. Speech signals have been recorded at a sampling frequency of 48 kHz. The recordings were made in an isolated booth by means of a digital audio tape recorder (Sony TCD D8) and a head-mounted microphone (AKG C41WL). The recordings have been transferred from the DAT recorder to computer hard disk via a digital-to-digital interface. Silent intervals before and after each recording have been removed by manual segmentation.

2.3. MLP Classifier. In the last years, neural networks are among the popular signal-processing technologies. In speech processing, neural networks supply as pattern classifiers and as nonlinear adaptive filters. The most popular example of neural network in many tasks of pattern recognition is the multilayer perceptron (MLP) [13]. In the MLP which has one layer or more, the neurons of each layer are interconnected with each other by weights. The activation function of each layer of neurons is a specific mathematical function that allows the neuron to generate an output for the next layer. This activation function is calculated based on the sum of the product between the input vector and the synaptic weights of each unit. Generally, the MLP is trained using the descent gradient method [13].

The trained neural networks are an essential step that can allow the system to learn the prospective interaction between voice quality indices and their corresponding classes. Also it can provide an output representing the definite category for each of voice class indices, whereas the testing step is used to verify the classification ability of the proposed neural networks and thus deduce the CCR of the used speech feature.

Figure 2 gives you an idea about the general schematic of a neural network (MLP) and to an artificial neuron.

2.4. Feature Extraction. The speech signal has many acoustic features which reflect the pathological voices characteristics. In the research domain of classification of pathological and normal voices, the importance of feature extraction is how to extract and select the most pertinent speech features with which most voice pathologies could be identified. Different parameters where chosen at the input of the neural networks such as speech rate, energy, pitch, formant, Linear Prediction Coefficients (LPC), Linear Prediction Cepstrum Coefficients (LPCC), Mel-Frequency Cepstrum Coefficients (MFCCs) and their derivative. The type of each parameter depends on its method of extraction. In this study, our objective is to introduce new speech features that are more robust in classification of pathological and normal voices in noisy environments. We propose a robust speech feature which is based on the combination of the Teager-Kaiser energy cepstrum and an auditory (Gammatone) filter bank (Figure 3).

We investigate the robustness and compare the performance of the proposed GTECC features to that of MFCCs and MTECCs by artificially introducing different levels of white noise to the speech signal and then computing their correct classification rate.

As illustrated in Figure 3, it can be seen that one of the main dissimilarity between MTECC and GTECC is the set of filters used in the extraction. In fact, triangular filter bank equally spaced in the Mel scale frequency axis is used to extract MTECCs features, while in GTECC, the Gammatone filter bank are used. For instance, on the computing of MTECC and GTECC features, the speech energy is estimated through Teager-Kaiser energy operator (TEO).

The MFCC features are a parameters family that may be deducted either using a parametric approach resulting from linear predictive coefficients or using a nonparametric approach based on the Fast Fourier Transform (FFT). In our study we use the nonparametric approach because it allows modelling of the effects induced by the presence of pathology over the excitation (vocal folds) and the system (vocal tract). In recent literature, the MFCC features are mostly used for speech recognition and it presents an excellent performance in this task. Their success occurs from the use of perceptually based Mel-spaced filter bank processing of the Fourier transform and the particular robustness and flexibility that can be achieved using cepstral analysis [18].

Consequently, we derive the filter bank values by catercornered; we multiply the K triangular filter bank weighting function by the NFFT magnitude coefficients and then we collect each filter triangle results.

In order to reflect the human hearing logarithmic compression, we usually take the log of the filter bank output. As indicated by Figure 4, the spacing of the triangle filters

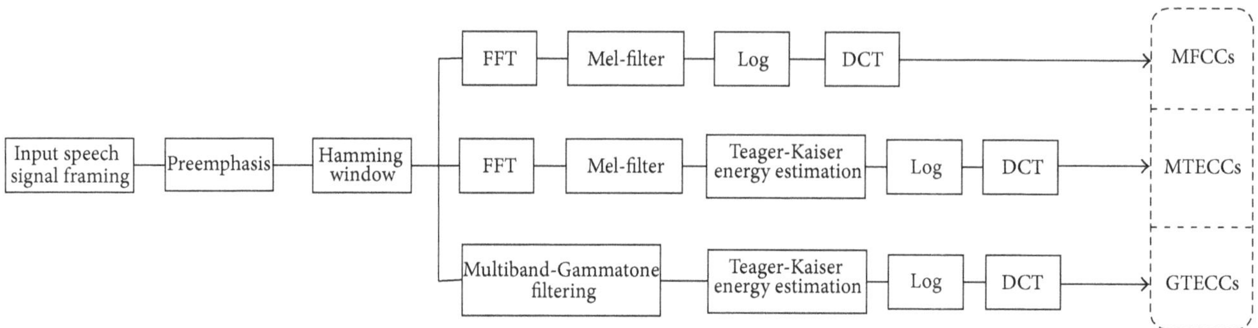

FIGURE 3: Block diagrams of the extraction of MFCC, MTECC, and GTECC features.

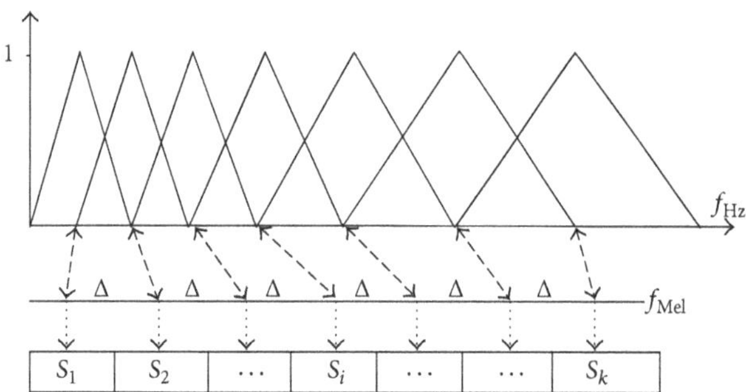

FIGURE 4: Principle of Mel scale filter bank.

bank centres occurs according to the Mel scale defined by the following [19, 20]:

$$f_{\mathrm{Mel}} = 2595\log_{10}\left(1+\frac{f_{\mathrm{Hz}}}{700}\right). \tag{1}$$

Finally and based on the Discrete Cosine Transform (DCT), we apply the cepstral analysis which consists in converting the log filter bank spectral values into cepstral coefficients as shown in

$$C_n = \sum_{i=1}^{K}\log_{10}(S_i)\cos\left[n\left(i-\frac{1}{2}\right)\frac{\pi}{K}\right], \tag{2}$$

where $i = (1,2,\ldots,K)$, K represents the number of the Mel bands in the Mel scale, $n = (1,2,\ldots,N)$, N being the number of MFCCs extracteds and S_i is the short-time Fourier transform (STFT) of the input discrete signal.

2.4.1. Teager-Kaiser Energy Operator (TEO). As shown by Figure 5, the mechanic oscillator with mass "m" and spring constant "K" is equivalent to electrical oscillator consists by a serial "LC" circuit.

This oscillator (Figure 5) is generally used either for generating signals at a particular frequency f_0 or picking out a signal at a particular frequency f_0 from a more complex signal [21, 22]:

$$\omega = \sqrt{\frac{1}{LC}} = \frac{1}{\sqrt{LC}}, \quad \omega = 2\pi f_0. \tag{3}$$

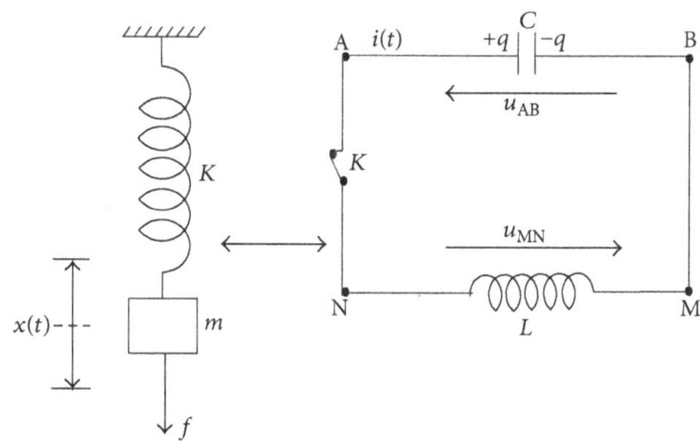

FIGURE 5: Electrical and mechanical resonant oscillators.

The dynamics of this system are described as follows:

$$\frac{d^2q}{dt^2} + \frac{1}{LC}q = 0. \tag{4}$$

The solution of this equation consists of a signal $q(t)$ defined by

$$q(t) = Q\cos(\omega t+\varphi) = Q\cos(\phi(t)). \tag{5}$$

The system's total energy E is the sum of the electrical (capacitive) energy and magnetic (inductive) energy which is given by

$$E = \frac{1}{2}\frac{q^2}{C} + \frac{1}{2}L\left(\frac{dq}{dt}\right)^2 \Longrightarrow E = \frac{1}{2}L\omega^2Q^2 \tag{6}$$

such as $\omega = d\phi(t)/dt$.

Based on this analysis, the Teager-Kaiser operator ψ is defined as follows:

$$\psi[q(t)] = \left(\frac{dq(t)}{dt}\right)^2 - q(t)\cdot\frac{d^2q(t)}{dt^2}. \tag{7}$$

In the approximate discrete form of this operator we discredited the time t in n points and then

$$\psi_d[q(n)] = q^2(n) - q(n+1)\cdot q(n-1). \tag{8}$$

In some cases it is made known that the speech signal can be modeled as a linear combination of AM-FM signals. Then the speech signal can be expressed as follows:

$$s(t) = a(t)\cos(\omega_0 t + \varphi) = a(t)\cos(\phi(t)) = a(t)\cos\left[\int_0^t \omega_i(\tau)\,d\tau + \phi(0)\right], \tag{9}$$

where $a(t)$ is the amplitude signal depending on a time and $\omega_i(t)$ is the instantaneous frequency defined by $\omega_i = d\phi(t)/dt$.

Once applying the TEO to the speech signal give up

$$\psi[s(t)] \simeq \left(a(t)\cdot\frac{d\phi(t)}{dt}\right)^2. \tag{10}$$

Herein, it is shown that TEO can track the modulation energy and identify the instantaneous amplitude and frequency. Motivated by this fact and in order to compute the real signal energy, we will use the TEO model as an alternative of using the commonly used instantaneous energy that only takes into account the "s^2" of the signal's source. The idea of using TEO is motivated by advantage of the modulation energy tracking capability of this technique. Indeed, the Teager-Kaiser estimated energy incorporates both amplitude and frequency information. It is hoped that additional information of the estimated energy will lead to an improvement of the accuracy of the automatic identification of pathological voices [23].

2.4.2. Auditory Filter Bank (Gammatone). In auditory modelling, the digital filter bank is one of the most fundamental concepts that resemble the characteristics of the basilar membrane. In the inner ear's cochlea, each band-pass filter modeled response of part of the basilar membrane to some localized frequency information of the speech signals. Human auditory processing is based on a set of density frequency asymmetric filters used to estimate the activity of each frequency band. The bandwidth of asymmetrical filters is quantified using the notion of the Equivalent Rectangular Bandwidth (ERB). The Gammatone function that represents the impulse response of each filter has the following temporal form [9–11]:

$$g(t) = At^{n-1}\exp(-2\pi b\mathrm{ERB}(f_c)t)\cos(2\pi f_c t), \tag{11}$$

where A, b, n are the Gammatone filter design parameters and f_c is the center frequency of the filter. Figure 6 shows the Gammatone function corresponding to a cochlea filter at order 4, centred at the frequency 1000 Hz and with bandwidth of 125 Hz.

In the Gammatone filter bank, the bandwidth of each filer is established according to the auditory critical band related to its centre frequency. Particularly, the filter's ERB is defined in Hz as in (12) and this is when we specified the magnitude of a filter's frequency response $|H(f)|$ and the filter's maximum gain $|H(f_{\max})|$ at the frequency $f_{\max}$:

$$\mathrm{ERB} = \frac{\int |H(f)|^2}{|H(f_{\max})|^2}. \tag{12}$$

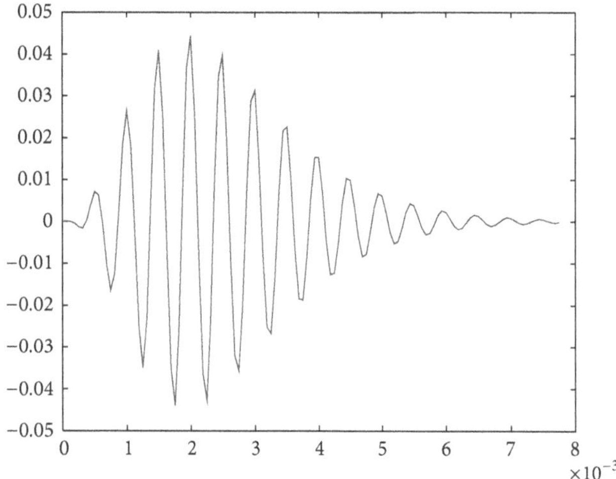

FIGURE 6: Gammatone function of the cochlear filter.

The ERB is the equivalent bandwidth of an orthogonal filter with constant gain $|H(f_{\max})|$ and energy equal to the original filter's energy. Such as the filter's energy is defined as the integral of the filter's frequency response squared. Based on the human physiology states, it is revealed in the current research [23] that the auditory filter bandwidths are given by the following ERB(f) function:

$$\mathrm{ERB}(f) = 6{,}23\left(\frac{f}{1000}\right)^2 + 93{,}39\left(\frac{f}{1000}\right) + 28{,}52, \tag{13}$$

where f is the filter center frequency expressed in Hz.

Using the critical Bark frequency scale, the filter insertion is equidistant as follows:

$$\mathrm{Bark}(f) = \frac{26{,}81 f}{f + 3920} - 0{,}53. \tag{14}$$

Being given the sampling frequency of the signal, the frequency f must verify the condition $0 \le f \le F_s/2$.

Regarding (11) and taking the values $b = 1{,}019$ and $n = 4$ of auditory filters [24], as a result, the filter frequency response $G(\omega)$ is specified by

$$G(\omega) = \frac{A}{2}\frac{6}{(2\pi b\,\mathrm{ERB}(f_c) + j(\omega - \omega_c))^4} + \frac{A}{2}\frac{6}{(2\pi b\,\mathrm{ERB}(f_c) + j(\omega + \omega_c))^4}. \tag{15}$$

Taking into consideration that $|H(\omega_c)| = 1$, the filter gain A is situate and is equal to

$$A = \frac{1}{\sum_{k=1}^{N} t^{n-1}\exp(-2\pi b\ \mathrm{ERB}(f_c)t)}, \tag{16}$$

where N is the sample number of the impulse response.

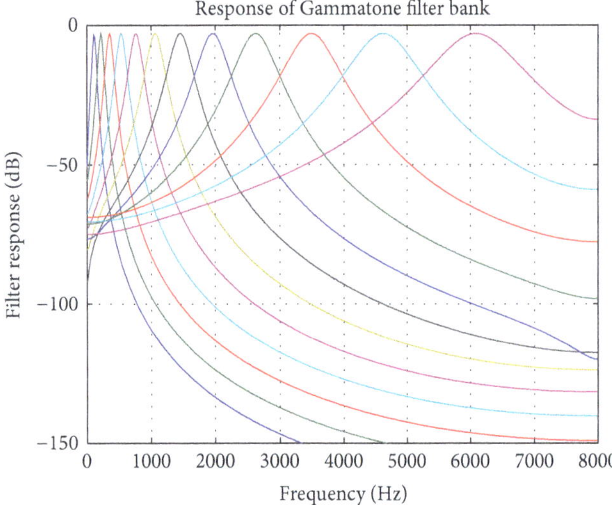

FIGURE 7: Gammatone filter bank with 25 filters.

In anther study [25], the authors discussed two parameters to create a family of Gammatone filter banks. These parameters are the filter bank density (number of filters) and the filters bandwidth parameter denoted F which is a multiplier parameter ($F * \mathrm{ERB}(f)$). The results provided show that both parameters are important for robust speech recognition. Best results are obtained for F around 1.5 and for 30 filters.

Figure 7 shows an example of the Gammatone filter bank with 25 filters and with $1.5 * \mathrm{ERB}(f)$.

3. Experiments and Results

3.1. Experimental Selection. We investigate the robustness of auditory TECCs (GTECCs) in noise by artificial addition of various levels of white noise to the speech signal and computing the correct classification rate (CCR) for each of MFCCs, MTECCs, and GTECCs features. The results are obtained using the databases described previously and based on the general classification algorithm shown in Figure 1 and on the block diagram of feature extraction shown in Figure 3. Concerning the development of the multilayer perceptron (MLP) and accordingly, the number of input layer nodes represents the number of voice quality features, while the single output layer nodes represent the two different class categories (pathological or normal). Many experimental investigations are conducted. The selected number of voice features is 13 MFCCs or TECCs. For the extraction of Gammatone Teager Energy Cepstrum Coefficients (GTECCs) we truncate the cepstrum coefficients to keep the first 13 coefficients similarly to the "standard" MFCC front end. The respective number of hidden nodes that provided the optimal result is 10 hidden nodes. Therefore, the architecture of the network is 13-10-1. The target mean square error (MSE) is fixed to 0.0001 after 5000 iterations. We have created the "Noisy Database" by adding white noise to the speech databases, respectively, at SNR levels of 0 dB, 5 dB, 10 dB, and 15 dB. We performed the CCR for each feature which is the average of two separate values corresponding to the two speech databases experiments. The 75% of the speech database was used to the training process, while the 25% was used to the validation (test) process.

3.2. Results and Discussion. The performance of each voice feature is performed using the correct classification rate (CCR). The speech database is mixed of pathological and normal voices, so the CCR is defined as follows:

$$\mathrm{CCR} = \frac{\mathrm{CCR}_{(\mathrm{Normal})} + \mathrm{CCR}_{(\mathrm{Pathological})}}{2}, \qquad (17)$$

where

$$\mathrm{CCR}_{(\mathrm{Normal})} = \frac{\text{Number of correct classification normal voices}}{\text{Total number of normal voices}} * 100,$$

$$\mathrm{CCR}_{(\mathrm{Path})} = \frac{\text{Number of correct classification pathological voices}}{\text{Total number of normal voices}} * 100. \qquad (18)$$

The signal-to-noise ratio (SNR) is defined as

$$\mathrm{SNR} = 10\log\frac{\sum_{n=0}^{M-1} s^2[n]}{\sum_{n=0}^{M-1} n^2[n]} = 10\log\frac{\sigma_s^2}{\sigma_n^2}, \qquad (19)$$

where $s^2[n]$ and $n^2[n]$ are, respectively, the speech and noise samples of analysed signal segment. Furthermore, σ_s and σ_n are, respectively, the power in the signal or noise frame.

Table 1 recapitulates the experiment results. It gives the CCRs for each voice feature: MFCC, MTECC, and GTECC at clean or noisy condition. In the noisy environment, the speech signal is admixed with white noise for different SNR levels.

3.3. Theoretical Framework

3.3.1. Mel-Frequency Cepstrum Coefficients (MFCCs). Table 1 presents the performance of three voice features in presence of various levels of additive noise. We note that the GTECC features that are extracted using the Gammatone filter bank exhibit the best CCR. Also, it is observable that the performance of the MFCC features decreases when the SNR decreases too, that is, when the speech signal becoming more noisy. Similarly, the performance of MTECC shows a decrease, but it is a relatively small decrease, whereas the GTECC features have the overall highest recognition rate throughout all SNR levels. These results assert well the major interest of the Teager energy operator and of the auditory filter bank analysis. Figure 8 is a graphical representation of Table 1 results.

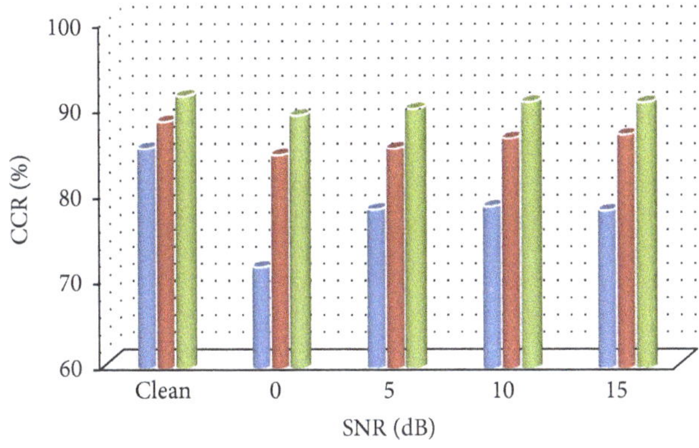

FIGURE 8: Feature performance in clean and noisy condition.

TABLE 1: Feature performance in clean and noisy condition.

SNR (dB)	Clean	15	10	5	0
MFCC					
CCR_{Norm}	86.76	80.52	80.22	78.59	72.06
CCR_{Path}	84.13	77.32	77.36	78.29	71.43
CCR	85.45	78. 29	78.79	78.44	71.74
MTECC					
CCR_{Norm}	88.24	86.76	87.50	86.76	85.29
CCR_{Path}	88.89	87.30	85.71	84.13	84.13
CCR	88.56	87.03	86.61	85.45	84.71
GTECC					
CCR_{Norm}	92.65	92.65	91.18	91.18	89.71
CCR_{Path}	90.48	88.89	90.48	88.89	88.89
CCR	91.56	90.77	90.83	90.03	89.30

4. Conclusion

In this paper, we concentrated on the implementation of an automatic classification of pathological and normal voices system able to worke in noisy environments. This system uses Teager energy cepstral features extracted from an audio signal after analysis by Gammatone filter bank. The proposed features (GTECCs) have been shown to be more robust than MFCCs in white noise environments for low SNR values. For clean conditions and white noise, the MTECCs performed similarly to the GTECCs. In noisy environment, the MFCCs have the lowest classification accuracy but in clean condition there is no big difference with respect to TECC features. The increased robustness of GTECCs is due to both the auditory filter bank design and the Teager energy estimation. In fact, the Gammatone filter bank with filters placed according to the Bark scale and with bandwidths given by the $\mathrm{ERB}(f)$ is a good approximation of the human auditory system. Also, the TEO presents a demodulation-like operation and the envelope of the spectrum produces more robust features.

Acknowledgments

The authors would like to thank the anonymous reviewers for their valuable suggestions and comments. The authors would also like to acknowledge the assistance in data collection of the ENT Department of the Rabta Hospital in Tunis, Tunisia. They also would like to express their appreciation to Professor Francis GRENEZ and to all the members of "Signals and Waves" Laboratory (LIST), Faculty of Engineering, Free University of Brussels, for their invaluable collaborations and for the availability of the voice database.

References

[1] P. Yu, M. Ouaknine, J. Revis, and A. Giovanni, "Objective voice analysis for dysphonic patients: a multiparametric protocol including acoustic and aerodynamic measurements," *Journal of Voice*, vol. 15, no. 4, pp. 529–542, 2001.

[2] B. Boyanov and S. Hadjitodorov, "Acoustic analysis of pathological voices: a voice analysis system for the screening and laryngeal diseases," *IEEE Engineering in Medicine and Biology Magazine*, vol. 16, no. 4, pp. 74–82, 1997.

[3] V. Parsa and D. G. Jamieson, "Identification of pathological voices using glottal noise measures," *Journal of Speech, Language, and Hearing Research*, vol. 43, no. 2, pp. 469–485, 2000.

[4] T. Dubuisson, T. Dutoit, B. Gosselin, and M. Remacle, "On the use of the correlation between acoustic descriptors for the normal/Pathological voices discrimination," *EURASIP Journal on Advances in Signal Processing*, vol. 2009, Article ID 173967, 2009.

[5] C. d'Alessandro, B. Bozkurt, B. Doval et al., "Phase-based methods for voice source analysis," in *Advances in Nonlinear Speech Processing*, pp. 1–27, 2007.

[6] B. Boashash, "Estimating and interpreting the instantaneous frequency of a signal. I. Fundamentals," *Proceedings of the IEEE*, vol. 80, no. 4, pp. 520–538, 1992.

[7] L. Liu, J. He, and G. Palm, "Effects of phase on the perception of intervocalic stop consonants," *Speech Communication*, vol. 22, no. 4, pp. 403–417, 1997.

[8] R. M. Hegde, H. A. Murthy, and V. R. R. Gadde, "Significance of the modified group delay feature in speech recognition," *IEEE Transactions on Audio, Speech and Language Processing*, vol. 15, no. 1, pp. 190–202, 2007.

[9] O. Ghitza, "Auditory models and human performance in tasks related to speech coding and speech recognition," *IEEE Transactions on Speech and Audio Processing*, vol. 2, no. 1, pp. 115–132, 1994.

[10] B. R. Glasberg and B. C. J. Moore, "Derivation of auditory filter shapes from notched-noise data," *Hearing Research*, vol. 47, no. 1-2, pp. 103–138, 1990.

[11] T. Irino and R. D. Patterson, "A time-domain, level-dependent auditory filter: the gammachirp," *Journal of the Acoustical Society of America*, vol. 101, no. 1, pp. 412–419, 1997.

[12] J. Kortelainen and K. Noponen, *Neural Networks*, Intelligent Systems, 2005.

[13] C. M. Bishop, *Neural Networks for Pattern Recognition*, Clarendon Press, Oxford, UK, 1996.

[14] A. Cherif, "Pitch detection and formant extraction of Arabic speech processing," *Journal of Applied Acoustics*, 2001.

[15] L. Salhi, M. Talbi, S. Abid, and A. Cherif, "Performance of wavelet analysis and neural networks for pathological voices

identification," *International Journal of Electronics*, vol. 98, no. 7–9, pp. 1129–1140, 2011.

[16] A. Kacha, F. Grenez, and J. Schoentgen, "Multiband frame-based acoustic cues of vocal dysperiodicities in disordered connected speech," *Biomedical Signal Processing and Control*, vol. 1, no. 2, pp. 137–143, 2006.

[17] F. Bettens, F. Grenez, and J. Schoentgen, "Estimation of vocal dysperiodicities in disordered connected speech by means of distant-sample bidirectional linear predictive analysis," *Journal of the Acoustical Society of America*, vol. 117, no. 1, pp. 328–337, 2005.

[18] S. B. Davis and P. Mermelstein, "Comparison of parametric representation for monosyllabic word recognition in continuously spoken sentences," *IEEE Transactions on Acoustics, Speech, and Signal Processing*, vol. 28, no. 4, pp. 357–366, 1980.

[19] A. Potamianos and P. Maragos, "Time-frequency distributions for automatic speech recognition," *IEEE Transactions on Speech and Audio Processing*, vol. 9, no. 3, pp. 196–200, 2001.

[20] K. S. R. Murty and B. Yegnanarayana, "Combining evidence from residual phase and MFCC features for speaker recognition," *IEEE Signal Processing Letters*, vol. 13, no. 1, pp. 52–55, 2006.

[21] F. Jabloun, A. E. Çetin, and E. Erzin, "Teager energy based feature parameters for speech recognition in car noise," *IEEE Signal Processing Letters*, vol. 6, no. 10, pp. 259–261, 1999.

[22] J. F. Kaiser, "Some observations on vocal tract operation from a fluid flow point of view," in *Vocal Fold Physiology: Biomechanics, Acoustics, and Phonatory Control*, 1983.

[23] H. A. Patil and K. K. Parhi, "Novel variable length teager energy based features for person recognition from their hum," in *Proceedings of the IEEE International Conference on Acoustics Speech and Signal Processing (ICASSP '10)*, pp. 4526–44529, March 2010.

[24] J. F. Kaiser, "On a simple algorithm to calculate the 'energy' of a signal," in *Proceedings of IEEE International Conference on Acoustics, Speech and Signal Processing (ICASSP '90)*, pp. 381–384, Albuquerque, NM, USA, April 1990.

[25] D. Dimitriadis, P. Maragos, and A. Potamianos, "Auditory teager energy cepstrum coefficients for robust speech recognition," in *Proceedings of the 9th European Conference on Speech Communication and Technology*, pp. 3013–3016, Lisbon, Portugal, September 2005.

8

QoS and Energy Aware Cooperative Routing Protocol for Wildfire Monitoring Wireless Sensor Networks

Mohamed Maalej, Sofiane Cherif, and Hichem Besbes

Engineering School of Telecommunications of Tunis (Sup'Com), City of Communications Technologies, El Ghazala 2083, Ariana, Tunisia

Correspondence should be addressed to Sofiane Cherif; sofiane.cherif@supcom.rnu.tn

Academic Editors: A. E. Cetin, I. Korpeoglu, B. U. Toreyin, and S. Verstockt

Wireless sensor networks (WSN) are presented as proper solution for wildfire monitoring. However, this application requires a design of WSN taking into account the network lifetime and the shadowing effect generated by the trees in the forest environment. Cooperative communication is a promising solution for WSN which uses, at each hop, the resources of multiple nodes to transmit its data. Thus, by sharing resources between nodes, the transmission quality is enhanced. In this paper, we use the technique of reinforcement learning by opponent modeling, optimizing a cooperative communication protocol based on RSSI and node energy consumption in a competitive context (RSSI/energy-CC), that is, an energy and quality-of-service aware-based cooperative communication routing protocol. Simulation results show that the proposed algorithm performs well in terms of network lifetime, packet delay, and energy consumption.

1. Introduction

The automatic monitoring of wildfire generally supports multimodal observations. This is due to the extent of the areas to be covered and the difficulty of detecting fire. In fact, most fire detection techniques, for example, based on the video, suffer from false alarms. The use of wireless sensor networks (WSNs) can improve the quality of the detection and consequently the reduction of the false alarm. WSN can be easily deployed and do not require special auxiliary installation. They are mainly used to control buildings, houses, or archaeological sites in the forest.

However, the forest environment presents the problem of wide covered areas requiring the transmission of a large amount of information through the network with the risk of significant energy consumption and hence limiting the lifetime of the network. Particularly, energy parameter is crucial for the wildfire application. This is due to the complexity of maintenance of the sensors and the substitution of dead batteries due to the difficulty of access to these sensors placed generally in large covered areas. The second problem which arises in this type of environment is the fading effect due to the presence of trees leading to an important shadowing phenomenon.

To solve these problems, we propose a new methodology to design and optimize WSN based on both energy conservation and consideration of the quality of transmission for choosing the routing protocol.

Cooperative communication is a promising solution for enhancing WSN lifetime. In recent works, this concept has been proposed to exploit the spatial diversity gains in wireless networks [1–3]. Data aggregation in WSN often uses multihop transmission techniques. At each hop, the network relies on only one sensor. This often results in a significant decrease in the energy of some sensors and thus limits the lifetime of the network while a large number of sensors are still in working condition. The main idea of cooperative communication consists in relying, at each hop, on the resources of multiple nodes or relays (called cooperative nodes) to transmit data from one sensor to another, instead of using only one sensor as relay. Thus, by sharing resources between nodes, the transmission quality is enhanced.

It is also obvious that the use of a cooperative scheme improves the reliability of communication in case of fire propagation. Indeed, the presence of several relays for each possible hop ensures the further communication of information and therefore the possibility of detection and tracking of potential wildfire.

Thus, cooperative mechanism is the key to the performance of cooperative communication protocols. However, it is challenging to find the optimal cooperative policies in dynamic WSN, where reinforcement learning (RL) algorithms can be used to find the optimal control policy without the need of centralized control.

Recently, a cooperative communication protocol for quality-of-service (QoS) provisioning has been proposed and named MRL-CC, a multiagent reinforcement learning-based cooperative communication routing algorithm [1]. The RL concept consists in considering the cooperative nodes as multiple agents learning their optimal policy through experiences and rewards. MRL-CC has been based on internode distance and packet delay to enhance the QoS metrics. However, it does not care about energy consumption and network lifetime which are important components for energy efficiency.

In this paper, we design cooperative communication routing protocol based on both energy consumption and QoS. The QoS is measured by the absolute received signal strength indicator (RSSI). To integrate these two parameters in the routing protocol, we use a competitive/opponent mechanism implemented at each node by the multiagent reinforcement-learning (MRL) algorithm. Our proposed algorithm (RSSI/energy-CC) is also an energy and QoS aware routing protocol since it ensures better performance in terms of end-to-end delay and packet loss rate, taking into account the consumed energy through the network.

The rest of the paper is organized as follows. Section 2 describes the RL algorithm and the design and implementation of MRL-CC algorithm and our algorithm, the RSSI/energy-CC. The performance analysis is presented in Section 3. Finally, Section 4 concludes the paper and gives future research discussions.

2. Cooperative Communication in WSN Using Reinforcement Learning

In this section, the background information on RL is provided. Then, we give an overview about the architecture and design issues of our concept of cooperative communication in WSN. Then, we describe the architecture and design issues of MRL-CC, a cooperative communication algorithm using RL. After that, we explain the architecture of new algorithm, RSSI/energy-CC, taking into account both QoS and energy consumption.

2.1. Reinforcement Learning. RL provides a framework in which an agent can learn control policies based on experiences and rewards. In the standard RL model, an agent is connected to its environment via perception and action, as shown in Figure 1. On each step of interaction, the agent receives as an input, i, some indication of the current state, s, of the environment; the agent then chooses an action, a, to generate as an output. The action changes the state

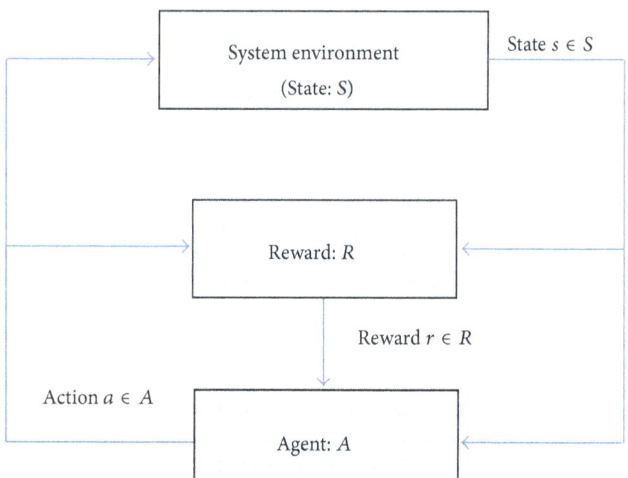

FIGURE 1: Reinforcement learning model.

of the environment, and the value of the state transition is communicated to the agent through a scalar RL signal, r. Depending on its behavior, the agent should choose actions that tend to increase the long-term sum of values of the reinforcement signal [4].

The main idea of RL is to strengthen the good behaviors of the agent while weakening the bad behaviors through rewards given by the environment.

The environment of the agent is described by a Markov decision process (MDP). An MDP models an agent acting in an environment with a tuple (S, A, P, R), where S is a set of states and A denotes a set of actions. $P(s' \mid s, a)$ is the transition model that describes the probability of entering state $s' \in S$ after executing action $a \in A$ at state $s \in S$. $R(s, a, s')$ is the reward obtained when the agent executes a at s and enter s'. The goal of solving an MDP is to find an optimal policy, $\pi : S \mapsto A$, that maps states to actions such that the cumulative reward is maximized [4].

Multiagent systems (MASs) are systems showing that multiple agents are connected to the environment and that they may take actions to change the state of the environment. The generalization of the Markov decision process to the multiagent case is the stochastic game (SG) [5].

In MAS case, each agent assumes itself as the only one that can change the state of the environment and does not consider the interactions between itself and other agents. Therefore, the state transitions are the result of the joint action of all agents, $\mathbf{a} = [a_1, \ldots, a_n]$, where $n$is the number of agents. Consequently, the rewards for each agent R_i, $i = 1, \ldots, n$, also depends on the joint action. The policies $\pi_i : S \mapsto A$ form together the joint policy $\mathbf{\Pi}$.

If $R_1 = \cdots = R_n$, all the agents have the same goal (to maximize the same expected return), and the SG is fully cooperative. If $n = 2$ and $R_1 = -R_2$, the two agents have opposite goals, and the SG is fully competitive. Mixed games are stochastic games that are neither fully cooperative nor fully competitive.

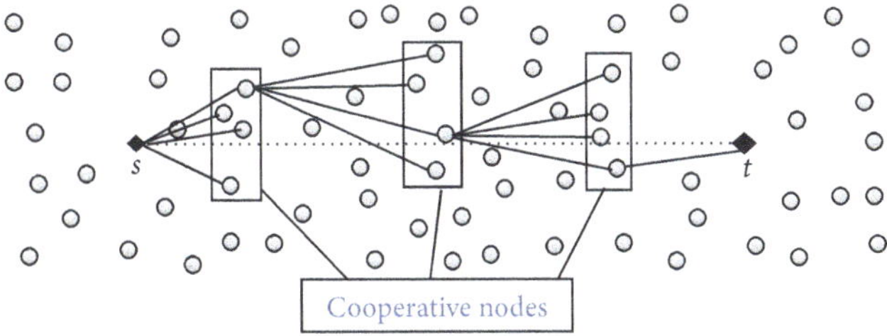

FIGURE 2: Multihop mesh cooperative structure for data dissemination in WSNs.

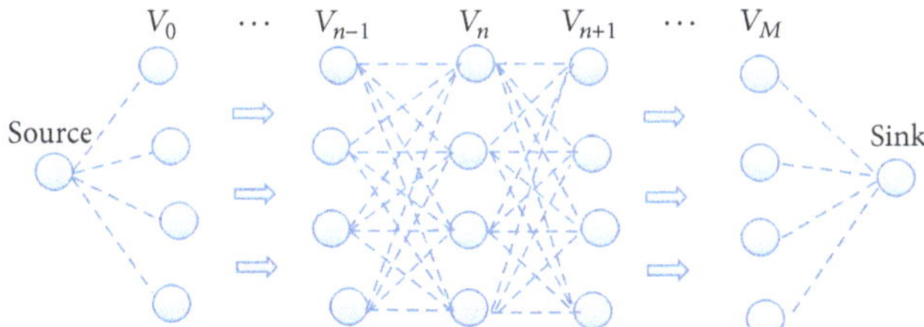

FIGURE 3: Cooperation between adjacent groups of cooperative nodes.

2.2. Cooperative Communication Concept in WSN

2.2.1. Adopted Architecture. For reliable data dissemination in WSNs, we use a multihop mesh cooperative structure. It consists in forming groups of cooperative nodes (denoted as CN) between the source node and the sink node. The data packets originated from a source node are forwarded towards the sink by these CN groups (Figure 2) using a multihop transmission. When a data packet is received by a CN group, a node from that group will be elected to broadcast the data packet to the adjacent CN group. The other nodes of that CN group will help in the packet forwarding in case the elected node fails in data packet transmission or in case the packet is corrupted.

Therefore, we can show the group of nodes connected to each other in a multihop mesh cooperative structure in Figure 3. In fact, the set of nth cooperative group (denoted by V_n) is connected with V_{n-1} and V_{n+1}, which are one hop farther and closer towards the sink than V_n, respectively, that is, each node in V_n is connected with all nodes in V_{n-1} and V_{n+1}.

To construct a multihop mesh cooperative structure, a set of nodes, termed as reference nodes (denoted as RN), between the source node and the sink node is first selected. After that, a set of nodes around each RN will be selected as CN, and thus a multihop mesh cooperative structure is constructed in this phase [6].

2.2.2. WSN Modeling with RL. From the point of view of RL, we can consider a WSN as multiagent system. In fact, sensor nodes can be considered as agents interacting with the environment which can be represented for node $i \in V_n$ as follows.

(i) *State*: the CN groups are modeled to be the environment states:

$$s_n = \{k\}, \quad \text{where } k \in \{\ldots, V_{n-1}, V_n, V_{n+1}, \ldots\}. \tag{1}$$

(ii) *Action*: an agent can operate one of these two actions:

a_f: forwarding of the packet from V_n to V_{n+1},
a_m: monitoring the forwarded packet;
so: $A = \{a_f, a_m\}$.

In our study, we have considered two approaches. The first approach is proposed in [1] where the RL strategy (policy, behaviors, and rewards) for the sensor nodes considers the packet delay and the packet loss rate. This technique has been called the MRL-CC algorithm. The goal of MRL-CC is to enhance packet delay and packet loss rate. The second approach is treated in our work in [7] where the RL strategy is based on the link quality between sensor nodes and their amount of energy consumption. Our strategy goal is to enhance energy efficiency and lifetime of the WSN, that is, to reduce network energy consumption and to maximize network lifetime.

2.3. Multiagent Reinforcement Learning-Based Cooperative Communication Routing Algorithm (MRL-CC)

2.3.1. MRL-CC Implementation. Node election in the CN group is based on a multiagent RL algorithm, performing a fully cooperative task using a "Q-learning" algorithm. The strategy is described as follows.

(i) *Behavior*: each node maintains Q-values of itself and its cooperative partners which reflect the qualities (transmission delay, packet delivery ratio) of the available routes to the sink.

(ii) *Policy*: when a packet is received by the nodes in a CN group, each node will compare its own Q-value with those of other nodes in the CN group; the node which determines that it has the highest Q-value will be elected to forward the data packet to the adjacent CN group towards the sink. The other cooperative nodes will monitor the packet transmission at the next hop.

(iii) *Reward*: the reward function is defined as follows:

$$r_i = \frac{\left(\left(d_{V_n,\text{sink}} - d_{V_{n+1},\text{sink}}\right)/d_{V_n,\text{sink}}\right)}{\left(\left(T_{V_{n+1}} - T_{V_n}\right)/T_{rmn}\right)}, \tag{2a}$$

$$r_i = -\frac{T_{rf}}{T_{rmn}}. \tag{2b}$$

Equation (2a) is used to calculate the reward when the packet forwarding is successful, where $d_{V_n,\text{sink}}$ is the average distance between V_n and the sink, which can be calculated as

$$d_{V_n,\text{sink}} = \frac{1}{N_{V_n}} \sum_{i \in V_n} d_{i,\text{sink}} \tag{3}$$

where N_{V_n} is the number of cooperative nodes in V_n, $T_{V_{n+1}}$ and T_{V_n} are the packet forwarding time at V_{n+1} and V_n, respectively; T_{rmn} is the maximum amount of time that can be elapsed in the remaining path to the sink to meet the QoS requirements on end-to-end delay. The positive reward reflects the quality of the packet forwarding.

Equation (2b) is used to calculate the reward when the packet forwarding fails; T_{rf} is the packet reforwarding timer used for failed forwarding packets. The negative reward reflects the delay caused by the unsuccessful packet transmission from V_n to V_{n+1}.

(i) *Q-value update*: in MRL-CC, for 1-hop forwarding, at iteration t, node $i \in V_n$ forwards a packet to V_{n+1}, and then $j \in V_{n+1}$ is elected to continue packet forwarding. Therefore, node i updates its Q-value as

$$\begin{aligned} Q_i^{t+1}\left(s_i^t, a_i^t\right) &= (1-\alpha)\, Q_i^t\left(s_i^t, a_i^t\right) \\ &+ \alpha \left(r_i^{t+1}\left(s_i^{t+1}\right) + \gamma\omega(i,j) \max_{a_j \in A} Q_j^t\left(s_j^t, a_j^t\right) \right. \\ &\left. + \gamma \sum_{\substack{i' \in V_n \\ i' \neq i}} \omega\left(i, i'\right) \max_{a_{i'} \in A} Q_{i'}^t\left(s_{i'}^t, a_{i'}^t\right) \right), \end{aligned} \tag{4}$$

where $\gamma \in [0,1]$ is the discount factor, $\alpha \in [0,1]$ is the learning rate parameter and $\omega(i,j)$ and $\omega(i,i')$ are, respectively, factors that weigh the maximum Q-value for node j in V_{n+1} and the maximum Q-value of node i' (neighbor of node i) in V_n.

Equation (4) shows that the Q-value of node i is a weighed sum of the Q-value of node i at the previous state, the action's immediate reward, the maximum Q-value of j which is elected as the forwarding node in V_{n+1} at the next hop, and the Q-values of all of i's cooperative partners in V_n.

Note that in the initialization phase, each node is assigned with an initial Q-value. For node $i \in V_n$, its initial Q-value (denoted as Q_i^{ini}) is calculated based on the relative distance (compared with its cooperative partners in V_n) from node i to the nodes in V_{n+1}, as shown in the following:

$$Q_i^{\text{ini}} = \frac{d_{V_n,V_{n+1}}}{d_{i,V_{n+1}}}, \tag{5}$$

where $d_{V_n,V_{n+1}}$ is the average distance between V_n and V_{n+1}, which can be calculated as

$$d_{V_n,V_{n+1}} = \frac{1}{N_{V_n}} \sum_{i \in V_n} d_{i,V_{n+1}}, \tag{6}$$

where N_{V_n} is the number of cooperative nodes in V_n.

The average distance between node i and V_{n+1}, denoted by $d_{i,V_{n+1}}$, can be calculated as

$$d_{i,V_{n+1}} = \frac{1}{N_{V_{n+1}}} \sum_{j \in V_{n+1}} d_{i,j}. \tag{7}$$

2.3.2. Interpretation. We can conclude that MRL-CC algorithm is considering each CN group as one single node because it is performing a fully cooperative task. In fact, all nodes of one CN group get the same positive/negative reward after each transmission procedure. The value of that reward represents the quality of packet forwarding in terms of delay and packet loss rate. Besides, the Q-values of the cooperative nodes are initially based on average distance. Therefore, by electing a node with the highest Q-value, we also understand that the policy adopted in MRL-CC is based on node election with the shortest distance and the lowest packet delay. Thus, MRL-CC ensures communication reliability. However, it has no information about energy consumption that can be a useful parameter to be considered in RL.

2.4. WSN Modeling with Reinforcement Learning in RSSI/Energy-CC Algorithm

2.4.1. Main Idea. Nodes in a CN group will be considered as opponents to each other, so that, each node will maintain a Q-value which reflects the payoff that would have been received if that node selected the action a_f and the other nodes jointly selected the action a_m. After that, the node with the highest total payoff will be elected to forward the data packet to the next CN group towards the sink.

For the rewarding procedure, there are two cases.

(i) Transmission succeeded: the Q-values of each node will be updated according to its energy consumption compared to its neighbors in its CN group.

(ii) Transmission failed: the Q-value of the node that failed to forward the data packet will be updated with a negative reward, whereas for the other nodes, their Q-value will be updated according to an indication about their signal quality.

In our work, we have chosen to use the RSSI as an available indication about signal quality for each packet received at a sensor node.

2.4.2. RSSI/Energy-CC Algorithm Strategy. Node election in the CN group is based on a multiagent RL algorithm, performing a fully competitive task using an "opponent modeling" algorithm [8]. The strategy is described as follows.

(i) *Policy*: node election, for packet forwarding, for the node with the best link quality and the lowest energy consumption, or a tradeoff between the two criteria.

(ii) *Behavior*: each node maintains Q-values which reflects the payoff that would have been received if that node selected the forwarding action a_f and another node in its CN group selected the monitoring action a_m.

(iii) *Reward*: Each time a packet is forwarded, all the nodes will receive immediate rewards from the environment, which represent a tradeoff about energy consumption and quality of the received signal.

2.4.3. Algorithm Initialization Phase. In the initialization phase, each node is assigned with an initial value regarding its opponents in V_n. The initial payoff of node $i \in V_n$ compared to its neighbor i' is the Q-value calculated based on its absolute RSSI in dBm measured from the next cooperative group V_{n+1}. The Q-value is defined as follows:

$$Q_{i',i}^{\text{ini}} = \frac{\text{RSSI}_{i,V_{n+1}} - \text{RSSI}_{i',V_{n+1}}}{\text{RSSI}_{V_n,V_{n+1}}}, \quad (8)$$

where $\text{RSSI}_{V_n,V_{n+1}}$ is the average RSSI between V_n and V_{n+1}, which can be calculated as

$$\text{RSSI}_{V_n,V_{n+1}} = \frac{1}{N_{V_n}} \sum_{i \in V_n} \text{RSSI}_{i,V_{n+1}}, \quad (9)$$

where N_{V_n} is the number of cooperative nodes in V_n.

The average RSSI between node i and V_{n+1}, $\text{RSSI}_{i,V_{n+1}}$, can be calculated as

$$\text{RSSI}_{i,V_{n+1}} = \frac{1}{N_{V_{n+1}}} \sum_{j \in V_{n+1}} \text{RSSI}_{i,j}. \quad (10)$$

2.4.4. Data Dissemination Phase. When a data packet is received by a CN group V_n, each node will compare its own total payoff, regarding all its opponents, with those of other cooperative nodes.

The node which determines that it has the highest total payoff will forward the data packet to V_{n+1}, and other nodes in V_n will deduce whether the packet forwarding is successful or not, by overhearing the packet transmission from V_{n+1} to V_{n+2}.

(i) *Q-value update*: the updating of Q-value iterates at each node in each forwarding procedure. For 1-hop forwarding, at iteration t, node $i \in V_n$ forwards a packet to V_{n+1} and nodes i'; neighbors of i in V_n monitor the packet forwarding. Then, $j \in V_{n+1}$ is elected to continue packet forwarding. Therefore, node $i$updates its Q-values as

$$\begin{aligned} &Q_{i',i}^{t+1}\left(s_i^t, a_f^t, a_m^t\right) \\ &= (1-\alpha)\, Q_{i',i}^{t}\left(s_i^t, a_f^t, a_m^t\right) \\ &\quad + \alpha\left(r_i^{t+1}\left(s_i^{t+1}\right) + \gamma \cdot \omega_{s_i^t} V\left(s_i^t\right) + \gamma\omega_{s_j^t} V\left(s_j^t\right)\right), \end{aligned} \quad (11)$$

where $\omega_{s_i^t}$ and $\omega_{s_j^t}$ are, respectively, factors that weigh the total payoff in V_n and V_{n+1} and $V(s^t)$ is the maximum payoff expressed by

$$V\left(s^t\right) = \max_{a_f^t} \sum_{a_m^t} \frac{C_{i'}^t\left(s^t, a_m^t\right)}{N\left(s\right)} Q_{i',i}^t\left(s^t, a_f^t, a_m^t\right), \quad (12)$$

where $C_{i'}^t(s^t, a_m^t)$ counts the number of times agent i observed agent i' taking action a_m in state s at packet t and $N(s)$ is the total counts for all agents taking action a_m in state s. Therefore, $C_{i'}^t(s^t, a_m^t)/N(s)$ is the probability in which the nodes other than i will select joint action a_m for packet t based on past experience.

So, for $i' \in V_n$ if agent i' chooses a_m action, then

$$\begin{aligned} C_{i'}^{t+1}\left(s^t, a_m^t\right) &= C_{i'}^t\left(s^t, a_m^t\right) + 1, \\ N(s) &= N(s) + 1. \end{aligned} \quad (13)$$

Equation (11) shows that the Q-value of node i is a weighed sum of the Q-value of node i at the previous state, the action's immediate reward and the maximum payoff of the group V_{n+1} and the maximum payoff of the group V_n.

(i) *Reward function*: the reward function is defined as follows:

$$r_i = \frac{\left(\left(\sum_{j \in V_n} E_j / N_{V_n}\right) - E_i\right)}{\left(\left(\sum_{j \in V_n} E_j / N_{V_n}\right) - \min_{j \in V_n} E_j\right)}, \quad (14a)$$

$$r_i = \frac{\text{RSSI}_{i,V_{n+1}}}{\text{RSSI}_{V_n,V_{n+1}}} - \sigma \cdot N_{V_n}. \quad (14b)$$

Equation (14a) is used to calculate the reward when the packet forwarding is successful, where E_i represents the consumed energy for node i of the group V_n. So, nodes with less energy consumption will receive positive rewards, and nodes with more energy consumption will receive negative rewards.

Equation (14b) is used to calculate the reward when the packet forwarding fails. The parameter σ takes 1 for the node that failed to forward data packet, whereas for the other nodes, it takes 0. So, the forwarding-node will receive a negative reward. The other nodes in V_n will receive positive reward according to their RSSI values.

In the opponent modeling case, all nodes in V_n are acting in a fully competitive task. So, the total sum of the attributed rewards to all cooperative nodes is zero.

After a certain number of iterations, nodes in V_n are able to use the learned policy to take appropriate actions.

2.5. Complexity Analysis. As noticed in the previous subsections, RL algorithms are composed of two main phases:

(i) updating phase of the Q-values for each agent;

(ii) node election for data forwarding.

For the Q-learning algorithm, the updating phase is realized through (4). The algorithm complexity concerning the Q-value updating is then equal to N^2.

For the node election phase, the node with the highest Q-value is elected for data forwarding:

$$a_f = \arg\max_i Q_i. \quad (15)$$

So, the algorithm complexity concerning node election equals N. Therefore, the algorithm complexity of the Q-learning algorithm equals to $N + N^2$.

For the opponent modeling algorithm, the updating phase is realized through (11). The algorithm complexity concerning the Q-value updating is then equal to $N \cdot (N - 1)$.

For the node election phase, the node with the highest payoff is elected for data forwarding:

$$a_f = \arg\max_i \sum_{a_m^t} \frac{C_{i'}\left(s^t, a_m^t\right)}{N(s)} Q_{i',i}^t\left(s^t, a_f^t, a_m^t\right). \quad (16)$$

So, the algorithm complexity concerning node election equals N^2. Therefore, the algorithm complexity of the Q-learning algorithm equals $2N^2 - N$.

3. Performance Evaluation

3.1. Simulation Environment. For performance evaluation, we use TOSSIM simulation platform in order to evaluate parameters of interest such as energy consumption. TOSSIM is a discrete event simulator for TinyOS sensor networks that builds directly from the same TinyOS code written for the actual motes.

We simulate different topologies, sizes of WSN, and channel environment parameters (path loss and shadowing effects). The sink node is also placed in different positions. Simulation results concern network lifetime, packet delay (average delay to the sink, percentage of delayed packets, and percentage of lost packets), and energy consumption (network energy consumption and maximal energy consumption per node). Performance of RSSI/energy-CC algorithm is compared each time to MRL-CC algorithm.

The application of wildfire requires special measurement and transmission of temperature. Other parameters may be useful as moisture but are not considered in this paper. The amount of information transmitted is therefore likely to be low data rate. The area to cover, the forest, can be of different shapes. It can even be sparse. In this paper, we consider two different deployment architectures: uniform deployment and circular deployment.

In the forest environment, the transmission of information between different sensors can be significantly affected by the presence of trees. To evaluate the effect of this distortion on the quality of the proposed approach, we have also simulated the network in the presence of shadowing effect modeling this type of fading.

In Table 1, we give the parameters fixed for simulating the different versions of the algorithms.

3.2. Simulation Results

3.2.1. Uniform Deployment. We simulate a WSN where 81 sensor nodes are uniformly distributed in a 80 m × 80 m area (distance between 2 successive nodes is 10 m). The sink node is placed according to three different topologies (Figure 4).

TABLE 1: Simulation parameters.

Packet delivery	every 200 milliseconds
Packet size	17 Bytes
Reforwarding time	10 ms
Communication range	30 m
Initial battery energy	2 Li-ion AA batteries
Path-loss exponent	2
Shadowing standard deviation	2 dB
MAC object	CSMA protocol
Node used for simulation	Mica2 platform

TABLE 2: Network lifetime (in days) till the first node dies.

Network architecture	A	B	C
MRL-CC	52	69	205
RSSI/energy CC	58	77	232

(a) Packet Delay Analysis. We compute in Figure 5 the average delay to the sink, percentage of delayed packets, and percentage of lost packets.

The simulation results show that for noncooperative algorithm, the percentage of lost packets is huge compared to the MRL-CC algorithm and the RSSI/energy CC algorithm. However, in terms of percentage of delayed packets and average delay to the sink, the RSSI/energy-CC algorithm is lower than the MRL-CC algorithm. This is due to the fact that RSSI/energy-CC algorithm relies on the average link quality between the CN groups, which is performing at the same time in a competitive context. This competitive task allows a CN group to elect the node with the best RSSI for packet transmission.

(b) Energy Consumption in a Cooperative Node Group. Figure 6 presents the selected CN groups for data transmission from node 4 to the sink node (topology B is considered).

We display the residual battery energy for each selected CN group in Figure 7, and we compare energy consumption behavior between the MRL-CC algorithm and the RSSI/energy-CC algorithm.

Figure 7 shows that the behavior of energy consumption for each CN group is different when comparing MRL-CC algorithm and RSSI/energy-CC algorithm. For nodes which belong to the same CN group, the residual energy is more balanced for the RSSI/energy-CC algorithm. Thus, energy consumption is saved for each node in each CN group.

(c) WSN Lifetime. Network lifetime is defined as the time when the first node's battery is out of energy. For our case, we have compared the MRL-CC algorithm to the RSSI/energy-CC algorithm, computing at the same time the total energy consumed in the WSN (in J). Results are given in Table 2.

We also present in Table 3 the maximal lifetime during which all sensors can transmit to the sink node.

We can notice from Tables 2 and 3 that network lifetime is enhanced when comparing MRL-CC algorithm to

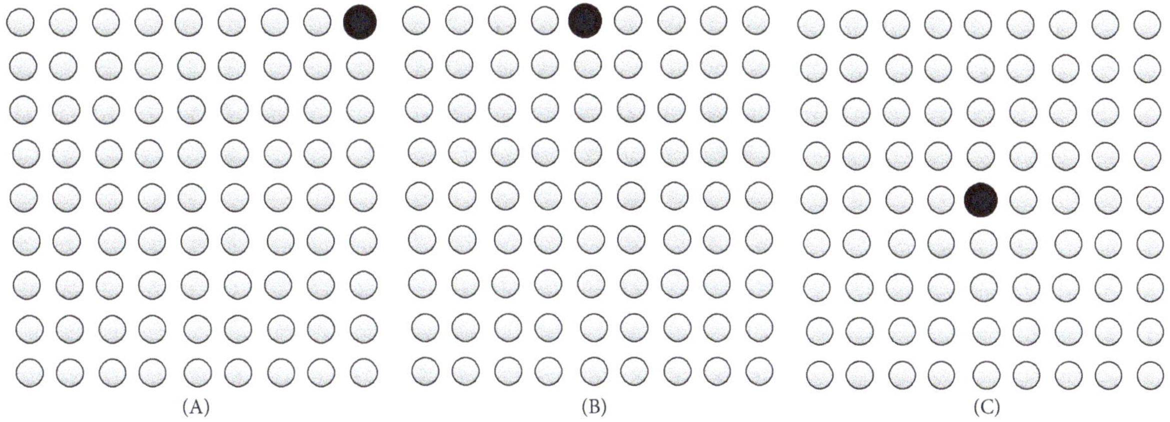

FIGURE 4: Sink node (in black) placement for topologies (A), (B), and (C).

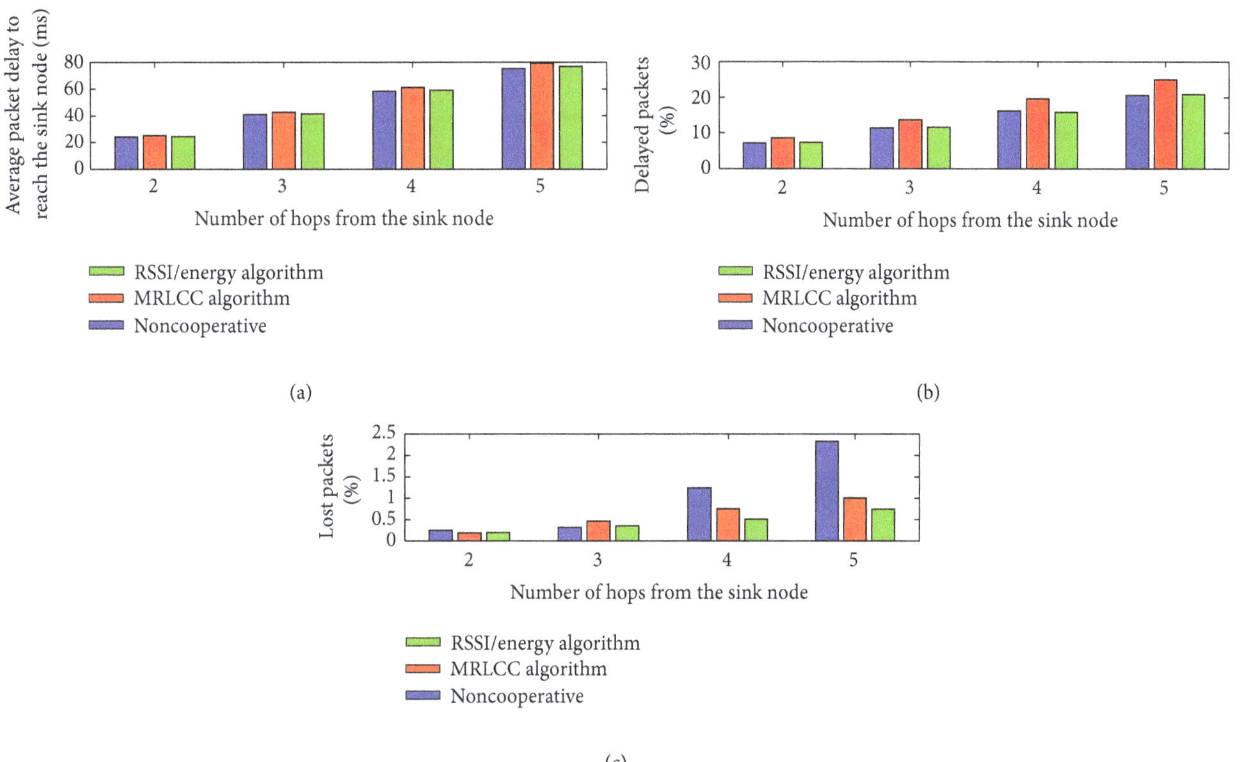

FIGURE 5: Average delay to the sink, percentage of delayed packets, and percentage of lost packets by averaging on the number of nodes being away with the same number of hops from the sink node.

TABLE 3: Network lifetime (in days) till the WSN cannot transmit to the sink node.

Network architecture	A	B	C
MRL-CC	100	178	251
RSSI/energy CC	101	187	275

RSSI/energy-CC algorithm. This enhancement is certainly due to some energy savings in the network.

(d) WSN Energy Consumption. We first investigate energy consumption in the whole network. A comparison between the different network architectures for the two algorithms is presented in Figure 8.

Comparing network architectures, we conclude that C has the lowest energy consumption compared to A and B. So, network lifetime for C is the longest.

Simulation results also show that when comparing network energy consumption between the two algorithms for the same network architecture, network energy consumption is saved for the RSSI/energy CC algorithm compared to the MRL-CC algorithm. This is because the RSSI is considered for the decision of the node election for packet forwarding.

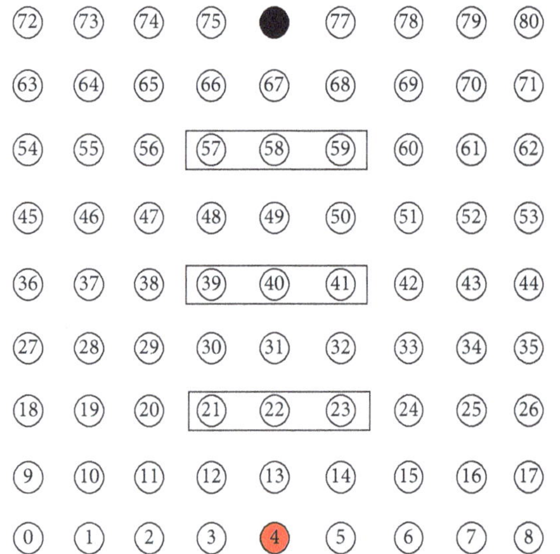

FIGURE 6: Selected CN groups for data transmission from source node 4 (in red) to sink node 76 (in black).

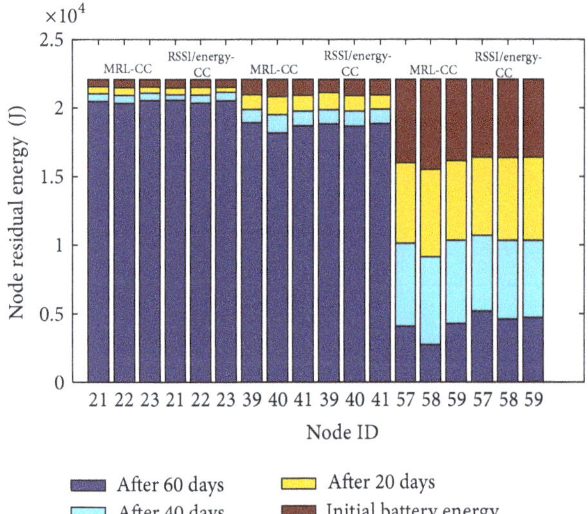

FIGURE 7: Energy consumption comparison for each selected CN group between MRL-CC algorithm and RSSI/energy-CC algorithm.

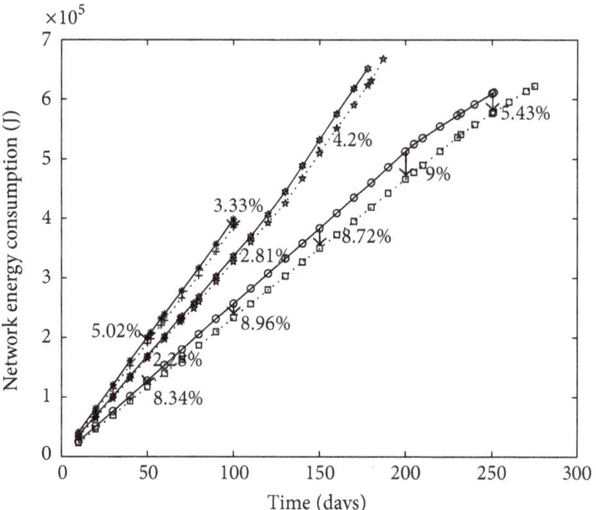

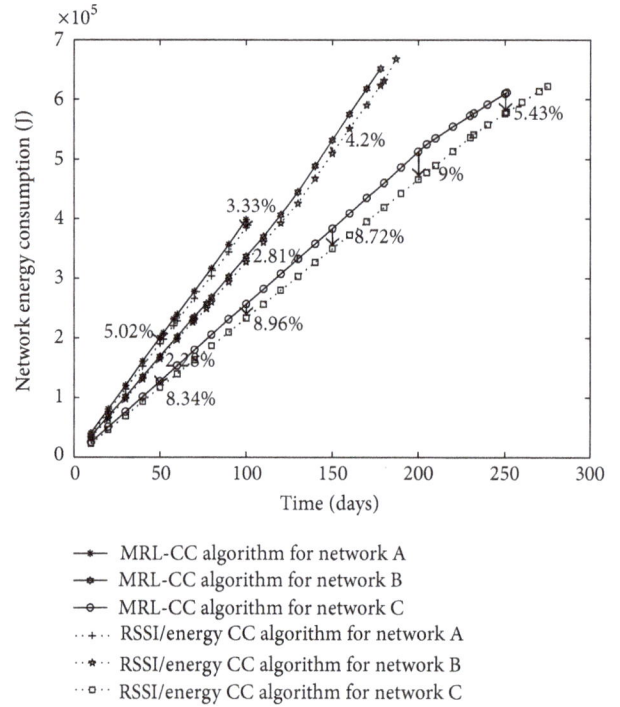

FIGURE 8: Network energy consumption, comparison between network architectures for MRL-CC and E/RSSI CC algorithm.

TABLE 4: Network lifetime (in days) till the first node dies.

Network architecture	9 × 9	13 × 13	21 × 21
MRL-CC	166	138	126
RSSI/energy CC	192	160	146

TABLE 5: Network lifetime (in days) till the WSN can not transmit to the sink node.

Network architecture	9 × 9	13 × 13	21 × 21
MRL-CC	219	206	205
RSSI/energy CC	231	219	205

Network energy consumption is saved from 3.33% to 5.19% for network A, from 2.28% to 6.23% for network B, and from 5.38% to 9.76% for network C.

At the same time, we compare the maximum energy consumption per node in the network, for the two algorithms. For each architecture, we obtain the charts presented in Figure 9.

The simulation results show that the maximum energy consumption per node is reduced for the RSSI/energy CC algorithm compared to MRL-CC algorithm. This is due to taking into account the energy consumption for the cooperative group before making the decision for node election. The maximal energy consumption is saved from 9.56% to 10.6% for network A, from 12.5% to 13.23% for network B, and from 10.79% to 14.76% for network C.

So, we can conclude that network lifetime enhancement is due to the enhancement of node's lifetime with maximal energy consumption.

In a second analysis of energy consumption, we propose to show results for extended grid networks where the sink is placed in the center (alike to topology C). Results about lifetime are shown in Tables 4 and 5.

We can notice from those tables that network lifetime is also enhanced for the RSSI/energy-CC algorithm.

We also display results about network energy consumption in Figure 10, and the maximum energy consumption per node in the network in Figure 11.

Comparing network architecture, we conclude that 9 × 9 network has the lowest energy consumption compared to 13 × 13 and 21 × 21 networks. So, network lifetime for 9 × 9 network is the longest. Simulation results, in Figure 10, also show that when comparing network energy consumption between the two algorithms for the same network architecture, the

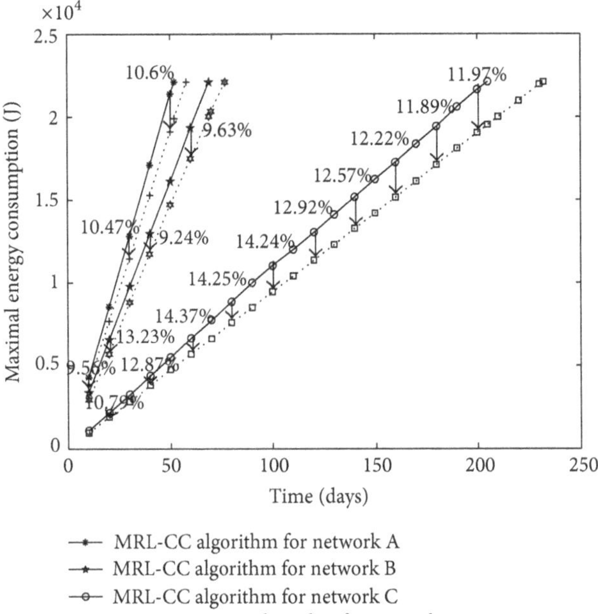

FIGURE 9: Maximal energy consumption in the whole WSN, comparison between MRL-CC and E/RSSI-CC algorithms for different network architectures.

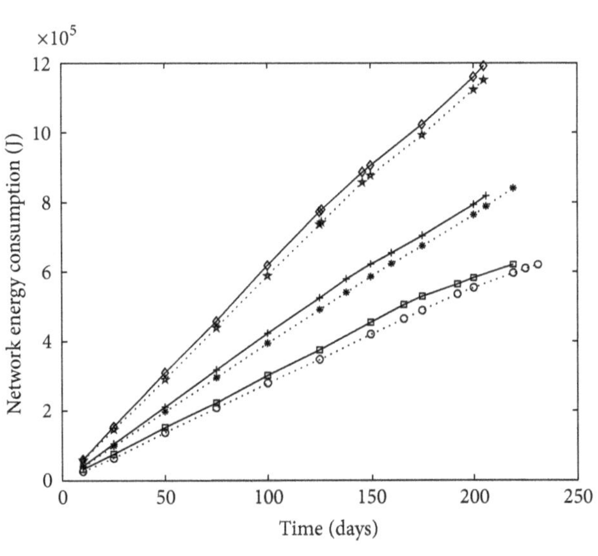

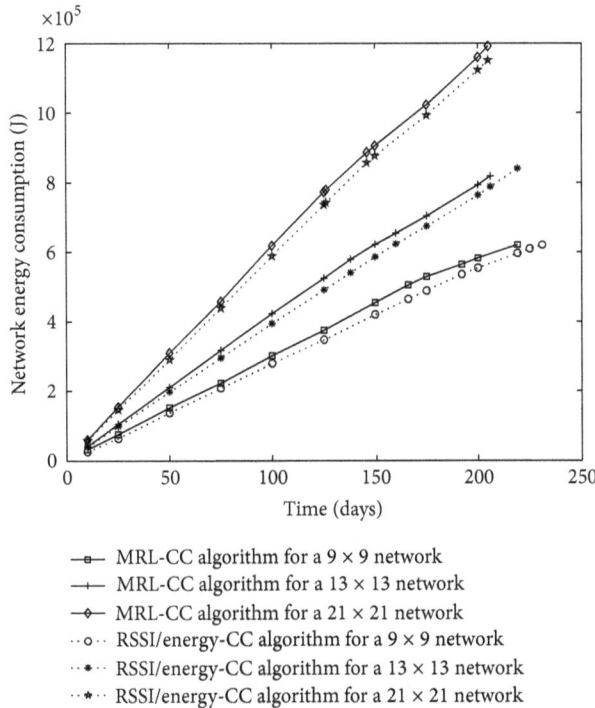

FIGURE 10: Network energy consumption, comparison between network architectures for MRL-CC and E/RSSI CC algorithms.

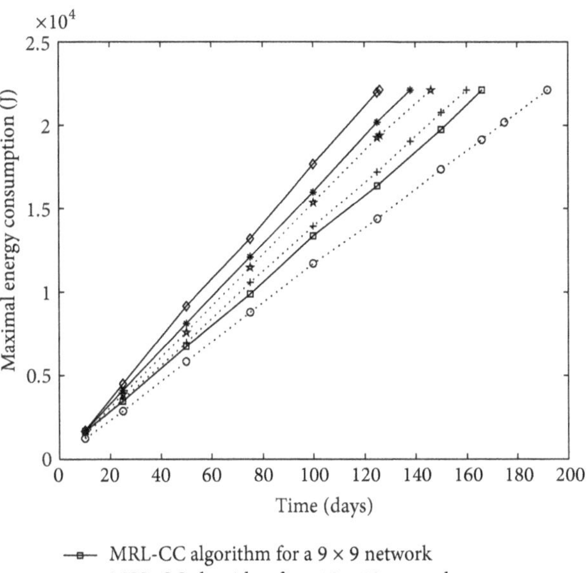

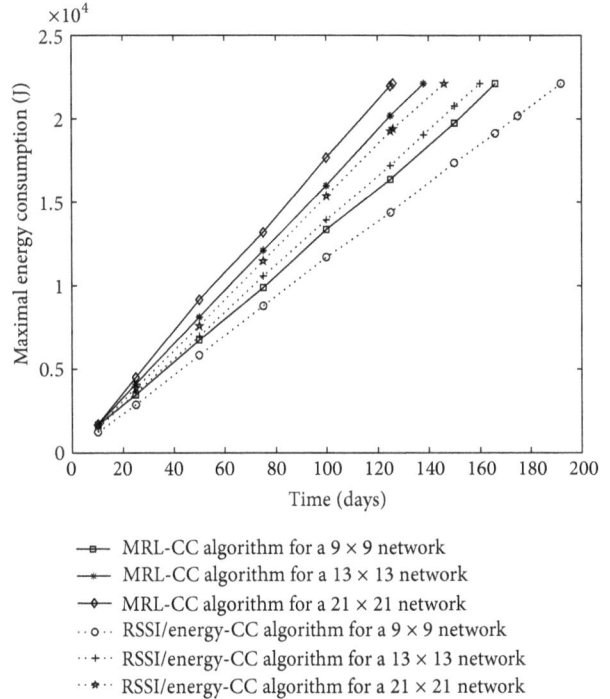

FIGURE 11: Maximal energy consumption in the whole WSN, comparison between MRL-CC and E/RSSI CC algorithms for different network architectures.

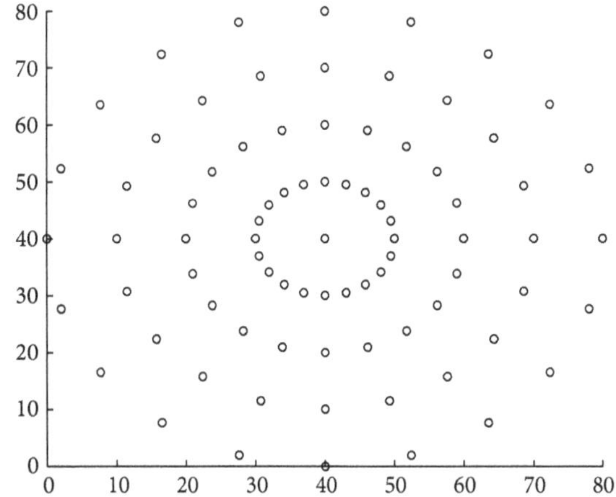

FIGURE 12: WSN topology in circles; sink node is at the center.

network energy consumption is saved for the RSSI/energy CC algorithm compared to the MRL-CC algorithm. Network energy consumption is saved up to 9.49% for 9 × 9 network, up to 6.78% for 13 × 13 network, and up to 6.08% for 21 × 21 network.

In Figure 11, the simulation results show that the maximum energy per node is reduced for the RSSI/energy CC algorithm compared to MRL-CC algorithm. Thus, the maximal energy consumption is saved up to 17.17% for 9 × 9 network, up to 14.12% for 13 × 13 network, and up to 14.01% for 21 × 21 network.

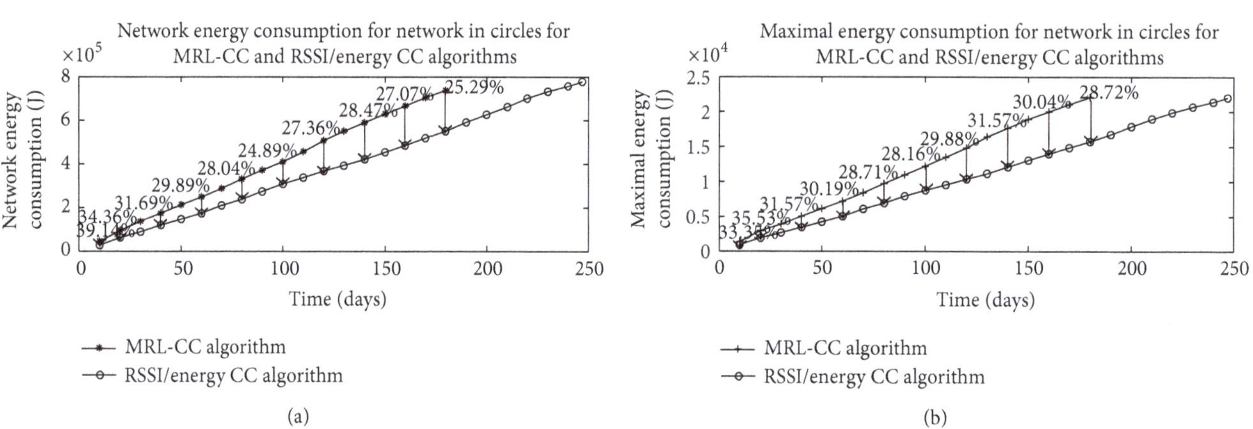

FIGURE 13: Network energy consumption and maximal energy consumption for network in form of circles, for MRL-CC algorithm and RSSI/energy CC algorithm.

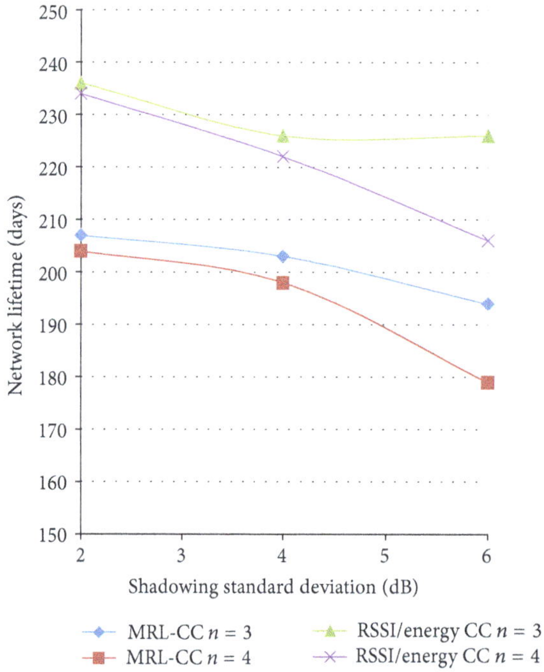

FIGURE 14: Network lifetime (architecture C) for different number of path losses, n, and different shadowing standard deviation, for the MRL-CC and the RSSI/energy algorithms.

3.2.2. Energy Consumption for Circular Topology. We also simulated our algorithms in the form of circles presented in Figure 12. The distance between circles is 10 meters.

Energy simulations for the network in circles for the two algorithms are presented in Figure 13.

The network lifetime for MRL-CC algorithm is 180 days. However, for the RSSI/energy CC algorithm, the network lifetime is 247 days. The gain in network lifetime is very valuable due to the special network topology. Network energy consumption savings go from 24.69% up to 39.14%. Also, for maximal energy consumption, savings are going from 28.16% up to 35.53%.

3.2.3. Shadowing and Path-Loss Effect. We propose to use the network architecture C (uniform deployment) to simulate the network lifetime when path-loss number takes the values: n = 3 and 4, and shadowing deviation takes the values: σ = 2, 4 and 6 dB. Simulation results are shown in Figure 14.

It is obviously clear that the network lifetime is reduced when the path-loss value increases and when the shadowing deviation increases. This result is both for the MRL-CC and the RSSI/energy CC algorithms. From that figure, we can also conclude that the RSSI/energy CC algorithm performs better than the MRL-CC algorithm in terms of network lifetime.

4. Conclusions

To help automatic monitoring of wildfire, we propose in this paper to deploy WSN. To design and optimize the routing protocol used for data aggregation in this network, we propose a new algorithm: the RSSI/energy-CC. This algorithm corresponds to the reinforcement learning optimization approach taking into account energy consumption and link quality measured by the RSSI, performing in a competitive task.

Simulations had shown that this algorithm is efficient in terms of percentage of lost packets, network energy consumption, maximal energy consumption per node, and network lifetime.

In future research, we will consider both the case of multiple sinks in the WSN in order to better process network energy consumption and better enhance the network lifetime and sparse deployment which describes better the forest environment.

Acknowledgment

The research leading to these results has received funding from the European Community's Seventh Framework Programme (FP7-ENV-2009-1) under Grant agreement no. FP7-ENV-244088 "FIRESENSE—Fire Detection and Management through a Multi-Sensor Network for the Protection of Cultural Heritage Areas from the Risk of Fire and Extreme Weather."

References

[1] X. Liang, M. Chen, Y. Xiao, I. Balasingham, and V. C. M. Leung, "MRL-CC: a novel cooperative communication protocol for QoS provisioning in wireless sensor networks," *International Journal of Sensor Networks*, vol. 8, no. 2, pp. 98–108, 2010.

[2] C.-K. Tham and J.-C. Renaud, "Multi-agent systems on sensor networks: a distributed reinforcement learning approach," in *Proceedings of the Intelligent Sensors, Sensor Networks and Information Processing Conference*, pp. 423–429, Melbourne, Australia, December 2005.

[3] A. Nosratinia, T. E. Hunter, and A. Hedayat, "Cooperative communication in wireless networks," *IEEE Communications Magazine*, vol. 42, no. 10, pp. 74–80, 2004.

[4] L. P. Kaelbling, M. L. Littman, and A. W. Moore, "Reinforcement learning: a survey," *Journal of Artificial Intelligence Research*, vol. 4, pp. 237–285, 1996.

[5] L. Buşoniu, R. Babuška, and B. De Schutter, "A comprehensive survey of multiagent reinforcement learning," *IEEE Transactions on Systems, Man and Cybernetics C*, vol. 38, no. 2, pp. 156–172, 2008.

[6] M. Chen, T. Kwon, S. Mao, Y. Yuan, and V. Leung, "Reliable and energy-efficient routing protocol in dense wireless sensor networks," *International Journal on Sensor Networks*, vol. 4, no. 12, pp. 104–117, 2008.

[7] M. Maalej, H. Besbes, and S. Cherif, "A cooperative communication protocol for saving energy consumption in WSNs," in *Proceedings of the IEEE 3rd International Conference on Communications and Networking (ComNet '12)*, pp. 1–5, Hammamet, Tunisia, 2012.

[8] W. Uther and M. Veloso, "Adversarial reinforcement learning," in *Proceedings of the AAAI Fall Symposium on Model Directed Autonomous Systems*, 1997.

9

Identification of Input Nonlinear Control Autoregressive Systems Using Fractional Signal Processing Approach

Naveed Ishtiaq Chaudhary,[1] Muhammad Asif Zahoor Raja,[2] Junaid Ali Khan,[2] and Muhammad Saeed Aslam[3]

[1] *Department of Electronic Engineering, International Islamic University, Islamabad 44000, Pakistan*
[2] *Department of Electrical Engineering, COMSATS Institute of Information Technology, Attock Campus, Attock 43600, Pakistan*
[3] *Pakistan Institute of Engineering and Applied Sciences, Nilore, Islamabad 45650, Pakistan*

Correspondence should be addressed to Muhammad Asif Zahoor Raja; muhammad.asif@ciit-attock.edu.pk

Academic Editors: M. F. G. Penedo and A. Ruano

A novel algorithm is developed based on fractional signal processing approach for parameter estimation of input nonlinear control autoregressive (INCAR) models. The design scheme consists of parameterization of INCAR systems to obtain linear-in-parameter models and to use fractional least mean square algorithm (FLMS) for adaptation of unknown parameter vectors. The performance analyses of the proposed scheme are carried out with third-order Volterra least mean square (VLMS) and kernel least mean square (KLMS) algorithms based on convergence to the true values of INCAR systems. It is found that the proposed FLMS algorithm provides most accurate and convergent results than those of VLMS and KLMS under different scenarios and by taking the low-to-high signal-to-noise ratio.

1. Introduction

Parameter estimation methods have been applied in many important applications arising in applied science and engineering including linear and nonlinear system identification, signal processing, and adaptive control [1–9]. Nonlinear systems are generally categorized into input, output, feedback, and hybrid, that is, combination of input and output nonlinear systems. Many nonlinear systems are modeled with Hammerstein model, a class of input nonlinear systems that consists of static nonlinear blocks followed by linear dynamical subsystems [10, 11]. Such models have been broadly used in diverse fields such as nonlinear filtering [12], biological systems [13], actuator saturations [14], chemical processes [15], audiovisual processing [16], and signal analysis [17].

A lot of interest has been shown by the research community for parameter estimation of Hammerstein nonlinear controlled autoregression models also known as input nonlinear controlled auto-regression (INCAR) systems. For instance, Ding and Chen have developed a least square based iterative procedure and an adaptive extended version of the least square algorithm for Hammerstein autoregressive moving average with exogenous inputs (ARMAX) system [18], Ding et al. also present an auxiliary model using recursive least square algorithm for Hammerstein output error systems [19], and Fan et al. have developed the least square identification algorithm for Hammerstein nonlinear autoregressive with exogenous inputs (ARX) models, while Wang and Ding have developed the extended stochastic gradient algorithm for Hammerstein-Wiener ARMAX models. As per authors' literature survey adaptive or recursive algorithms based on fractional signal processing approach like fractional least mean square algorithm (FLMS) and its normalized version have not been exploited in this domain.

The application of fractional signal processing has been arising in many fields of science and technology including modeling of fractional Brownian motion [20], description of fractional damping [21], charge estimation of lead acid battery through identification of fractional systems [22], which differintegration [23], and Identifying a transfer function from a frequency response[24] etc. Fundamental description, subject terms, importance, and history of fractional signal process can be seen in [25, 26]. Wealth of information about fractional signal processing is also available in special issues of renewed journals [27, 28]. Fractional time integral approach to image structure denoising [29] and design for the adjustable fractional order differentiator [30] are other illustrative recent applications of these approaches. These are also motivation factors for the authors to explore applications of fractional signal processing specially in the area of Hammerstein nonlinear systems.

In this paper, adaptive algorithm based on fractional least mean square (FLMS) approach is applied for parameter estimation of INCAR model to find unknown parameter vector. The FLMS algorithm with different step size parameters is applied to two examples of INCAR model, and performance of the proposed scheme is analyzed for different scenarios of signal-to-noise ratios. The optimization problem is also adaptive with Volterra LMS and recently proposed kernel LMS, and comparison of the results is made with FLMS algorithm for each case of both examples.

The organization of the paper is as follows; in Section 2 the description of the problem based on INCAR model is presented. In Section 3, proposed adaptive algorithms are described. Results of detailed simulations are given in Section 4 alone with necessary discussion. We conclude our finding in the last sections along with few future research directions in this domain.

2. Input Nonlinear Control Autoregressive Systems

In this section, the brief description of input nonlinear control autoregressive (INCAR) systems is presented.

Let us consider the following governing equation of INCAR model as [18, 31]

$$P(z)\,y(t) = Q(z)\,\overline{u}(t) + v(t), \tag{1}$$

here $y(t)$ represents the output of system, $v(t)$ is the disturbance noise, $\overline{u}(t)$ is output of nonlinear block and is given as a nonlinear function of m known basis $(f_1, f_2, \ldots, f_m)$ of the system input $u(t)$ as

$$\begin{aligned}\overline{u}(t) &= f(x(t)) \\ &= a_1 f_1(u(t)) + a_2 f_2(u(t)) + \cdots + a_m f_m((t)),\end{aligned} \tag{2}$$

where $\mathbf{A} = [a_1, a_2, \ldots, a_m]^T \in \mathbb{R}^m$ is the vector of constants, $P(z)$ and $Q(z)$ are known polynomials and given in term of unit backward shift operator $z^{-1}[z^{-1}y(t) = y(t-1)]$, as

$$\begin{aligned}P(z) &= 1 + p_1 z^{-1} + p_2 z^{-2} + \cdots + p_n z^{-n}, \\ Q(z) &= q_1 z^{-1} + q_2 z^{-2} + q_3 z^{-3} + \cdots + q_n z^{-n},\end{aligned} \tag{3}$$

where $\mathbf{p} = [p_1, p_2, \ldots, p_n]^T \in \mathbb{R}^n$ and $\mathbf{q} = [q_1, q_2, \ldots, q_n]^T \in \mathbb{R}^n$ are the constants coefficient vectors. Rearranging equation (1) one has

$$y(t) = [1 - P(z)]\,y(t) + Q(z)\,\overline{u}(t) + v(t) \tag{4}$$

Using (3) in (4) one has

$$\begin{aligned}y(t) = &-\sum_{i=1}^{n} p_i y(t-i) + \sum_{i=1}^{n}\sum_{j=1}^{m} q_i a_j f_j u(t-i) + v(t) \\ = &-\sum_{i=1}^{n} (p_i y(t-i)) + q_1 a_1 f_1(u(t-1)) \\ &+ q_1 a_2 f_2(u(t-1)) + \cdots + q_1 a_m f_m(u(t-1)) \\ &+ q_2 a_1 f_1(u(t-2)) + q_2 a_2 f_2(u(t-2)) + \cdots \\ &+ q_2 a_m f_m(u(t-2)) + \cdots + q_n a_1 f_1(u(t-n)) \\ &+ q_n a_2 f_2(u(t-n)) + \cdots \\ &+ q_n a_m f_m(u(t-n)) + v(t) \\ = &\ \varphi^T(t)\,\boldsymbol{\theta} + v(t),\end{aligned} \tag{5}$$

where the parameter vector $\boldsymbol{\theta}$ and information vector $\varphi(t)$ are defined as

$$\begin{aligned}\boldsymbol{\theta} &= [\mathbf{p}^T, q_1\mathbf{a}^T, q_2\mathbf{a}^T, \ldots, q_m\mathbf{a}^T]^T \in \mathbb{R}^{n_0}, \quad n_0 = n + mn, \\ \varphi(t) &= \left[\varphi_0^T(t), \varphi_1^T(t), \varphi_2^T(t), \ldots, \varphi_m^T(t)\right]^T \in \mathbb{R}^{n_0}, \\ &\qquad n_0 = n + mn, \\ \varphi_0(t) &= [-y(t-1), -y(t-2), \ldots, y(t-n)]^T \in \mathbb{R}^n, \\ \varphi_j(t) &= \left[f_j(u(t-1)), f_j(u(t-2)), \ldots, f_j(u(t-n))\right]^T \in \mathbb{R}^n, \\ &\qquad j = 1, 2, \ldots, m.\end{aligned} \tag{6}$$

Equation (5) represents the linear-in-parameters identification model for Hammerstein control autoregressive systems using parameterization. The detail studies of input nonlinear systems, interested reader are referred to [32].

3. Methodologies for Parameter Estimation of INCAR Model

In this section, brief introductory material is presented for proposed adaptive algorithms for identification of INCAR model given in Section 5. Three recursive algorithms are used: optimization of the model including fractional least mean square (FLMS), Volterra least mean square (VLMS), and kernel least mean square (KLMS).

3.1. Fractional Least Mean Square (FLMS) Algorithm. FLMS belongs to the class of nonlinear adaptive algorithms which is introduced by Zahoor and Qureshi [33] in their work of identification of autoregressive (AR) systems. Since origination of FLMS algorithm, it has been utilized immensely in various problems effectively such as dual-channel speech enhancement [34, 35], acoustic echo cancellation [36], and performance analysis of Bessel beamformers [37, 38]. Our intention is thin study to use FLMS with a different order for parameter estimation of INCAR systems.

The cost function for adaptive algorithm like FLMS is given as

$$j(n) = E\left[|e(n)|^2\right], \tag{7}$$

where

$$e(n) = d(n) - y(n) \tag{8}$$

$e(n)$ represents the difference between desired $d(n)$ and $y(n)$ filter response, $u(n)$ is the input to the filter, and μ is the step size parameter.

Normally, the filter weight update equation for least mean square (LMS) algorithm is written as

$$w_k(n+1) = w_k(n) - \mu\frac{\partial j(n)}{\partial w_k}, \quad k = 0, 1, 2, \ldots, M-1, \tag{9}$$

where M is the number of tap weight and $w_k(n)$ indicates the kth filter weight at n time index. The final weight updated equation for LMS algorithm [39] is given in vector form as

$$\mathbf{w}(n+1) = \mathbf{w}(n) + \mu[\mathbf{u}(n)e(n)]. \tag{10}$$

Accordingly, for FLMS algorithm, filter weight update equation for kth tap weight is written with inclusion of fractional term as

$$w_k(n+1) = w_k(n) - \mu\frac{\partial C(n)}{\partial w_k} - \mu_{\mathrm{fr}}\frac{\partial^{\mathrm{fr}} C(n)}{\partial w_k^{\mathrm{fr}}}, \tag{11}$$

where fr represents the fractional order which is generally taken as real value between 0 and 1, and μ_{fr} is fractional step size parameter. The final weight updated equation for kth tap in case of FLMS algorithm is written as [33]

$$\begin{aligned} w_k(n+1) &= w_k(n) + \mu e(n)u(n-k) \\ &\quad + \mu_{\mathrm{fr}} e(n)u(n-k)\frac{1}{\Gamma(2-\mathrm{fr})}w_k^{1-\mathrm{fr}}(n). \end{aligned} \tag{12}$$

The detailed derivation of (12) can be seen in [33, 40].

3.2. Third-Order Volterra Least Mean Square (VLMS) Algorithm. In this section, brief description of third-order Volterra is presented. Volterra model is widely used in many applications of nonlinear systems including system identification, echo cancellation, acoustic noise control, and nonlinear channel equalization and is also used in transmission channels to compensate the nonlinear effects [41–43].

The governing mathematical relations for Volterra series for a causal discrete time nonlinear system having input $u[n]$ and output $y[n]$ are introduced by Schetzen, in 1980, and given as [42, 44]

$$y[n] = \sum_{r=1}^{N}\sum_{k_1=0}^{M-1}\cdots\sum_{k_r=0}^{M-1} w_r[k_1,\ldots,k_r]\,u[n-k_1]\cdots u[n-k_r], \tag{13}$$

where N represents the degree of nonlinearity in the model, M is the filter memory, $w_r[k_1,\ldots,k_r]$ is the rth-order Volterra kernel. By taking $N = 3$ in (13), the input-output expression for third-order Volterra filter is given as

$$\begin{aligned} y[n] &= w_0 + \sum_{k_1=0}^{M-1} w_1[k_1]\,u[n-k_1] \\ &\quad + \sum_{k_1=0}^{M-1}\sum_{k_2=0}^{M-1} w_2[k_1,k_2]\,u[n-k_1]\,u[n-k_1] \\ &\quad + \sum_{k_1=0}^{M-1}\sum_{k_2=0}^{M-1}\sum_{k_3=0}^{M-1} w_3[k_1,k_2,k_3] \\ &\qquad \times u[n-k_1]\,u[n-k_2]\,u[n-k_3], \end{aligned} \tag{14}$$

here $w_3[k_1,k_2,k_3]$ is the third-order Volterra kernel of the system. In case of symmetric kernels having memory M, then coefficient $M(M+1)(M+2)/6$ is required for third-order kernel [44]. For the third degree of nonlinearity with memory M, the volterra kernel coefficient vector $\mathbf{W}$ is given as:

$$W_k^{(3)T} = \left[w_k^3[0,0,0]\,w_k^3[0,0,1]\cdots w_k^3[M-1,M-1,M-1]\right]. \tag{15}$$

The corresponding input vector $\mathbf{U}$ for $M = 3$ is written as

$$\begin{aligned} U^{(3)T} = {} & \left[u^3[n]\,u^2[n]\,u[n-1]\cdots u[n]\,u^2[n-2]\,u^3\right. \\ & \left.\times[-1]\cdots u[n-1]\,u^2[n-2]\,u^3[n]\right]. \end{aligned} \tag{16}$$

The weights update equation for third-order VLMS is given as

$$W_{k+1}^{(3)} = W_k^{(3)} + \mu e_k U_k^{(3)}, \tag{17}$$

where e_k is the error and μ is the step size parameter. For the detail description of VLMS, interested readers are referred to [44].

3.3. Kernel LMS (KLMS) Algorithm. Pokharel et al. have developed the least mean square (LMS) adaptive algorithm in kernel feature space known in the literature as kernel least mean square (KLMS) algorithm [45]. The basic idea of KLMS algorithm is to transform the data from the input space to a high-dimensional feature space. The importance, fundamental theory, the definition of mathematical term, and applications can be seen in [46–49].

The KLMS algorithm is a modified version of LMS with introduction of kernel feature space, and its weight updating equation is written as

$$\boldsymbol{\omega}(n+1) = \boldsymbol{\omega}(n) + 2\mu e(n)\Phi(\mathbf{u}(n)), \quad (18)$$

where $e(n)$ represents the error term similar to (8) but for KLMS, filter output y is computed as

$$y(n) = \langle \boldsymbol{\omega}(n), \Phi(\mathbf{u}(n))\rangle, \quad (19)$$

here $\langle\cdot,\cdot\rangle$ represents inner product in the kernel Hilbert space and Φ is a mapping which transforms input vector $\mathbf{u}(n)$ to high-dimensional kernel feature space such that

$$\begin{aligned}\langle \Phi(\mathbf{u}(j)), \Phi(\mathbf{u}(n))\rangle &= \langle \kappa(\cdot,\mathbf{u}(i)), \kappa(\cdot,\mathbf{u}(n))\rangle \\ &= \kappa(\mathbf{u}(j), \mathbf{u}(n)),\end{aligned} \quad (20)$$

where $\Phi(\mathbf{u}(n)) = \kappa(\cdot,\mathbf{u}(n))$ defines the Hilbert space associated with the kernel and can be taken as a nonlinear transformation from the input to feature space. Using (20) in (19) gives

$$y(n) = \mu\sum_{j=0}^{n-1} e(j)\kappa(\mathbf{u}(j), \mathbf{u}(n)). \quad (21)$$

Equation (21) is called the KLMS algorithm and further detail about the procedure for the derivation of the algorithm is given in [45, 46].

In this study we will only consider most widely used Mercer kernel which is given by translation invariant radial basis (Gaussian) kernel as

$$\kappa(\mathbf{u},\mathbf{v}) = \exp\left(-\frac{\|\mathbf{u}-\mathbf{v}\|^2}{\sigma^2}\right). \quad (22)$$

4. Simulations and Results

In this section, results of simulations are presented for two case studies of INCAR model using proposed FLMS, VLMS, and KLMS algorithms. The parameter estimation is carried in both studies by taking different levels of signal-to-noise ratio (SNR) and with various step size μ parameters. Moreover, FLMS operates based on different values of fractional orders.

4.1. Case Study 1. The INCAR model for this case is taken as follows:

$$\begin{aligned}P(z)\,y(t) &= Q(z)\,\bar{u}(t) + v(t),\\ P(z) &= 1 + p_1z^{-1} + p_2z^{-2} = 1 + 1.35z^{-1} - 0.75z^{-2},\\ Q(z) &= q_1z^{-1} + q_2z^{-2} = z^{-1} + 1.68z^{-2},\\ \bar{u}(t) &= f(u(t)) = a_1u(t) + a_2u^2(t) + a_3u^3(t)\\ &= u(t) + 0.50u^2(t) + 0.20u^3(t),\\ \boldsymbol{\theta} &= [\theta_1,\theta_2,\theta_3,\theta_4,\theta_5,\theta_6,\theta_7,\theta_8]^{\mathrm{T}}\\ &= [p_1,p_2,a_1,a_2,a_3,q_2a_1,q_2a_2,q_2a_3]^{\mathrm{T}}\\ &= [1.35,-0.75,1.00,0.50,0.20,1.68,0.84,0.336]^T.\end{aligned} \quad (23)$$

In numerical experimentation, the input $u(t)$ is taken as persistent excitation signal sequence with zero mean and unit variance, and $v(t)$ is taken as a white noise sequence with zero mean and constant variance. Before applying the design methodology, a figure of merit or fitness function is developed based on estimation error as

$$\varepsilon = \frac{\|\mathbf{w}(n)-\boldsymbol{\theta}\|}{\|\boldsymbol{\theta}\|}, \quad (24)$$

where $\mathbf{w}(n)$ is vector of adaptive parameter for INCAR model based on nth iteration of the algorithm and vector for the true or desired values is represented by $\boldsymbol{\theta}$. Now the requirement is to find weight vector $\mathbf{w}$ such that the value of fitness function given in (24) approaches zero, and, consequently, the $\mathbf{w}$ approaches $\boldsymbol{\theta}$.

The proposed adaptive algorithms based on FLMS, VLMS, and KLMS are applied to find the optimal weight vector $\mathbf{w}$ for INCAR system using sufficient large number of iteration, that is, $n = 20000$. Two types of step size variation strategy are adopted for each algorithm. Firstly, up to $n = 1000$ iterations, the larger values of step size parameter are taken, that is, $\mu = 10^{-04}$, for fast convergence and for remaining iteration smaller value of step size is used, that is, $\mu = 10^{-08}$, for the stability. Secondly, initially the step size is taken as $\mu = 10^{-03}$and later on $\mu = 10^{-05}$ for $n > 1000$. The design schemes are evaluated for INCAR models based on four different levels of signal-to-noise ratio, that is, 30 dB, 20 dB, 10 dB, and 3 dB. The iterative results of each algorithm against the values of merit function are shown graphically in Figures 1 and 2 for first and second strategy of μ, respectively, for all four variants of SNR. It is found that for higher values of SNR and lower values of step size, all the three algorithms are convergent but the accuracy and convergence of the FLMS algorithm are much better than those of VLMS and KLMS. Moreover, with the increase in step size VLMS algorithm diverges, while efficiency of both KLMS and FLMS algorithms increases and remains convergent.

The design parameters of INCAR model obtained with adaptation procedure of VLMS, KLMS, and FLMS_1 for fr = 0.5 and FLMS_2 for fr = 0.75 are listed in Tables 1, 2, 3, and 4 for SNR = 30 dB, 20 dB, 10 dB, and 3 dB, respectively, for both step size strategies. The values of mean square error (MSE) from true parameters of INCAR model are calculated and its values are also tabulated in Tables 1, 2, 3, and 4 for each algorithm. The values of absolute error (AE) for each element of the design parameter are calculated from reference value of INCAR model and results are presented in Figure 3 for each variant of SNR and for both step size strategies. In order to broaden the small difference in the values, results are plotted on semilog scale. It is seen from the results presented that for high SNR values, like 30 dB, the values of MSE for FLMS_1 and FLMS_2 are of the order 10^{-07} to 10^{-06} and for low SNR values like 3 dB the values of MSE are around 10^{-04} to 10^{-05} for FLMS algorithm. Moreover, for increased values of step size, that is, $\mu \in (10^{-03}, 10^{-05})$, the VLMS algorithm is not providing the convergent results while both KLMS and FLMS give accurate results. The MSE and AE values of KLMS algorithm are considerably inferior to FLMS

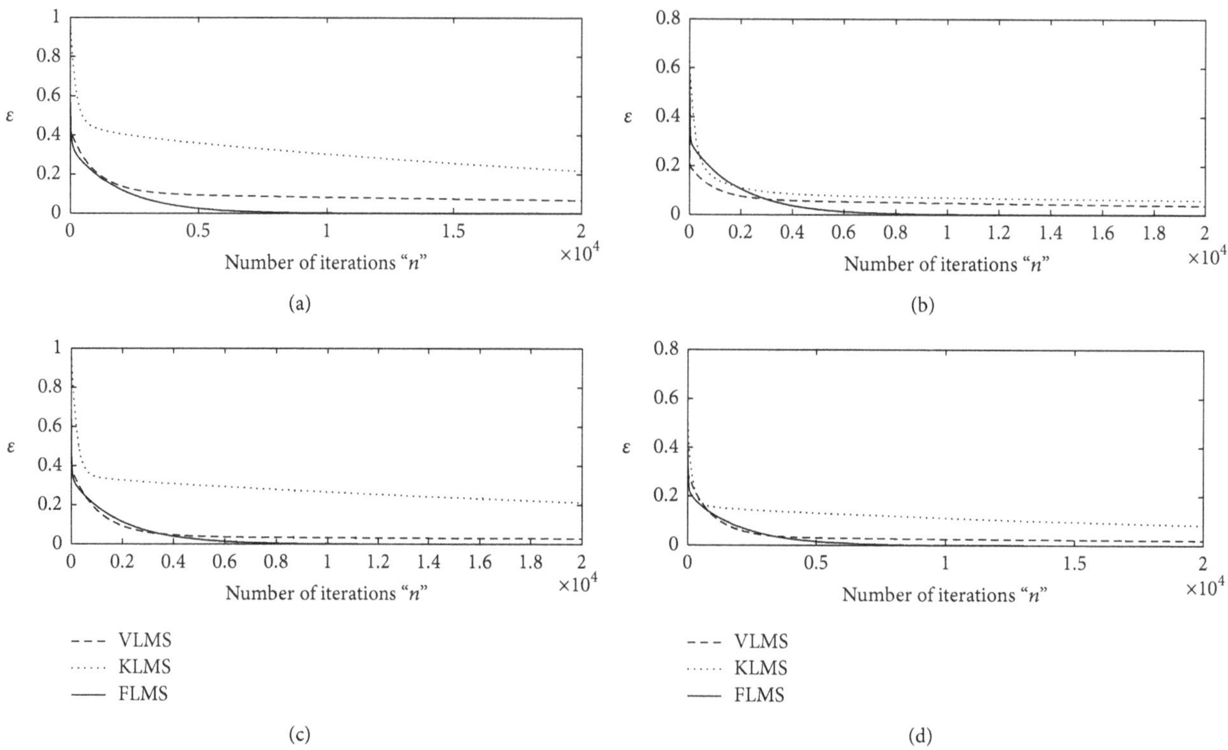

FIGURE 1: Iterative adaptation of merit function by VLMS, KLMS, and FLMS for fr = 0.5 algorithms for $\mu \in (10^{-04}, 10^{-08})$; (a) for SNR = 30 dB, (b) for SNR = 20 dB, (c) for SNR = 10, and (d) for SNR = 3 dB.

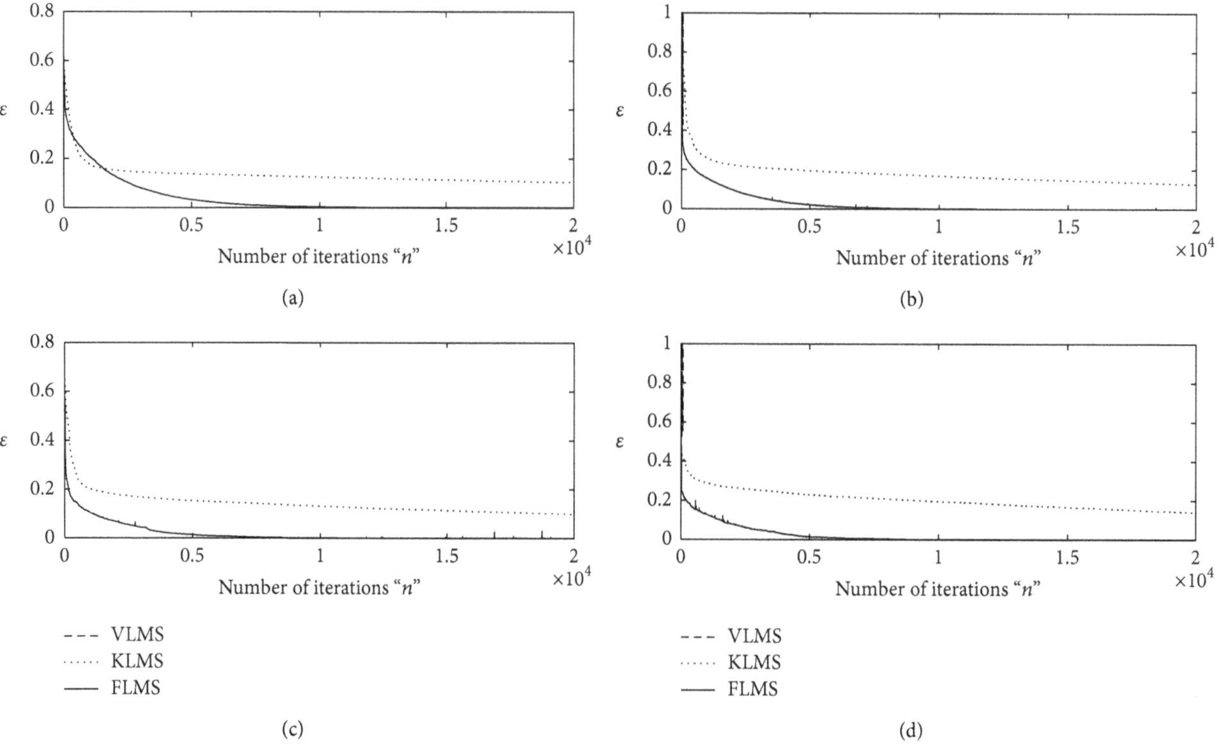

FIGURE 2: Iterative adaptation of merit function by VLMS, KLMS, and FLMS for fr = 0.5 algorithms for $\mu \in (10^{-03}, 10^{-05})$; (a) for SNR = 30 dB, (b) for SNR = 20 dB, (c) for SNR = 10, and (d) for SNR = 3 dB.

TABLE 1: Comparison of proposed results against true values of INCAR model for 30 dB SNR.

μ	Method	Design parameters								MSE
		p_1	p_2	a_1	a_2	a_3	q_2a_1	q_2a_2	q_2a_3	
$(10^{-04}, 10^{-08})$	VLMS	1.339192	−0.738613	0.783251	0.722700	0.143211	1.614600	0.948003	0.311984	$1.46E-02$
	KLMS	1.347467	−0.751995	1.185194	0.341222	0.235790	1.435438	1.067026	0.282885	$2.19E-02$
	$FLMS_1$	1.350010	−0.749510	0.999882	0.499921	0.200304	1.679315	0.839550	0.336252	$1.36E-07$
	$FLMS_2$	1.350001	−0.749505	1.000112	0.499909	0.200210	1.679258	0.839549	0.336301	$1.44E-07$
$(10^{-03}, 10^{-05})$	VLMS	NaN	NaN	NaN	NaN	NaN	NaN	NaN	NaN	NaN
	KLMS	1.353513	−0.746410	1.210424	0.385133	0.139330	1.317242	0.989680	0.541315	$3.22E-02$
	$FLMS_1$	1.349938	−0.748981	0.997433	0.502193	0.201849	1.676682	0.839442	0.335640	$3.41E-06$
	$FLMS_2$	1.349940	−0.748991	0.997454	0.502326	0.201821	1.676806	0.839434	0.335511	$3.37E-06$
True values		1.350000	−0.750000	1.000000	0.500000	0.200000	1.680000	0.840000	0.33600	

TABLE 2: Comparison of proposed results against true values of INCAR model for 20 dB SNR.

μ	Method	Design parameters								MSE
		p_1	p_2	a_1	a_2	a_3	q_2a_1	q_2a_2	q_2a_3	
$(10^{-04}, 10^{-08})$	VLMS	1.347467	−0.751995	1.185194	0.341222	0.235790	1.435438	1.067026	0.282885	$2.19E-02$
	KLMS	1.352035	−0.745072	0.923361	0.244504	0.577959	1.563722	0.660224	0.673292	$4.67E-02$
	$FLMS_1$	1.351852	−0.750247	0.998773	0.499946	0.201665	1.679453	0.841320	0.336062	$1.23E-06$
	$FLMS_2$	1.351805	−0.750225	0.998823	0.499995	0.201691	1.679341	0.841338	0.336139	$1.22E-06$
$(10^{-03}, 10^{-05})$	VLMS	NaN	NaN	NaN	NaN	NaN	NaN	NaN	NaN	NaN
	KLMS	1.353428	−0.745015	1.141632	0.298110	0.300711	1.325873	1.163538	0.362182	$3.77E-02$
	$FLMS_1$	1.345754	−0.753166	0.996107	0.496821	0.201403	1.679157	0.838343	0.338860	$8.36E-06$
	$FLMS_2$	1.345854	−0.753357	0.996484	0.496661	0.201533	1.679222	0.838346	0.338957	$8.30E-06$
True values		1.350000	−0.750000	1.000000	0.500000	0.200000	1.680000	0.840000	0.33600	0

TABLE 3: Comparison of proposed results against true values of INCAR model for 10 dB SNR.

μ	Method	Design parameters								MSE
		p_1	p_2	a_1	a_2	a_3	q_2a_1	q_2a_2	q_2a_3	
$(10^{-04}, 10^{-08})$	VLMS	1.350924	−0.748145	0.772281	0.747924	0.138729	1.481738	1.066460	0.281078	$2.63E-02$
	KLMS	1.355082	−0.736640	0.657397	0.631919	0.529419	0.932575	1.184584	0.819516	$1.44E-01$
	$FLMS_1$	1.348795	−0.753541	0.999514	0.505212	0.194240	1.687344	0.839732	0.332946	$1.72E-05$
	$FLMS_2$	1.348876	−0.753596	0.999636	0.505522	0.193962	1.686987	0.839622	0.333021	$1.74E-05$
$(10^{-03}, 10^{-05})$	VLMS	NaN	NaN	NaN	NaN	NaN	NaN	NaN	NaN	NaN
	KLMS	1.339625	−0.751687	1.065474	0.302732	0.375939	1.634616	0.705183	0.515233	$3.88E-02$
	$FLMS_1$	1.346058	−0.732482	0.977546	0.494359	0.193604	1.691872	0.857421	0.345628	$1.80E-04$
	$FLMS_2$	1.346365	−0.732517	0.977584	0.494478	0.192037	1.688032	0.857678	0.347713	$1.79E-04$
True values		1.350000	−0.750000	1.000000	0.500000	0.200000	1.680000	0.840000	0.33600	

TABLE 4: Comparison of proposed results against true values of INCAR model for 3 dB SNR.

μ	Method	Design parameters								MSE
		p_1	p_2	a_1	a_2	a_3	q_2a_1	q_2a_2	q_2a_3	
$(10^{-04}, 10^{-08})$	VLMS	1.349988	−0.749965	1.004847	0.536095	0.187753	1.244750	1.281211	0.230035	$4.96E-02$
	KLMS	1.354804	−0.739470	1.021873	0.808820	−0.040308	0.858857	1.095944	0.934113	$1.56E-01$
	$FLMS_1$	1.340508	−0.735628	1.003795	0.503585	0.206571	1.684274	0.845499	0.346739	$6.64E-05$
	$FLMS_2$	1.340415	−0.735617	1.003511	0.503718	0.207224	1.683801	0.845381	0.347265	$6.84E-05$
$(10^{-03}, 10^{-05})$	VLMS	NaN	NaN	NaN	NaN	NaN	NaN	NaN	NaN	NaN
	KLMS	1.354922	−0.759970	1.011317	0.632627	0.038110	1.601097	0.584400	0.668535	$3.93E-02$
	$FLMS_1$	1.351284	−0.725432	1.008120	0.504721	0.201932	1.694754	0.805207	0.332677	$2.67E-04$
	$FLMS_2$	1.351821	−0.726599	1.015656	0.504366	0.198380	1.696545	0.805365	0.331465	$2.89E-04$
True values		1.350000	−0.750000	1.000000	0.500000	0.200000	1.680000	0.840000	0.33600	

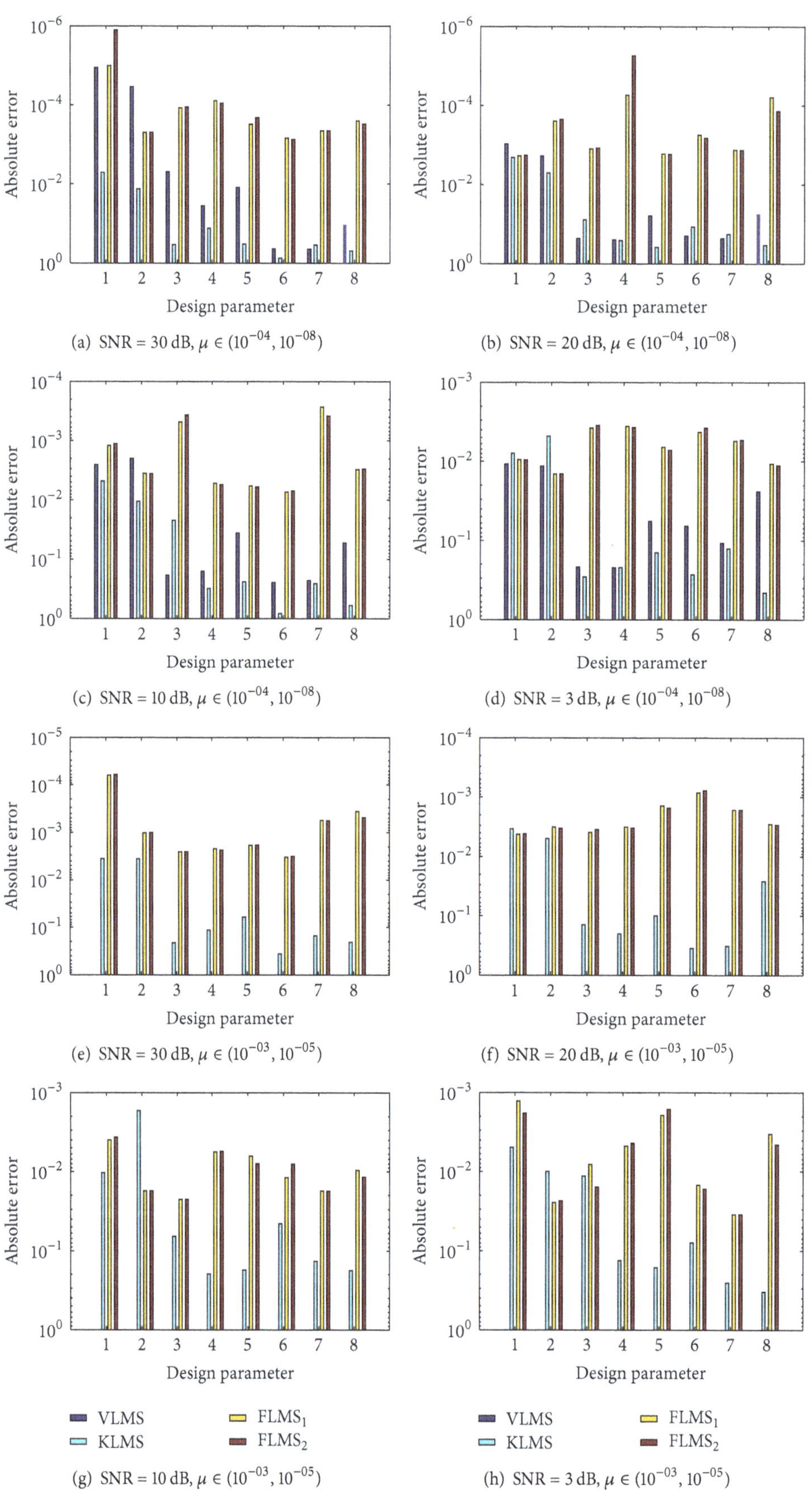

(a) SNR = 30 dB, $\mu \in (10^{-04}, 10^{-08})$

(b) SNR = 20 dB, $\mu \in (10^{-04}, 10^{-08})$

(c) SNR = 10 dB, $\mu \in (10^{-04}, 10^{-08})$

(d) SNR = 3 dB, $\mu \in (10^{-04}, 10^{-08})$

(e) SNR = 30 dB, $\mu \in (10^{-03}, 10^{-05})$

(f) SNR = 20 dB, $\mu \in (10^{-03}, 10^{-05})$

(g) SNR = 10 dB, $\mu \in (10^{-03}, 10^{-05})$

(h) SNR = 3 dB, $\mu \in (10^{-03}, 10^{-05})$

FIGURE 3: Comparison on the basis of absolute error from true values for INCAR model in case study 1.

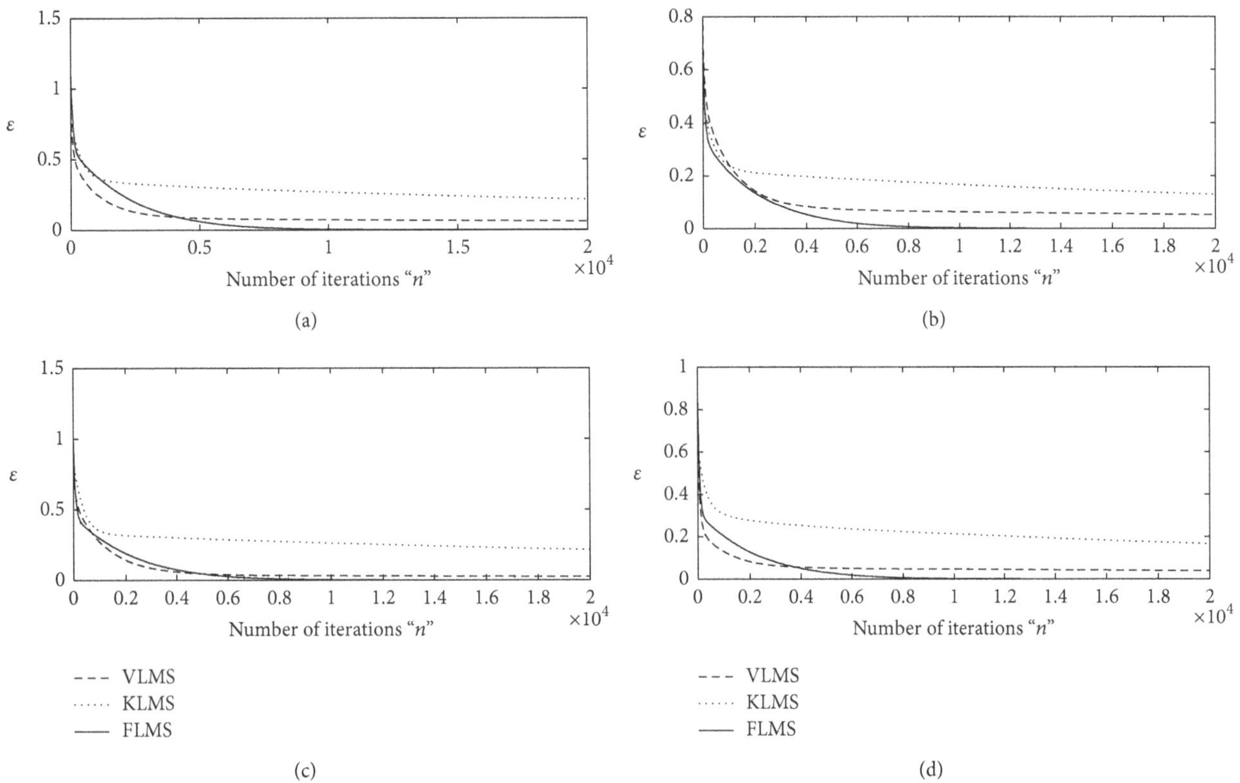

FIGURE 4: Iterative adaptation of merit function by VLMS, KLMS, and FLMS for fr = 0.5 algorithm for $\mu \in (10^{-04}, 10^{-08})$; (a) for SNR = 30 dB, (b) for SNR = 20 dB, (c) for SNR = 10, and (d) for SNR = 3 dB.

algorithm. Generally, it is observed that with decrease in the values of step size parameter, the stability of the algorithm is observed but needs more computational budget to achieve better results.

4.2. Case Study 2. Another INCAR system has been taken in this case as

$$\begin{aligned} P(z)\,y(t) &= Q(z)\,\bar{u}(t) + v(t), \\ P(z) &= 1 + p_1 z^{-1} + p_2 z^{-2} = 1 + 1.35z^{-1} - 0.75z^{-2}, \\ Q(z) &= q_1 z^{-1} + q_2 z^{-2} = z^{-1} + 1.68z^{-2}, \\ \bar{u}(t) &= f(u(t)) = a_1 u(t) + a_2 u^2(t) + a_3 u^3(t) \\ &= u(t) + 0.50u^2(t) - 0.20u^3(t), \\ \boldsymbol{\theta} &= [\theta_1, \theta_2, \theta_3, \theta_4, \theta_5, \theta_6, \theta_7, \theta_8]^{\mathrm{T}} \\ &= [p_1, p_2, a_1, a_2, a_3, q_2 a_1, q_2 a_2, q_2 a_3]^{\mathrm{T}} \\ &= [1.35, -0.75, 1.00, 0.50, -0.20, 1.68, 0.84, -0.336]^{T}. \end{aligned} \quad (25)$$

The numerical experimentation for this case has been performed on a similar pattern as in the previous case study. The proposed schemes based on FLMS, VLMS, and KLMS methods are applied to find vector of design parameters $\mathbf{w}$ for INCAR system using sufficient large number of iterations, that is, $n = 20000$. The same types of step size variation strategy and variants of SNR are used for each algorithm in this case as described in the last example. The iterative results of each algorithm against the values of merit function are plotted in Figures 4 and 5 for first and second strategy of μ, respectively, for all four variants of SNR. The vector for design parameters of INCAR systems optimized with VLMS, KLMS, and FLMS_1 for fr = 0.5 and FLMS_2 for fr = 0.75 are tabulated in Tables 5, 6, 7, and 8 for SNR = 30 dB, 20 dB, 10 dB, and 3 dB, respectively, for both step size strategies. The values of MSE and AE of the proposed schemes from true parameters of INCAR model are calculated and results are given in Tables 5, 6, 7, and 8 and Figure 6, respectively.

It is seen from the results presented that with high SNR values, that is, 30 dB, the values of MSE for FLMS_1 and FLMS_2 are around 10^{-06} to 10^{-07} while for low SNR values, that is, 3 dB, the values of MSE are around 10^{-04} to 10^{-05}. By increasing the values of step size, that is, $\mu \in (10^{-03}$ and $10^{-05})$, the VLMS algorithm is also giving the convergent results for this case, as well as both KLMS and FLMS provide accurate and convergent results. The MSE and AE values for the KLMS and VLMS algorithms for this case are also found to be inferior from FLMS algorithm. Moreover, it is found that with decrease in the values of step size parameter, the stability in the algorithm is observed but needs relatively more computations to get better results.

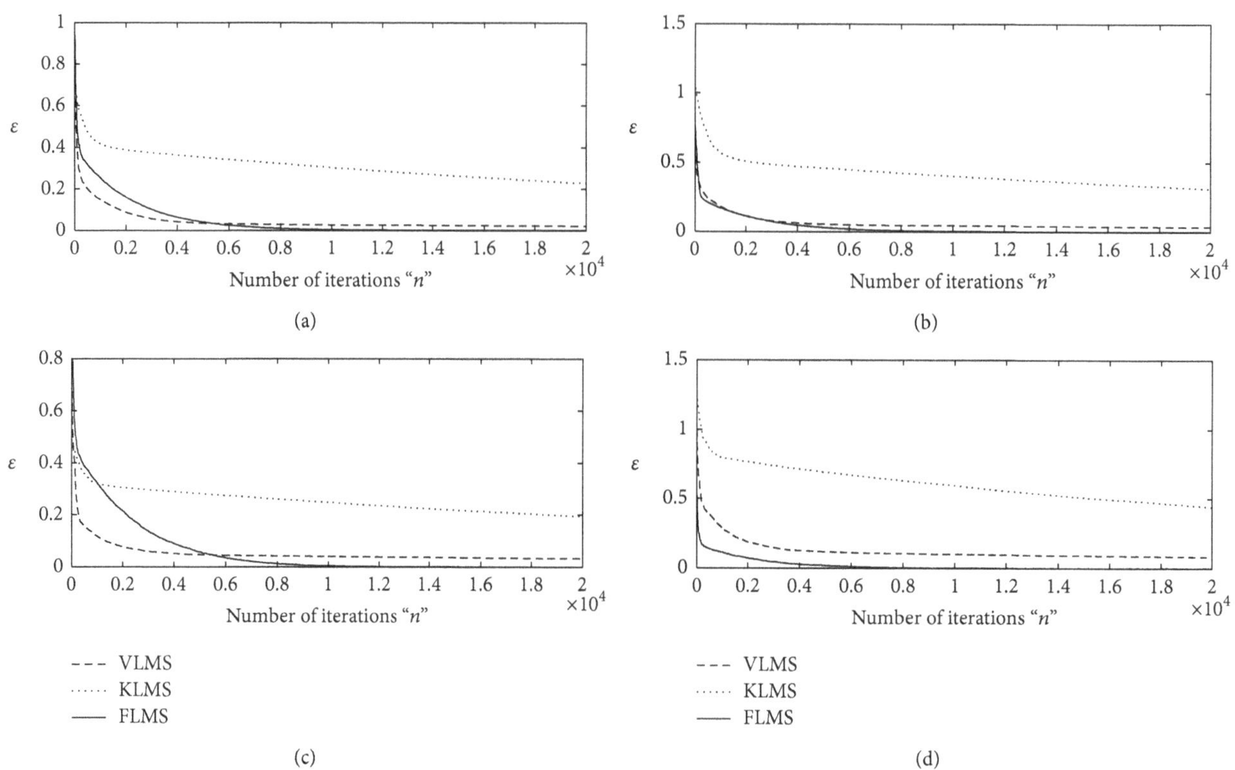

FIGURE 5: Iterative adaptation of merit function by VLMS, KLMS, and FLMS for fr = 0.5 algorithms for $\mu \in (10^{-03}, 10^{-05})$ (a) for SNR = 30 dB, (b) for SNR = 20 dB, (c) for SNR = 10, and (d) for SNR = 3 dB.

TABLE 5: Comparison of proposed results against true values of INCAR model for 30 dB SNR.

μ	Method	Design parameters								MSE
		p_1	p_2	a_1	a_2	a_3	q_2a_1	q_2a_2	q_2a_3	
$(10^{-04}, 10^{-08})$	VLMS	1.353089	−0.745850	0.939847	0.577771	−0.222600	1.428070	1.095959	−0.399576	$1.79E-02$
	KLMS	1.355430	−0.739217	0.694061	0.493618	0.180732	1.425650	0.558743	0.272328	$9.41E-02$
	$FLMS_1$	1.349448	−0.750309	0.998908	0.499814	−0.199658	1.679808	0.840671	−0.336060	$2.79E-07$
	$FLMS_2$	1.349481	−0.750319	0.999085	0.499812	−0.199776	1.680022	0.840677	−0.336151	$2.22E-07$
$(10^{-03}, 10^{-05})$	VLMS	1.347530	−0.747896	1.023045	0.527512	−0.177532	1.591528	0.921393	−0.372951	$2.20E-03$
	KLMS	1.351936	−0.745468	0.941774	0.405908	−0.002987	1.483573	0.737265	−0.011737	$2.57E-02$
	$FLMS_1$	1.347141	−0.749222	1.002343	0.501873	−0.200487	1.679787	0.840078	−0.337043	$2.39E-06$
	$FLMS_2$	1.347230	−0.749253	1.002398	0.501947	−0.200409	1.679837	0.840063	−0.337117	$2.40E-06$
True values		1.350000	−0.750000	1.000000	0.500000	−0.200000	1.680000	0.840000	−0.336000	

TABLE 6: Comparison of proposed results against true values of INCAR model for 20 dB SNR.

μ	Method	Design parameters								MSE
		p_1	p_2	a_1	a_2	a_3	q_2a_1	q_2a_2	q_2a_3	
$(10^{-04}, 10^{-08})$	VLMS	1.349454	−0.744427	1.111856	0.406555	−0.186142	1.348684	1.163043	−0.409249	$3.01E-02$
	KLMS	1.355797	−0.738500	0.777289	0.235371	0.385983	1.384078	0.601534	0.270259	$1.22E-01$
	$FLMS_1$	1.353185	−0.747387	0.997633	0.499185	−0.199222	1.681031	0.838613	−0.335203	$3.43E-06$
	$FLMS_2$	1.353038	−0.747304	0.997603	0.499176	−0.199241	1.680798	0.838613	−0.335052	$3.37E-06$
$(10^{-03}, 10^{-05})$	VLMS	1.344572	−0.743960	0.921542	0.588116	−0.222840	1.580521	0.967188	−0.370809	$5.22E-03$
	KLMS	1.351239	−0.745972	0.971052	0.517420	−0.160944	1.518107	0.630421	0.060205	$2.87E-02$
	$FLMS_1$	1.355256	−0.749658	1.000849	0.497428	−0.198812	1.683680	0.847624	−0.336714	$1.36E-05$
	$FLMS_2$	1.355106	−0.749508	1.000717	0.497390	−0.198819	1.683638	0.847686	−0.336603	$1.35E-05$
True values		1.350000	−0.750000	1.000000	0.500000	−0.200000	1.680000	0.840000	−0.336000	

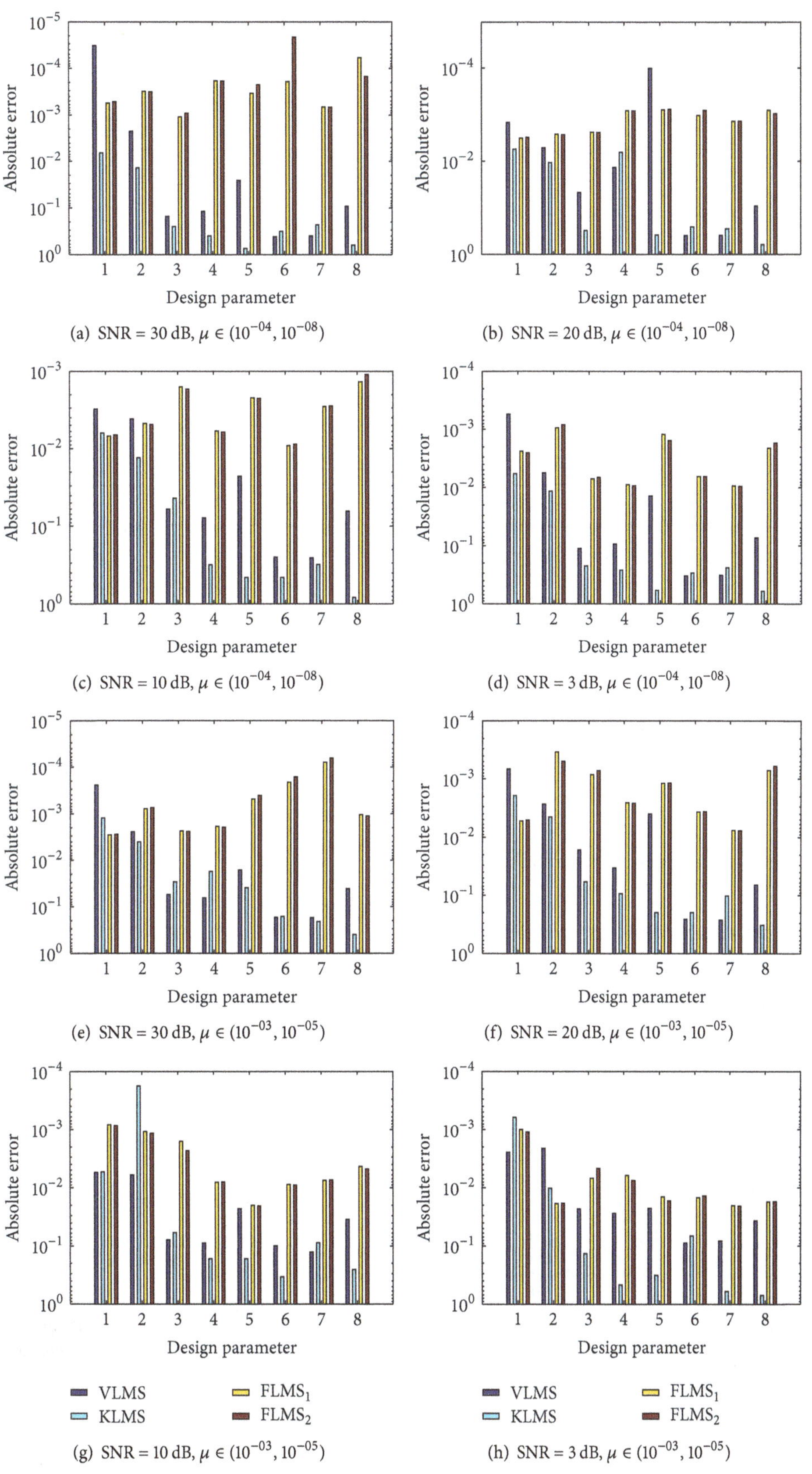

FIGURE 6: Comparison on the basis of absolute error from true values for INCAR model in case study 2.

TABLE 7: Comparison of proposed results against true values of INCAR model for 10 dB SNR.

μ	Method	Design parameters								MSE
		p_1	p_2	a_1	a_2	a_3	q_2a_1	q_2a_2	q_2a_3	
$(10^{-04}, 10^{-08})$	VLMS	1.351468	−0.744938	1.047010	0.486303	−0.200102	1.285267	1.225023	−0.427779	$3.94E-02$
	KLMS	1.356262	−0.736962	0.956832	0.185102	0.259909	1.220949	0.529005	0.495796	$1.64E-01$
	FLMS_1	1.356850	−0.745266	1.001581	0.494077	−0.202190	1.670903	0.837135	−0.334630	$2.56E-05$
	FLMS_2	1.356608	−0.745109	1.001695	0.493895	−0.202230	1.671239	0.837177	−0.334908	$2.48E-05$
$(10^{-03}, 10^{-05})$	VLMS	1.349757	−0.747542	0.945167	0.564599	−0.216172	1.511909	1.011739	−0.377157	$8.36E-03$
	KLMS	1.355306	−0.750178	1.058421	0.333366	−0.033855	1.341960	0.926972	−0.080043	$3.08E-02$
	FLMS_1	1.350819	−0.748919	0.998394	0.491951	−0.179965	1.688782	0.847405	−0.340293	$7.76E-05$
	FLMS_2	1.350853	−0.748833	0.997694	0.492029	−0.179449	1.688987	0.847312	−0.340771	$8.13E-05$
True values		1.350000	−0.750000	1.000000	0.500000	−0.200000	1.680000	0.840000	−0.336000	

TABLE 8: Comparison of proposed results against true values of INCAR model for 3 dB SNR.

μ	Method	Design parameters								MSE
		p_1	p_2	a_1	a_2	a_3	q_2a_1	q_2a_2	q_2a_3	
$(10^{-04}, 10^{-08})$	VLMS	1.350033	−0.747776	1.151411	0.382217	−0.174445	1.266046	1.237098	−0.428614	$4.69E-02$
	KLMS	1.356493	−0.736365	0.748964	0.105174	0.542420	1.362022	0.609956	0.290923	$1.65E-01$
	FLMS_1	1.352386	−0.750941	0.992991	0.491026	−0.198780	1.686456	0.849316	−0.338108	$3.38E-05$
	FLMS_2	1.352534	−0.750836	0.993316	0.490747	−0.198439	1.686445	0.849539	−0.337726	$3.44E-05$
$(10^{-03}, 10^{-05})$	VLMS	1.349324	−0.747288	0.983402	0.533690	−0.203984	1.423701	1.105389	−0.402156	$1.77E-02$
	KLMS	1.350614	−0.760252	1.136970	0.030726	0.119125	1.612852	0.232501	0.372932	$1.52E-01$
	FLMS_1	1.351001	−0.731407	0.993154	0.493843	−0.214263	1.694738	0.819746	−0.318643	$1.95E-04$
	FLMS_2	1.351107	−0.731604	0.995350	0.492513	−0.216675	1.693931	0.819381	−0.318775	$2.01E-04$
True values		1.350000	−0.750000	1.000000	0.500000	−0.200000	1.680000	0.840000	−0.336000	

5. Conclusion

On the basis of the simulation and results presented in the last section, the following conclusions are drawn.

(i) The adaptive algorithms based on fractional signal processing approach are used effectively for parameter estimation of input nonlinear control autoregressive (INCAR) models for both case studies.

(ii) The variation of step size strategies shows that for smaller and relatively larger value of step size parameter both order of fractional least mean square (FLMS) algorithms provide accurate and convergent results than those of VLMS and KLMS algorithms.

(iii) The variants of signal-to-noise ratio (SNR) in INCAR models show that the performance of all the algorithm decreases as SNR decreases from higher level to lower level, but FLMS algorithm still achieved the values for mean square error around 10^{-04} to 10^{-05} for even SNR = 3 dB.

(iv) Comparative studies between FLMS, VLMS, and KLMS algorithms for each variants of both case studies validate the correctness of the adaptive algorithms based on FLMS algorithm.

In future, one may look for heuristic computing techniques based on genetic algorithms, swarm intelligence, differential evolution, genetic programming, and memetic computing approaches, and so forth, for parameter estimation of INCAR models.

References

[1] M. R. Zakerzadeh, M. Firouzi, H. Sayyaadi, and S. B. Shouraki, "Hysteresis nonlinearity identification using new Preisach model-based artificial neural network approach," *Journal of Applied Mathematics*, vol. 2011, Article ID 458768, 22 pages, 2011.

[2] X. X. Li, H. Z. Guo, S. M. Wan, and F. Yang, "Inverse source identification by the modified regularization method on poisson equation," *Journal of Applied Mathematics*, vol. 2012, Article ID 971952, 13 pages, 2012.

[3] Y. Shi and H. Fang, "Kalman filter-based identification for systems with randomly missing measurements in a network environment," *International Journal of Control*, vol. 83, no. 3, pp. 538–551, 2010.

[4] Y. Liu, J. Sheng, and R. Ding, "Convergence of stochastic gradient estimation algorithm for multivariable ARX-like systems," *Computers and Mathematics with Applications*, vol. 59, no. 8, pp. 2615–2627, 2010.

[5] F. Ding, G. Liu, and X. P. Liu, "Parameter estimation with scarce measurements," *Automatica*, vol. 47, no. 8, pp. 1646–1655, 2011.

[6] J. Ding, F. Ding, X. P. Liu, and G. Liu, "Hierarchical least squares identification for linear SISO systems with dual-rate sampled-data," *IEEE Transactions on Automatic Control*, vol. 56, no. 11, pp. 2677–2683, 2011.

[7] Y. Liu, Y. Xiao, and X. Zhao, "Multi-innovation stochastic gradient algorithm for multiple-input single-output systems using the auxiliary model," *Applied Mathematics and Computation*, vol. 215, no. 4, pp. 1477–1483, 2009.

[8] J. Ding and F. Ding, "The residual based extended least squares identification method for dual-rate systems," *Computers and Mathematics with Applications*, vol. 56, no. 6, pp. 1479–1487, 2008.

[9] L. Han and F. Ding, "Identification for multirate multi-input systems using the multi-innovation identification theory," *Computers and Mathematics with Applications*, vol. 57, no. 9, pp. 1438–1449, 2009.

[10] F. Ding, Y. Shi, and T. Chen, "Gradient-based identification methods for hammerstein nonlinear ARMAX models," *Nonlinear Dynamics*, vol. 45, no. 1-2, pp. 31–43, 2006.

[11] F. Ding, T. Chen, and Z. Iwai, "Adaptive digital control of Hammerstein nonlinear systems with limited output sampling," *SIAM Journal on Control and Optimization*, vol. 45, no. 6, pp. 2257–2276, 2007.

[12] F. Ding and T. Chen, "Identification of Hammerstein nonlinear ARMAX systems," *Automatica*, vol. 41, no. 9, pp. 1479–1489, 2005.

[13] I. W. Hunter and M. J. Korenberg, "The identification of nonlinear biological systems: wiener and Hammerstein cascade models," *Biological Cybernetics*, vol. 55, no. 2-3, pp. 135–144, 1986.

[14] Y. Y. Cao and Z. Lin, "Robust stability analysis and fuzzy-scheduling control for nonlinear systems subject to actuator saturation," *IEEE Transactions on Fuzzy Systems*, vol. 11, no. 1, pp. 57–67, 2003.

[15] K. P. Fruzzetti, A. Palazoğlu, and K. A. McDonald, "Nonlinear model predictive control using Hammerstein models," *Journal of Process Control*, vol. 7, no. 1, pp. 31–41, 1997.

[16] S. Dupont and J. Luettin, "Audio-visual speech modeling for continuous speech recognition," *IEEE Transactions on Multimedia*, vol. 2, no. 3, pp. 141–151, 2000.

[17] M. Karimi-Ghartemani and M. R. Iravani, "A nonlinear adaptive filter for online signal analysis in power systems: applications," *IEEE Transactions on Power Delivery*, vol. 17, no. 2, pp. 617–622, 2002.

[18] F. Ding and T. Chen, "Identification of Hammerstein nonlinear ARMAX systems," *Automatica*, vol. 41, no. 9, pp. 1479–1489, 2005.

[19] F. Ding, Y. Shi, and T. Chen, "Auxiliary model-based least-squares identification methods for Hammerstein output-error systems," *Systems and Control Letters*, vol. 56, no. 5, pp. 373–380, 2007.

[20] B. B. Mandelbrot and J. W. Van Ness, "Fractional Brownian motions, fractional noises and applications," *SIAM Review*, vol. 10, no. 4, pp. 422–437, 1968.

[21] L. Gaul, P. Klein, and S. Kemple, "Damping description involving fractional operators," *Mechanical Systems and Signal Processing*, vol. 5, no. 2, pp. 81–88, 1991.

[22] J. Sabatier, M. Aoun, A. Oustaloup, G. Grégoire, F. Ragot, and P. Roy, "Fractional system identification for lead acid battery state of charge estimation," *Signal Processing*, vol. 86, no. 10, pp. 2645–2657, 2006.

[23] M. D. Ortigueira, J. A. Tenreiro Machado, and J. S. da Costa, "Which differintegration? [fractional calculus]," *IEE Proceedings: Vision Image and Signal Processing*, vol. 152, no. 6, pp. 846–850, 2005.

[24] D. Valério, M. D. Ortigueira, and J. Sá da Costa, "Identifying a transfer function from a frequency response," *ASME Journal of Computational and Nonlinear Dynamics*, vol. 3, no. 2, Article ID 021207, 7 pages, 2008.

[25] M. D. Ortigueira, "Introduction to fractional linear systems. Part 1: continuous-time case," *IEE Proceedings: Vision, Image and Signal Processing*, vol. 147, no. 1, pp. 62–70, 2000.

[26] M. D. Ortigueira, "Introduction to fractional linear systems. Part 2: discrete-time case," *IEE Proceedings: Vision, Image and Signal Processing*, vol. 147, no. 1, pp. 71–78, 2000.

[27] M. D. Ortigueira and J. A. Tenreiro Machado, "Fractional signal processing and applications," *Signal Processing*, vol. 83, no. 11, pp. 2285–2286, 2003.

[28] M. D. Ortigueira and J. A. T. Machado, "Fractional calculus applications in signals and systems," *Signal Processing*, vol. 86, no. 10, pp. 2503–2504, 2006.

[29] E. Cuesta, M. Kirane, and S. A. Malik, "Image structure preserving denoising using generalized fractional time integrals," *Signal Processing*, vol. 92, no. 2, pp. 553–563, 2012.

[30] C. C. Tseng and S. L. Lee, "Design of adjustable fractional order differentiator using expansion of ideal frequency response," *Signal Processing*, vol. 92, no. 2, pp. 498–508, 2012.

[31] W. Fan, F. Ding, and Y. Shi, "Parameter estimation for Hammerstein nonlinear controlled auto-regression models," in *Proceedings of the IEEE International Conference on Automation and Logistics (ICAL '07)*, pp. 1007–1012, August 2007.

[32] X. Weili, W. Fan, and R. Ding, "Least-squares parameter estimation algorithm for a class of input nonlinear systems," *Journal of Applied Mathematics*, vol. 2012, Article ID 684074, 14 pages, 2012.

[33] R. M. A. Zahoor and I. M. Qureshi, "A modified least mean square algorithm using fractional derivative and its application to system identification," *European Journal of Scientific Research*, vol. 35, no. 1, pp. 14–21, 2009.

[34] M. Geravanchizadeh and S. G. Osgouei, "Dual-channel speech enhancement using normalized fractional least-mean-squares algorithm," in *Proceedings of the 19th Iranian Conference on Electrical Engineering (ICEE '11)*, May 2011.

[35] S. G. Osgouei and M. Geravanchizadeh, "Speech enhancement using convex combination of fractional least-mean-squares algorithm," in *Proceedings of the 5th International Symposium on Telecommunications (IST '10)*, pp. 869–872, December 2010.

[36] D. S. Kumar and N. K. Rout, "FLMS algorithm for acoustic echo cancellation and its comparison with LMS," in *Proceedings of the IEEE 1st International Conference on Recent Advances in Information Technology (RAIT '12)*, pp. 852–856, March 2012.

[37] A. Pervez and M. Yasin, "Performance analysis of bessel beamformer and LMS algorithm for smart antenna array in mobile communication system," *Emerging Trends and Applications in Information Communication Technologies*, vol. 281, pp. 52–61, 2012.

[38] M. Yasin and P. Akhtar, "Performance analysis of bessel beamformer with LMS algorithm for smart antenna array," in *Proceeding of the International Conference on Open Source Systems and Technologies (ICOSST '12)*, pp. 1–5, December 2012.

[39] S. Haykin, *Adaptive Filter Theory (ISE)*, 2003.

[40] R. M. A. Zahoor, *Application of fractional calculus to engineering: a new computational approach [Ph.D. thesis]*, International Islamic University, Islamabad, Pakistan, 2011.

[41] B. Georgeta, *Nonlinear Systems Identification Using the Volterra Model*, University of Timisoara, 2005.

[42] A. Guérin, G. Faucon, and R. Le Bouquin-Jeannès, "Nonlinear acoustic echo cancellation based on volterra filters," *IEEE Transactions on Speech and Audio Processing*, vol. 11, no. 6, pp. 672–683, 2003.

[43] F. Küch and W. Kellermann, "Nonlinear line echo cancellation using a simplified second order Volterra filter," in *Proceedings of the IEEE International Conference on Acoustic, Speech and Signal Processing*, vol. 2, pp. 1117–1120, May 2002.

[44] B. Georgeta and C. Botoca, "Nonlinearities identification using the LMS Volterra filter," in *Proceedings of the WSEAS International Conference on Dynamical Systems and Control*, pp. 148–153, Communications Department, Faculty of Electronics and Telecommunications, Venice, Italy, November 2005.

[45] P. P. Pokharel, L. Weifeng, and J. C. Principe, "Kernel LMS," in *Proceedings of the IEEE International Conference on Acoustics, Speech and Signal Processing (ICASSP '07*, vol. 3, pp. 1421–1424, April 2007.

[46] W. Liu, P. P. Pokharel, and J. C. Principe, "The kernel least-mean-square algorithm," *IEEE Transactions on Signal Processing*, vol. 56, no. 2, pp. 543–554, 2008.

[47] A. Gunduz, J. P. Kwon, J. C. Sanchez, and J. C. Principe, "Decoding hand trajectories from ECoG recordings via kernel least-mean-square algorithm," in *Proceedings of the 4th International IEEE/EMBS Conference on Neural Engineering (NER '09)*, pp. 267–270, May 2009.

[48] B. Pantelis, S. Theodoridis, and M. Mavroforakis, "The augmented complex kernel LMS," *IEEE Transactions on Signal Processing*, vol. 60, no. 9, pp. 4962–4967, 2012.

[49] H. Bao and I. M. S. Panahi, "Active noise control based on kernel least-mean-square algorithm," in *Proceedings of the 43rd Asilomar Conference on Signals, Systems and Computers*, pp. 642–644, November 2009.

10

Design Methodology of a New Wavelet Basis Function for Fetal Phonocardiographic Signals

Vijay S. Chourasia[1] and Anil Kumar Tiwari[2]

[1] *Manoharbhai Patel Institute of Engineering & Technology, Kudwa, Gondia, Maharashtra 441 614, India*
[2] *Indian Institute of Technology, Rajasthan MBM College Campus, Old Residency Road, Ratanada, Jodhpur, Rajasthan 342 011, India*

Correspondence should be addressed to Vijay S. Chourasia; chourasiav@gmail.com

Academic Editors: Z.-L. Sun and V. Valimaki

Fetal phonocardiography (fPCG) based antenatal care system is economical and has a potential to use for long-term monitoring due to noninvasive nature of the system. The main limitation of this technique is that noise gets superimposed on the useful signal during its acquisition and transmission. Conventional filtering may result into loss of valuable diagnostic information from these signals. This calls for a robust, versatile, and adaptable denoising method applicable in different operative circumstances. In this work, a novel algorithm based on wavelet transform has been developed for denoising of fPCG signals. Successful implementation of wavelet theory in denoising is heavily dependent on selection of suitable wavelet basis function. This work introduces a new mother wavelet basis function for denoising of fPCG signals. The performance of newly developed wavelet is found to be better when compared with the existing wavelets. For this purpose, a two-channel filter bank, based on characteristics of fPCG signal, is designed. The resultant denoised fPCG signals retain the important diagnostic information contained in the original fPCG signal.

1. Introduction

Continuous and long-term fetal monitoring has become an essential approach for better accuracy in diagnosis [1]. Variations in fetal heart rate (FHR) provide up-to-date information about the fetal health status [2]. Doppler ultrasound based fetal cardiotocography (fCTG) is currently being used as a technique for recording and analysis of the FHR. In this technique, a graph of fetal heart rate (cardio-) and uterine contractions (-toco-) are recorded during pregnancy [3]. This technique cannot be used for long-term monitoring of the fetus due to several reasons [4] such as the cost and complexity of the monitoring instrument, ultrasound radiation exposure, and gel application.

Fetal phonocardiography, recording of vibroacoustic (fPCG) signals from the maternal abdominal surface, may become an important alternative to fCTG [5]. However, these fPCG signals are heavily contaminated by noise from various sources [6–11]. Hence to extract important diagnostic information such as FHR, frequency/pitch, intensity, timing, and energy of fetal heart sound, this technique requires matured signal processing strategies.

Researchers from biomedical signal processing community have applied several techniques for denoising of fPCG signals. Varady (2001) presented a wavelet-based denoising method for phonocardiographic signals using two-channel signal recording and an adaptive cross-channel coefficient thresholding technique [12]. Messer et al. (2001) attempted to answer about which wavelet families, levels of decomposition, and thresholding techniques best removes the noise in a PCG [13]. Jianfeng et al. [5] used a modified spectral subtraction algorithm to remove the unwanted stationary background noise from the noisy fetal heart sound. The AM/FM modulation technique is employed to make the fetal heart sound more audible so that both pregnant women and gynecologists can identify the rhythmic fetal heart beat sounds easily [5]. Mittra et al. [14] compared and analyzed the performance of various digital noise cancellation techniques for fetal heart sound. Their experimental results show that adaptive filtering using Recursive Least Squares (RLS) algorithm is an appropriate methodology for the fetal phonocardiographic signals denoising implementation [14]. Tosanguan et al. (2008) proposed a 2-channel signal processing technique, termed interference suppression via spectral comparison (ISSC) and aims at improving the quality of the recorded heart sound or the PCG data [15]. Zhao et al. (2009) developed a denoising method for heart sound signal using improved thresholding function in wavelet domain [16]. Chourasia et al. (2010) presented a technique for denoising of fPCG signals using wavelet transform [17]. In another work, Chourasia et al. (2012) also developed a methodology for removal of unwanted noise from the fPCG signal using non-negative matrix factorization [18].

The above-mentioned researchers have used conventional filtering techniques and wavelet transform for denoising the fPCG signals. The conventional techniques also impact on the useful signals and hence may result in loss of information of diagnostic importance. Additionally, the passed band may still contain noise. In view of these limitations of conventional methods of denoising, wavelet threshold denoising has been applied to denoise heart sound recordings. The wavelet based denoising method try to preserve the signal by operating only on those selected regions of the bandwidth that need filtering. This technique requires appropriate selection of wavelet family, level of decomposition, and the method to be used for calculation of threshold. The existing works are limited only to employ the existing wavelet family from the bank of wavelet transform.

This present work proposes a novel algorithm for denoising of fPCG signals using wavelet transform (WT). In this approach, a new wavelet family and its mother wavelet are developed. For this purpose, a quadratic mirror filter (QMF) bank [19] is designed based on the characteristics of the fPCG signals. This filter bank requires a low-pass and a high-pass filter in decomposition (analysis) phase and reconstruction (synthesis) phase. Appropriate denoising algorithm and thresholding rule have been selected and used with the developed mother wavelet. The developed wavelet family provides better results when compared with the existing wavelets. The obtained denoised fPCG signals retain the vital diagnostics information contained in the original signals.

The rest of the paper is organized as follows. A brief introduction about fetal phonocardiographic signals and wavelet theory has been provided in Sections 2 and 3, respectively. Section 4 discusses the selection of appropriate denoising algorithm and thresholding rule for fPCG signals. Design methodology of filter bank for new wavelet has been explained in Section 5. Finally, the experimental results are described in Section 6, followed by conclusion and discussion in Section 7.

2. Fetal Phonocardiographic Signals

Fetal phonocardiography is the recording of natural vibroacoustic signals from the maternal abdominal surface. The fPCG signal carries valuable information concerning the physiological state of unborn [20]. It is also capable of recognizing additional potential dysfunctional signals of the fetal heart such as those related to cardiac murmurs, split effect, and breathing movements, which can be detected from the analysis of abdominal sound (fPCG) signals. However these cardiac anomalies are impossible to detect with the widely accepted fCTG technique due to its principle of operation [21]. Additionally, phonocardiography provides a permanent and lasting record of fetal heart sounds which by comparison at a later stage may prove to be of great prognostic importance. The fPCG technique has been introduced earlier but overlooked by biomedical scientists and medical experts mainly due to its low signal-to-noise ratio (SNR) [22]. The fPCG signals are a linear summation of [23]

(i) fetal heart sound,

(ii) internal noise,

(iii) external noise.

The fetal heart sound is a signal produced by mechanical activity of the fetal heart. The fetal heart is basically divided into two pairs of chambers and has four valves: the mitral and tricuspid valves. In the fetal cardiac cycle, when the ventricles begin to contract, the blood attempts to flow back into the lower pressure atrial chambers. This reverse flow of blood is arrested by the shutting of the mitral and tricuspid valves, which produces the first heart sound (S1). Whenever the pressure in the ventricular chambers becomes too high for the pulmonary valves to withstand, they open, and the pressurized blood is rapidly ejected into the arteries. While the ventricles are being evacuated, the pressure of the remaining blood decreases with respect to that in the arteries. This pressure gradient causes the arterial blood to flow back into the ventricles. The pulmonary valves arrest this reverse flow by shutting, which gives rise to the second heart sound (S2) [24]. The frequency spectrum of fetal heart sound lies below 200 Hz. Figures 1(a) and 1(b) show a typical fetal phonocardiographic (fPCG) signal and its frequency spectrum, respectively.

The internal noise is a random signal caused due to maternal respiratory sounds, acoustic noise produced by the fetal movement, maternal digestive sound, maternal respiratory sound, maternal heart sound, and placental blood turbulence. These noises are of low amplitude with main frequency components from 0 to 25 Hz [4].

Similarly, the external noise is a combination of shear noise from movement of the sensor during recording and ambient noise originating from the environment such as sound produced by fan, air conditioner, and hue and cry of the nearby people. It is comparatively of high amplitude and frequency (100–20000 Hz) [4].

3. Theoretical Background

3.1. Wavelet Transform. The WT is a two-dimensional time-scale processing method for nonstationary signals with adequate scale values and shifting in time. The major advantage of the WT is that it has a varying window size, being broad at low frequencies and narrow at high frequencies, thus leading to best possible time-frequency resolution in all frequency ranges [25, 26]. The analysis of nonstationary signals requires proper location of transitions or discontinuities and identification of their long-term behavior. The WT represents a time

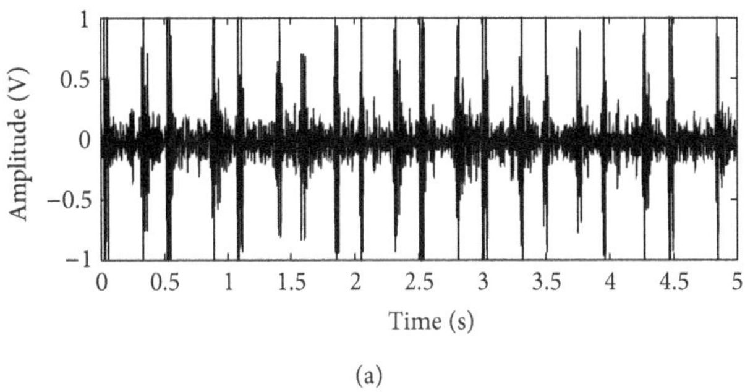

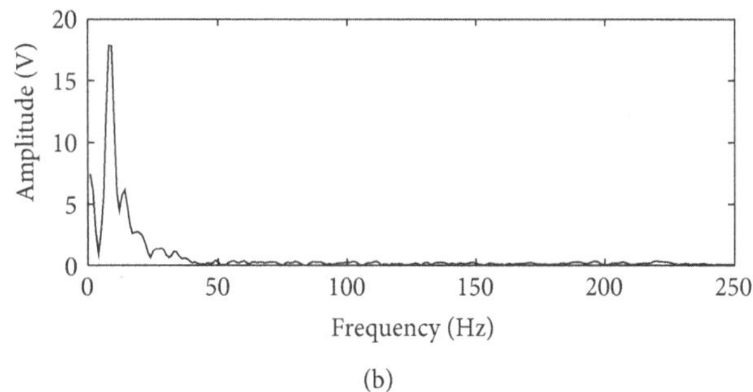

FIGURE 1: (a) Typical fPCG signal and (b) its frequency spectrum.

function in terms of simple and fixed building blocks derived from mother wavelet by translation and dilation operations. The translation is the shifting of mother wavelet along the time axis while dilation or scaling starches or compresses it.

The WT can be categorized as continuous wavelet transform (CWT) or discrete wavelet transform (DWT). The CWT is defined as the convolution between the original signal $s(t)$ and a wavelet $\psi(t)$ which can be calculated by

$$\begin{aligned} \mathrm{CWT}_{\psi}(a,b) &= \int_{-\infty}^{+\infty} s(t)\,\psi_{a,b}^{*}(t)\,dt \\ &= \frac{1}{\sqrt{a}}\int_{-\infty}^{+\infty} s(t)\,\psi^{*}\left(\frac{t-b}{a}\right)dt, \end{aligned} \tag{1}$$

where $\mathrm{CWT}_{\psi}(a,b)$ is a continuous wavelet transform, "$s(t)$" is a signal under study, "a" is a scale coefficient connected with stretching or compression of signal in time, "b" is a shift connected with time location, "$\psi_{a,b}^{*}(t)$" is a wavelet function or mother wavelet representing a wavelet family, "*" denotes the complex conjugation, and the factor $1/\sqrt{a}$ is used for energy normalization purposes so that the transformed signal will have the same energy at every scale. In CWT, the scaling parameter a and translation parameter b change with time continuously. Hence wavelet coefficients are calculated for every possible scale, which requires huge processing power and results in large amount of data.

The Discrete Wavelet Transform (DWT) coefficients are usually sampled from the CWT on a dyadic grid, choosing parameters of translation $b = k^{*}2^{-j}$ and scale $a = 2^{-j}$. Where $j, k \in Z$ a set of positive integers and $k = 0, 1, \ldots, n-1$, n represents the number of samples. These dilation and translation parameters are discretized leading to the DWT. After discretization, the wavelet function is defined as

$$\mathrm{DWT}_{\psi}(j,k) = \int_{-\infty}^{+\infty} s(t)\,\psi_{j,k}^{*}(t)\,dt. \tag{2}$$

Here $\psi_{j,k}^{*}(t)$ is the dilated and translated version of the wavelet function and given as

$$\psi_{j,k}(t) = 2^{j/2}\psi\left(2^{j}t-k\right), \tag{3}$$

where ψ is called as mother wavelet and $\psi_{j,k}$ is called as daughter wavelet. The level j determines how many wavelets are needed to cover the mother wavelet, and the number k determines the position of the wavelet and gives the indication of time. DWT analyzes the signal by decomposing it into its coarse and detail information, which is accomplished by using successive high-pass and low-pass filtering operations, on the basis of the following equations:

$$\begin{aligned} y_{\mathrm{high}}(k) &= \sum_{n} s(n)\cdot h(2k-n), \\ y_{\mathrm{low}}(k) &= \sum_{n} s(n)\cdot g(2k-n), \end{aligned} \tag{4}$$

where $y_{\mathrm{high}}(k)$ and $y_{\mathrm{low}}(k)$ are the outputs of the high-pass and low-pass filters with impulse response h and g, respectively, after downsampling by 2 [27, 28]. The coefficients of low-pass filter are called "approximation" ($c_{j,k}$) and coefficients from high-pass filter are called "detail" ($b_{j,k}$) wavelet coefficients. The detail coefficients are defined by the following equation:

$$b_{j,k} = \int s(t)\,\psi_{j,k}^{*}(t)\,dt, \tag{5}$$

where $\psi_{j,k}$ are wavelet functions given by

$$\psi_{j,k}(t) = \frac{1}{\sqrt{2^{j}}}\,\psi\left(\frac{t-k2^{j}}{2^{j}}\right). \tag{6}$$

Similarly, the approximate coefficients are

$$c_{j,k} = \int s(t)\,\Phi_{j,k}^{*}(t)\,dt, \tag{7}$$

where $\Phi_{j,k}$ are called scaling functions as follows:

$$\Phi_{j,k}(t) = \frac{1}{\sqrt{2^{j}}}\,\Phi\left(\frac{t-k2^{j}}{2^{j}}\right). \tag{8}$$

The discrete inverse transform is found by adding the translated, dilated wavelets, weighted by the coefficients:

$$f(t) = \sum_{j,k} b_{j,k}\psi_{j,k}(t). \tag{9}$$

The DWT gives a multiresolution description of a signal which is very useful in analyzing real-time signals [29].

The general wavelet denoising procedure is described in the following steps:

(i) decomposition of the fPCG signal using DWT to obtain the approximation and detail coefficients,

(ii) thresholding of these decomposed coefficients using an appropriate denoising algorithm,

(iii) reconstruction of the fPCG signal from these thresholded coefficients using the inverse transform (IDWT) [17].

3.2. Denoising Algorithm. The *denoising algorithm* uses statistical regression of noisy coefficients over time to obtain a nonparametric estimation of the reconstructed signal without noise. The thresholding algorithms commonly employed for denoising of the nonstationary signals are [30]

(i) universal threshold,

(ii) minimax threshold,

(iii) rigorous Stein's Unbiased Risk Estimate (SURE) threshold.

(i) Universal Threshold (Sqtwolog). The universal threshold denoising algorithm is a fixed threshold method which can be calculated by

$$\lambda = \sigma\sqrt{2\log(n)}, \tag{10}$$

where n denotes the length of the signal and σ is the standard deviation.

(ii) Minimax Threshold (Minimaxi). Minimax threshold also uses fixed threshold and it yields minimax performance for Mean Square Error (MSE) against an ideal procedures. This threshold level depends on the noise and signal relationships in the input data and it is given by $\lambda = \sigma\lambda_n$, where λ_n is determined by a minimax rule such that the maximum risk of estimation error across all locations of the data is minimized.

(iii) Rigorous SURE Threshold (Rigrsure). The denoising algorithms described previously use global thresholds. That is, the computed threshold is applied to all wavelet coefficients. The rigorous SURE threshold algorithm describes a scheme that uses a threshold value λ_j at each resolution level j of the wavelet coefficients. This algorithm is also known as SureShrink and uses the Stein's Unbiased Risk Estimate (SURE) criterion to get an unbiased estimate.

These denoising algorithms can be divided into linear and nonlinear methods. The linear method is independent of the size of empirical wavelet coefficients, and therefore the size of the coefficient by itself is not taken into account. It assumes that signal noise can be found mainly in fine scale coefficients and not in coarse scales. The nonlinear method is based on the idea that the signal noise can be found in every coefficient and is distributed over all scales.

3.3. Wavelet Thresholding Rules. Wavelet thresholding is a signal estimation technique that exploits the capabilities of wavelet transform for signal denoising [31]. This method has been researched extensively due to its effectiveness and simplicity. Any signal $s(t)$ can be represented by the summation of the original signal $x(t)$ and the noise $n(t)$ as follows:

$$s(t) = x(t) + n(t). \tag{11}$$

After performing the wavelet transform

$$S_{j,k} = X_{j,k} + N_{j,k}, \tag{12}$$

where $S_{j,k}$ is the kth wavelet coefficient in the scale j. There are two ways of thresholding with threshold λ; the shapes of these thresholding operators are illustrated in Figure 2.

Hard Thresholding (h). In hard thresholding, those wavelet coefficients with absolute values below or at the threshold level (λ) are affected only and they are replaced by zero value whereas others are kept unchanged

$$\widehat{Y}_{j,k}^{\text{hard}} = \begin{cases} Y_{j,k} & \text{for } \left|Y_{j,k}\right| > \lambda \\ 0 & \text{for } \left|Y_{j,k}\right| \le \lambda. \end{cases} \tag{13}$$

Soft Thresholding (s). In soft thresholding, coefficients above threshold level (λ) are also modified; they are reduced by particular value of the threshold

$$\widehat{Y}_{j,k}^{\text{soft}} = \begin{cases} Y_{j,k} - \lambda & \text{for } Y_{j,k} \ge \lambda \\ 0 & \text{for } \left|Y_{j,k}\right| < \lambda \\ Y_{j,k} + \lambda & \text{for } Y_{j,k} \le -\lambda. \end{cases} \tag{14}$$

Hard thresholding maintains the scale of the signal but introduces ringing and artifacts after reconstruction due to a discontinuity in the wavelet coefficients. Soft thresholding eliminates this discontinuity resulting in smoother signals but slightly decreases the magnitude of the reconstructed signal.

4. Selection of Denoising Algorithm and Thresholding Rule for fPCG Signals

The presented work also contributes in the selection of suitable algorithm and thresholding rule for denoising of fPCG signals. Based on the discussion in Sections 3.2 and 3.3, there are three denoising algorithms, namely, universal threshold, minimax threshold, and rigorous SURE threshold and two thresholding rules which can be employed in denoising of fPCG signals. These denoising algorithms and thresholding rules are practically implemented and their results are compared. The mean squared error (MSE) is used to evaluate the performance of all denoising algorithms and thresholding rules in denoising the fPCG signals. It can be obtained using following expression:

$$\text{MSE} = \frac{\sum_{i=1}^{n} (s - s_e)_i^2}{n}, \tag{15}$$

where n denotes the length of the signal, s represents the original signal and s_e is the estimated signal obtained from the denoised wavelet coefficients.

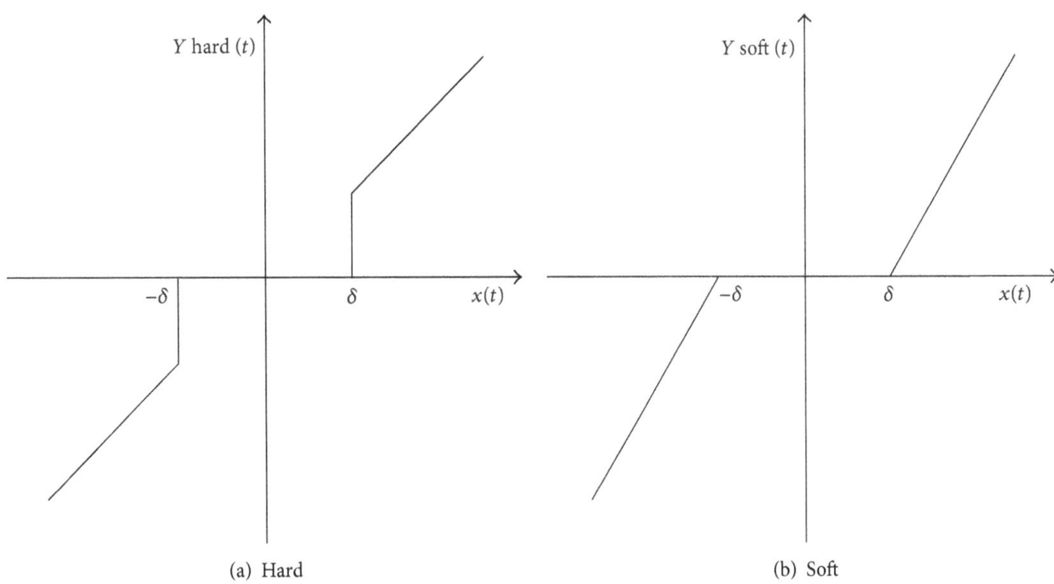

(a) Hard

(b) Soft

FIGURE 2: Wavelet thresholding.

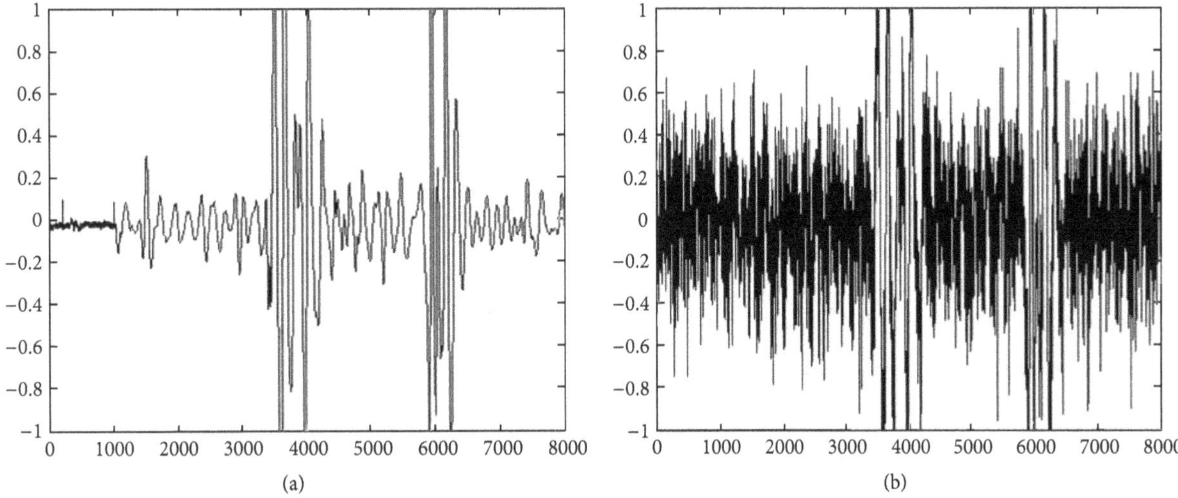

(a)

(b)

FIGURE 3: Waveform of (a) reference fPCG signal and (b) test fPCG signal with additive noise.

Figure 3(a) shows the waveform of one simulated fPCG signal as an example. This signal is used as a reference signal in the selection process of appropriate denoising algorithms and wavelet threshold. Figure 3(b) is a test signal generated by adding simulated stationary random noise in the original fPCG signal which is shown in Figure 3(a). The noise is generated by recording it, first, in actual conditions. After recording, the noise signal is simulated with similar average characteristics. The simulated noise is then mixed with the fPCG signal shown in Figure 3(a) and used as an input to evaluate the performance of the selection of appropriate denoising algorithm and thresholding rule.

The test signal so obtained is analyzed with DWT based multiresolution analysis. The signal is decomposed to five levels using fourth order Coiflets wavelet. This selection of mother wavelet is based on the fact that it possesses all the properties needed for analysis of the fPCG signals [29]. All the three algorithms with soft or hard thresholding rule are applied for denoising of the fPCG signal. The resultant waveforms from this implementation are shown in Figure 4, and the best estimations obtained are depicted in Table 1. In Figures 3 and 4, the x-axis represents number of samples and the y-axis represents relative amplitude.

Table 1 shows a comparison of three denoising algorithms with soft or hard thresholding rule. The rigorous SURE threshold algorithm with soft thresholding rule yields the best estimation with considerably smaller MSE as compared to the other algorithms for denoising of fPCG signals.

5. Design of Filter Bank for New Wavelet

Since it has been discussed earlier in Section 1 that the effective implementations of wavelet transform in denoising the fPCG signals requires appropriate wavelet basis function,

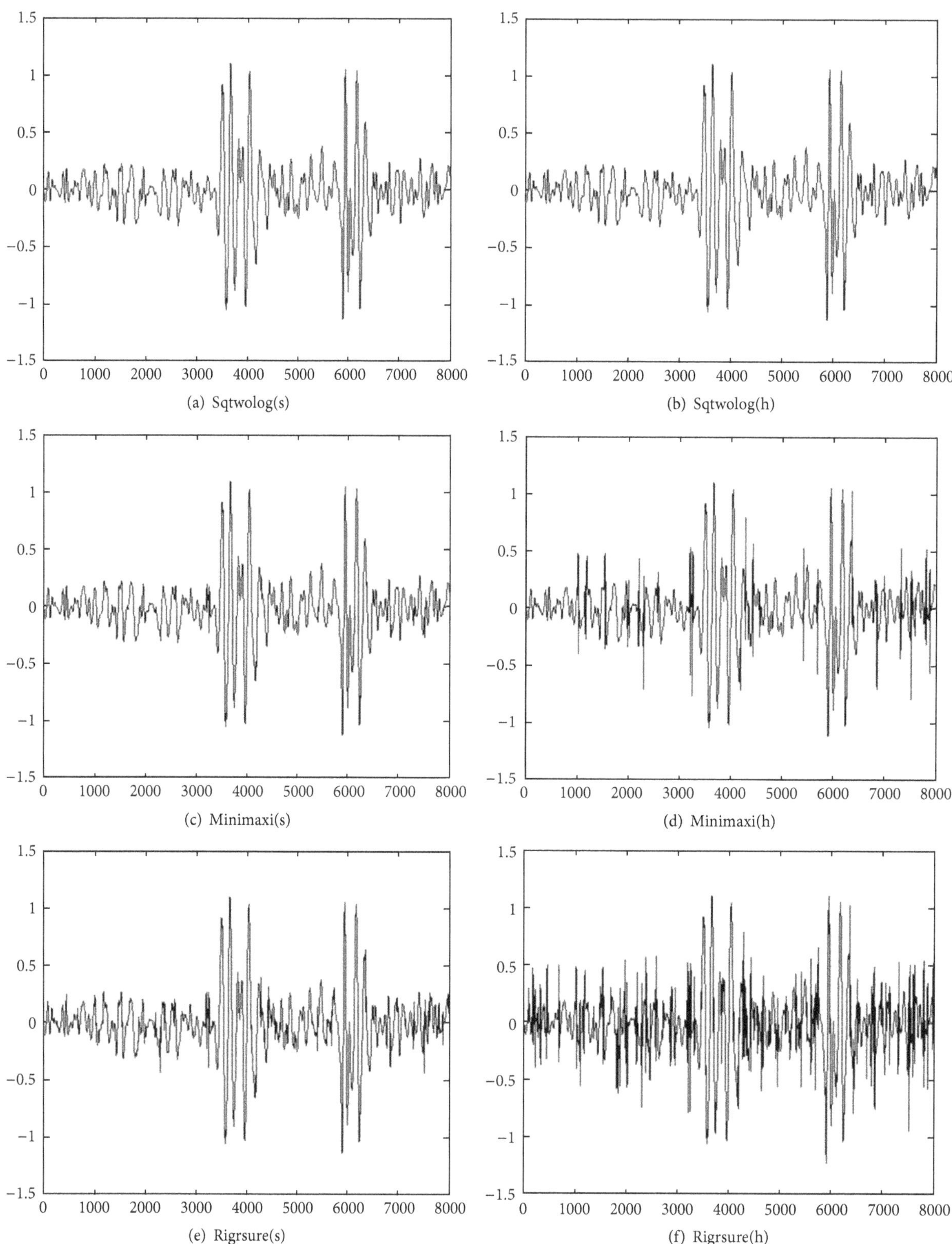

FIGURE 4: Denoised fPCG signal using different algorithms: (a) sqtwolog(s), (b) sqtwolog(h), (c) minimaxi(s), (d) minimaxi(h), (e) rigrsure(s), and (f) rigrsure(h).

Table 1: Comparison of denoising algorithms.

Denoising algorithm	Code	MSE
Universal threshold algorithm with soft thresholding rule	sqtwolog(s)	0.8331
Universal threshold algorithm with hard thresholding rule	sqtwolog(h)	0.843
Minimax threshold algorithm with soft thresholding rule	minimaxi(s)	0.6295
Minimax threshold algorithm with hard thresholding rule	minimaxi(h)	0.7892
Rigorous SURE threshold algorithm with soft thresholding rule	**rigrsure(s)**	**0.575**
Rigorous SURE threshold algorithm with hard thresholding rule	rigrsure(h)	0.8165

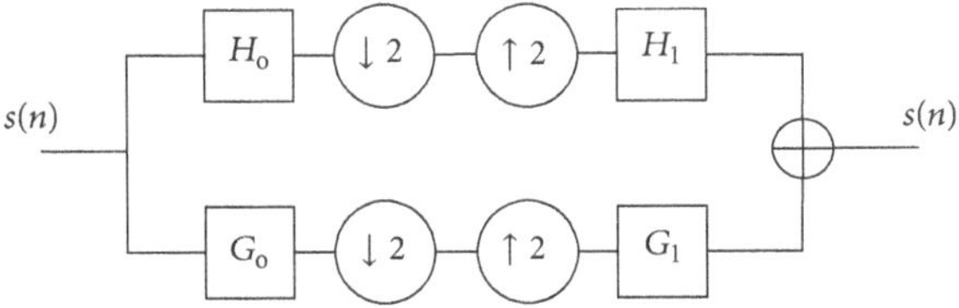

Figure 5: Two-channel filter bank.

we propose to design a new wavelet function which will be based on the characteristics of the fPCG signals and hence will improve the performance of its denoising. The main properties to be verified in designing a family of wavelet are existence of scaling function, symmetricity and compactly support of wavelet and scaling functions, availability of filter bank, orthogonality of filter bank, and smoothness of wavelet function. With these properties, the wavelet family may be called as suitable wavelet for the analysis of intended signal. This wavelet will lead to maximization of wavelet coefficient values to produce the highest local maxima of the signal in wavelet domain. It also produces the possibility of best characterization of frequency content of that signal [32].

In view of these considerations, we developed a new wavelet function which is orthogonal and named as "fetal." In orthogonal wavelet analysis, the number of convolutions at each scale is proportional to the width of the wavelet basis at that scale. This produces a wavelet spectrum that contains discrete "blocks" of wavelet power and is useful for signal processing as it gives the most compact representation of the signal [33]. Conversely, a nonorthogonal analysis is highly redundant at large scales, where the wavelet spectrum at adjacent times is highly correlated. The new wavelet also has a small number of coefficients in high-pass subbands and allows the signal singularities, transitions, and edges intact in the low-pass subband.

To synthesize new wavelet it requires a two-channel filter bank which has a low-pass and a high-pass filter in decomposition (analysis) phase and reconstruction (synthesis) phase. This two-channel multirate filter bank consists of filters which process the input signal at half of its original rate. A block diagram of two-channel filter bank is shown in Figure 5. The signal $s(n)$ gets filtered by G_0 and H_0 filters and downsampled. These filters are called analysis or decomposition filters. The output of these filters contains the signal at half rate. These are called subbands of the signal. Each subband can be further divided into smaller subbands using the same filter bank. After being processed, the signal is upsampled and filtered using G_1 and H_1 filters, which are called synthesis or reconstruction filters. For perfect reconstruction, in a two-channel filter bank, the downsampled signal that contains only the even samples is given to the first channel and the downsampled signal that contains only the odd samples is given to the second channel. This separation of signal into even and odd components is called polyphase representation of the signal.

Four important characteristics are taken into account while designing the filter bank: (i) perfect reconstruction, (ii) orthogonality of the filter bank and the underlying wavelet based multiresolution structure, (iii) flatness of the filters and vanishing moments in the wavelets, and (iv) smoothness of the wavelets. The output of a perfect reconstruction two-channel filter bank is [34]

$$Y(z) = \frac{1}{2}\left[G_0(z)G_1(z) + H_0(z)H_1(z)\right]X(z) + \frac{1}{2}\left[G_0(-z)G_1(z) + H_0(-z)H_1(z)\right]X(-z), \quad (16)$$

where $X(z)$, $G(z)$, and $H(z)$ are the z-transform of the input signal, analysis filters and synthesis filters respectively. For perfect reconstruction: firstly, the alias term $X(-z)$ must be zero, hence

$$G_0(-z)G_1(z) + H_0(-z)H_1(z) = 0. \quad (17)$$

This can be achieved by letting

$$G_1(z) = H_0(-z), \qquad H_1(z) = -G_0(-z). \quad (18)$$

Second, the distortion term must be constant or a pure delay time; that is,

$$G_0(z)G_1(z) + H_0(z)H_1(z) = 2z^{-l}, \quad (19)$$

where l denotes a time delay.

Equation (18) can also be written as

$$H_0(z) = G_1(-z), \qquad H_1(z) = -G_0(-z). \quad (20)$$

Substituting this in (19)

$$G_0(z)G_1(z) - G_1(-z)G_0(-z) = 2z^{-l}, \quad (21)$$

$$P_0(z) - P_0(-z) = 2z^{-l}, \quad (22)$$

where $P_0(z)$ denotes the product of two low-pass filters $G_0(z)$ and $G_1(z)$. Equation (21) indicates that all odd terms of

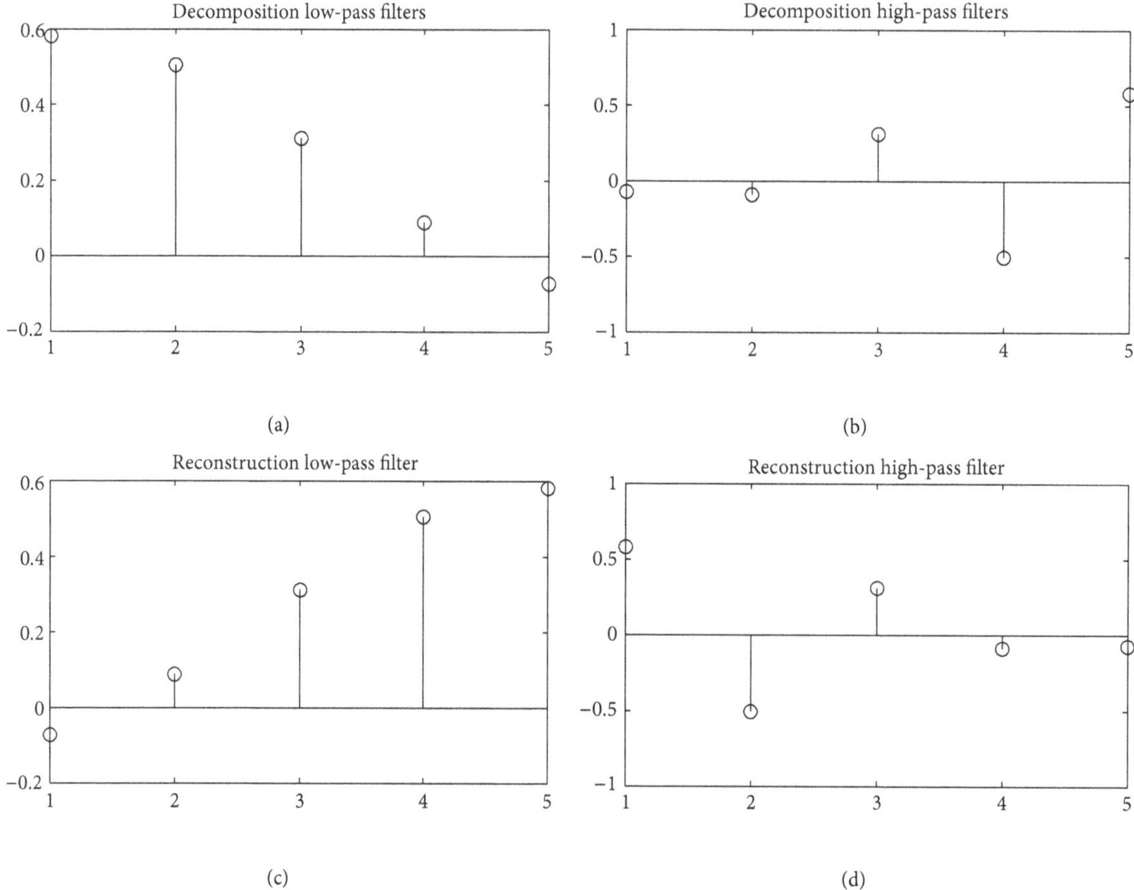

FIGURE 6: Impulse response for the reconstruction and decomposition filters of fetal.

product of two low-pass filters must be zero for order l, where l must be odd and even order terms are arbitrary [35]. Hence it can be written as

$$P_o(n) = \begin{cases} 0 & n \text{ odd}, \ n \neq l \\ 2 & n = l \\ \text{arbitrary} & n \text{ even.} \end{cases} \quad (23)$$

Hence it can be concluded that the design process of two-channel filter bank for a new wavelet is reduced into two steps as follows:

(1) design of filter $P_o(z)$ which satisfies (23),

(2) factorize $P_o(z)$ into $G_o(z)$ and $G_1(z)$.

In this work, the filter $P_o(z)$ is designed based on the characteristics of fPCG signals. The common requirements of this design are linear phase, minimum phase, and orthogonality of the filter. A fourth-order low-pass Butterworth filter is chosen because the transition width requirement is not stringent for the given cut-off frequency. This will also help in reducing the computational complexity. The Butterworth filter satisfies the conditions for perfect reconstruction. It has linear frequency response in the pass band as compared to Chebyshev Type I/Type II and elliptic filters. The filter $P_o(z)$ is factorized into $G_o(z)$ and $G_1(z)$, and then the coefficients of $H_o(z)$ and $H_1(z)$ are derived using (20) [32, 36]. Figure 6 shows the impulse response of the four filters computed for construction of filter bank of the new wavelet "fetal."

The wavelet and scaling functions are then derived from the coefficients of these filters using (6) and (8), respectively. Figure 7 shows the wavelet and scaling functions of the "fetal" wavelet.

With these wavelet and scaling functions, the wavelet and scaling coefficients for multiresolution analysis are obtained. The developed wavelet "fetal" is now ready to use. All the discrete analysis functions, including dwt, idwt, and wavedec. can operate on the new wavelet. Similarly, all the continuous analysis functions, including cwt, wscalogram, and the corresponding GUI tools, can also operate on the new wavelet. In this work, this new wavelet "fetal" is used for intended denoising of the fPCG signals.

6. Experimental Results

The newly developed wavelet "fetal" is used for denoising of real-time fetal heart sound signals which are of widely distinct nature and quality. A Simulink model is developed for computer based denoising of these signals [37]. Figure 8 shows the model for wavelet denoising of the fPCG signal

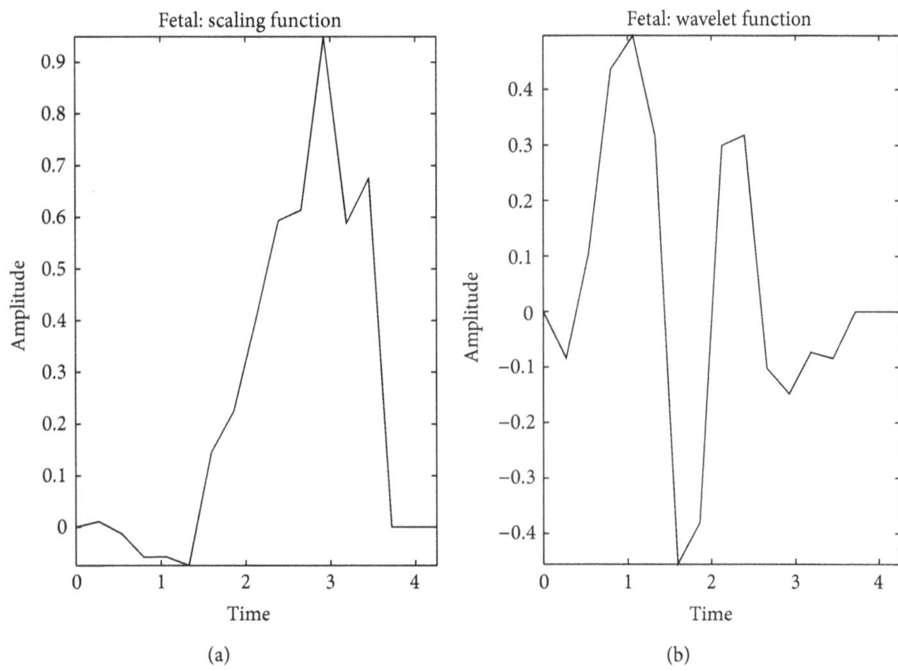

FIGURE 7: Wavelet and scaling function of "fetal" wavelet.

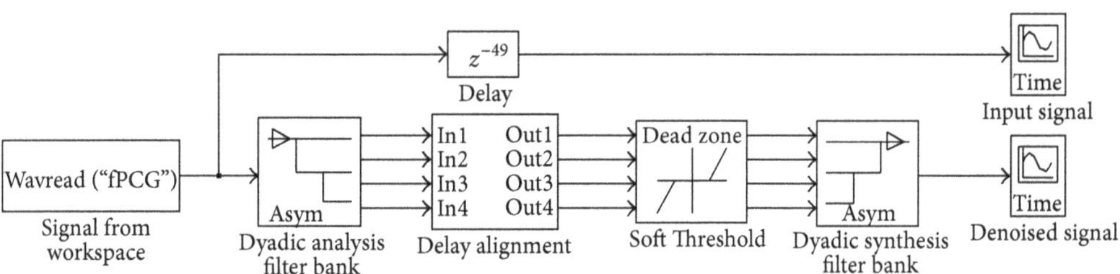

FIGURE 8: Simulink model for wavelet denoising.

through which the filters of developed wavelet family and selected threshold criterion are implemented.

In this model, the fPCG signals are fetched from the work space and applied to the Dyadic analysis filter bank. These signals carry fetal heart sound and a damped version of simulated maternal organs' sounds along with the external noise. The analysis filter bank decomposes the fPCG signals into a collection of subbands with smaller bandwidths and slower sample rates. This bank uses a series of high-pass and low-pass FIR filters to repeatedly divide the input frequency range. The fPCG signals are decomposed to the 3 levels by newly developed "fetal" wavelet with the analysis filter bank. These decomposed wavelet coefficients consist of details in the input fPCG signals. The denoising of fPCG signals is carried out by using selected algorithm (Rigorous SURE) and thresholding rule (Soft threshold). The synthesis filter bank reconstructs the signal decomposed by the analysis filter bank block. This bank takes in subbands of this signal and uses them to reconstruct the signal by using a series of high-pass and low-pass FIR filters. The reconstructed signals have a wider bandwidth and faster sample rate than the input subbands. The waveforms of input signal and denoised outputs are displayed through "Time Scope" blocks.

For performance evaluation of the developed wavelet, five test fPCG signals (S1–S5) are generated using a signal simulation module (SSM) [38]. These test signals are denoised using some existing competitive wavelets families and the developed wavelet. The competitive wavelets are the one that possess required properties of orthogonality and a scaling function Φ such as Daubechies, Symlets, and Coiflets. The results of this comparison are computed using MSE equation (15) and presented in Table 2. As explained earlier, all the three denoising algorithms with soft or hard threshold are used for denoising of the fPCG signals.

The results in Table 2 show that the developed wavelet "fetal" along with rigorous SURE algorithm and soft threshold provides the best performance for denoising the fPCG signals.

The developed wavelet family was also implemented for denoising of a real fPCG signal. This signal was recorded using a specially developed wireless data acquisition system [39]. The recording of fPCG signal was obtained in a quiet

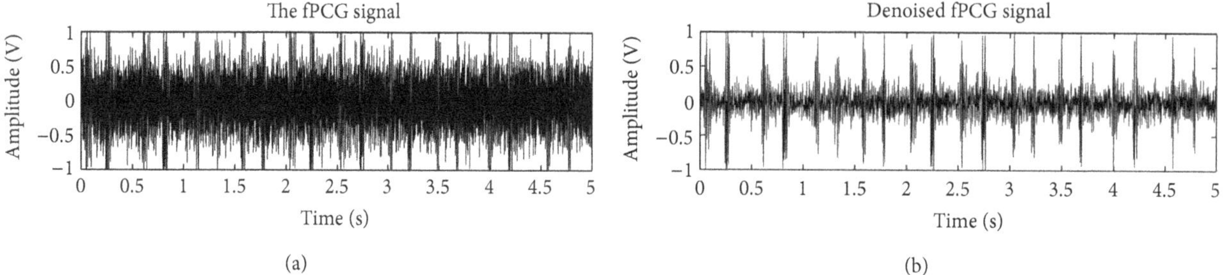

(a) (b)

FIGURE 9: Waveforms of a real fPCG signal and its denoised version.

TABLE 2: Comparison of different mother wavelets for denoising of fPCG signals in terms of MSE.

	Wavelet family									
			db5					coif4		
Algorithm↓ Signal→	S1	S2	S3	S4	S5	S1	S2	S3	S4	S5
(1) Sqtwolog(s)	0.8331	0.9124	1.099	0.3274	0.3328	0.5826	0.5744	1.1137	0.2615	0.2817
(2) Sqtwolog(h)	0.843	0.9124	1.099	0.3274	0.3328	0.5906	0.5666	1.1137	0.2615	0.2817
(3) Minimaxi(s)	0.6295	0.9073	1.099	0.3274	0.3328	0.5371	0.5716	1.1137	0.2712	0.2817
(4) Minimaxi(h)	0.7892	0.8953	1.099	0.7691	0.7779	0.8199	0.7972	1.1137	0.8162	0.8092
(5) Rigrsure(s)	0.575	0.8875	1.1001	0.4331	0.4396	0.4661	0.5531	1.081	0.2539	0.2669
(6) Rigrsure(h)	0.8165	0.8055	1.2945	0.7476	0.7561	1.0091	0.8633	1.2254	1.0062	0.8851
	Wavelet family									
			sym7					**Fetal (New)**		
Algorithm↓ Signal→	S1	S2	S3	S4	S5	S1	S2	S3	S4	S5
(1) Sqtwolog(s)	0.5608	0.6064	1.1245	0.2663	0.2769	0.5723	0.5766	1.1054	0.2712	0.2865
(2) Sqtwolog(h)	0.5629	0.5934	1.1245	0.2663	0.2769	0.6001	0.5613	1.1054	0.2712	0.2865
(3) Minimaxi(s)	0.5615	0.602	1.1245	0.2877	0.2837	0.5298	0.5743	1.1054	0.2712	0.2865
(4) Minimaxi(h)	0.836	0.8363	1.5005	0.8267	0.8224	0.8263	0.7889	1.1054	0.8133	0.8324
(5) **Rigrsure(s)**	0.5033	0.5979	1.1673	0.4122	0.4436	**0.4212**	**0.539**	**0.9998**	**0.2469**	**0.2518**
(6) Rigrsure(h)	0.8431	0.8394	1.5005	0.8369	0.8353	1.1776	0.8519	1.2173	1.1123	0.8732

room with the help of a medical expert and a trained nurse. A pregnant woman visiting for antenatal care was requested to contribute in the real-time testing of the developed system. The subject was having 34 weeks of gestation and singleton pregnancy. She was asked to lie in the supine position, with the head resting on a pillow. The phonocardiographic sensor was positioned on the abdominal surface of the mother and adjusted to acquire the maximum intensity of the signal. An abdominal belt was used to fasten the sensor. The signal was sampled with sampling frequency of 8000 Hz, 16 bit resolution and saved for further processing.

The recorded fPCG signal is now fetched from the memory through the model, as shown in Figure 8, for denoising. Figure 9 shows the waveforms of original and denoised version of this signal. In these waveforms, x-axis represents the time in seconds, whereas y-axis represents amplitude of signal in volts.

7. Discussion and Conclusion

The fPCG signals are of very low amplitude and contain poor signal-to-noise ratio. The main sources of noise are maternal biological activities, external noises such as sound produced by electrical appliances, movement of transducer, and so forth. These noises show overlapping spectra with the actual fPCG signals. Hence conventional noise removal techniques are not suitable for denoising of these signals. During denoising, care has to be practiced to preserve the features contained in the original signal. These preserved features are relevant and necessary for an appropriate diagnosis about fetal health.

In this paper design of a new wavelet basis function for denoising of fPCG signals has been carried out. The key features of newly designed family are its speed of convergence at infinity to 0, regularity, and orthogonality. It also has a small number of coefficients in high-pass subbands and allows the signal singularities, transitions, and edges intact in the low-pass subband. It has been found that the combination of optimal perfect reconstruction filter bank and appropriate denoising algorithm can improve the performance of denoising. The experimental results revealed suitability of the newly developed wavelet to be the most appropriate wavelet basis function for denoising of fPCG signals in terms of MSE. The resultant denoised fPCG signals will preserve physiological information of diagnostic importance regarding health status of the unborn.

Acknowledgments

The authors of this paper would like to thank the experts and technical staff of Government Women Hospital, Gondia (M.S.), India, for their kind support. They also thank the pregnant women who contributed in the clinical experimentation.

References

[1] V. S. Chourasia and A. K. Tiwari, "A review and comparative analysis of recent advancements in fetal monitoring techniques," *Critical Reviews in Biomedical Engineering*, vol. 36, no. 5-6, pp. 335–373, 2008.

[2] M. Murray, *Antepartal and Intrapartal Fetal Monitoring*, Springer, 3rd edition, 2007.

[3] J. Karin, M. Hirsch, O. Segal, and S. Akselrod, "Non invasive fetal ECG monitoring," in *Computers in Cardiology*, pp. 365–368, IEEE Computer Society Press, Bethesda, Md, USA, 1994.

[4] F. Kovács, M. Török, and I. Habermajer, "A rule-based phonocardiographic method for long-term fetal heart rate monitoring," *IEEE Transactions on Biomedical Engineering*, vol. 47, no. 1, pp. 124–130, 2000.

[5] C. Jianfeng, K. Phua, S. Ying, and L. Shue, "A portable phonocardiographic fetal heart rate monitor," in *Proceedings of the IEEE International Symposium on Circuits and Systems*, pp. 2141–2144, Island of Kos, Greece, May 2006.

[6] H. G. Goovaerts, O. Rompelman, and H. P. Van Geijn, "A transducer for detection of fetal breathing movements," *IEEE Transactions on Biomedical Engineering*, vol. 36, no. 4, pp. 471–478, 1989.

[7] J. T. E. McDonnell, "Knowledge-based interpretation of foetal phonocardiographic signals," *IEE Proceedings F*, vol. 137, no. 5, pp. 311–318, 1990.

[8] M. Godinez, A. Jimenez, R. Ortiz, and M. Pena, "On-line fetal heart rate monitor by Phonocardiography," in *Proceedings of the 25th Annual International Conference of the IEEE EMBS*, pp. 3141–3144, Cancun, Mexico, September 2003.

[9] R. Acharyya, N. L. Scott, P. Teal, E. Deuss, and J. Flierl, "Signal separation for non-invasive monitoring of foetal heartbeat," in *Proceedings of the 1st International Conference on BioMedical Engineering and Informatics*, pp. 497–501, May 2008.

[10] A. Jiménez-González and C. J. James, "Extracting sources from noisy abdominal phonograms: a single-channel blind source separation method," *Medical & Biological Engineering & Computing*, vol. 47, no. 6, pp. 655–664, 2009.

[11] F. Kovacs, C. Horváth, A. T. Balogh, and G. Hosszú, "Extended non-invasive fetal monitoring by detailed analysis of data measured with phonocardiography," *IEEE Transactions on Biomedical Engineering*, vol. 58, no. 1, pp. 64–70, 2011.

[12] P. Varady, "Wavelet-based adaptive denoising of phonocardiographic records," in *Proceedings of the 23rd Annual International Conference of the IEEE Engineering in Medicine and Biology Society*, vol. 2, pp. 1846–1849, 2001.

[13] S. R. Messer, J. Agzarian, and D. Abbott, "Optimal wavelet denoising for phonocardiograms," *Microelectronics Journal*, vol. 32, no. 12, pp. 931–941, 2001.

[14] A. K. Mittra, A. Shukla, and A. S. Zadgaonkar, "System simulation and comparative analysis of foetal heart sound de-noising techniques for advanced phonocardiography," *International Journal of Biomedical Engineering and Technology*, vol. 1, no. 1, pp. 73–85, 2007.

[15] T. Tosanguan, R. J. Dickinson, and E. M. Drakakis, "Modified spectral subtraction for de-noising heart sounds: interference suppression via spectral comparison," in *Proceedings of the IEEE-BIOCAS Biomedical Circuits and Systems Conference*, pp. 29–32, Baltimore, Md, USA, November 2008.

[16] X. M. Zhao and G. T. Cao, "A novel de-noising method for heart sound signal using improved thresholding function in wavelet domain," in *Proceedings of the International Conference on Future BioMedical Information Engineering*, pp. 65–68, Sanya, China, December 2009.

[17] V. S. Chourasia and A. K. Mittra, "Wavelet-based denoising of fetal phonocardiographic signals," *International Journal of Medical Engineering and Informatics*, vol. 2, no. 2, pp. 139–150, 2010.

[18] V. S. Chourasia, A. K. Tiwari, R. Gangopadhyay, and K. A. Akant, "Foetal phonocardiographic signal denoising based on non-negative matrix factorization," *Journal of Medical Engineering & Technology*, vol. 36, pp. 57–66, 2012.

[19] D. Esteban and C. Galand, "Application of quadrature mirror filters to split band voice coding schemes," in *Proceedings of the IEEE International Conference on Acoustics, Speech, and Signal Processing (ICASSP '77)*, pp. 191–195, May 1977.

[20] N. Colley, D. G. Talberg, and D. P. Southall, "Biophysical profile in the fetus from a phonocardiographic sensor," *European Journal of Obstetrics & Gynecology and Reproductive Biology*, vol. 23, no. 5-6, pp. 261–266, 1986.

[21] A. Di Lieto, M. De Falco, M. Campanile et al., "Regional and international prenatal telemedicine network for computerized antepartum cardiotocography," *Telemedicine and e-Health*, vol. 14, no. 1, pp. 49–54, 2008.

[22] B. H. Tan and M. Moghavvemi, "Real time analysis of fetal phonocardiography," in *Proceedings of the TENCON Proceedings*, vol. 2, pp. 135–140, Kuala Lumpur, Malaysia, September 2000.

[23] N. Colley, N. G. Abraham, P. Fayers, D. G. Talbert, W. L. Daviesa, and D. P. Southall, "The fetal phonogram: a measure of fetal activity," *Lancet*, vol. 327, no. 8487, pp. 931–935, 1986.

[24] D. G. Talbert, W. L. Davies, and F. Johnson, "Wide bandwidth fetal phonography using a sensor matched to the compliance of the mother's abdominal wall," *IEEE Transactions on Biomedical Engineering*, vol. 33, no. 2, pp. 175–181, 1986.

[25] X. C. Yin, P. Han, J. Zhang, F. Q. Zhang, and N. L. Wang, "Application of wavelet transform in signal denoising," in *Proceedings of the International Conference on Machine Learning and Cybernetics*, vol. 1, pp. 436–441, November 2003.

[26] I. Daubechies, *Ten Lectures on Wavelets*, Society for Industrial and Applied Mathematics (SIAM), New York, NY, USA, 1992.

[27] M. R. Raghuveer and A. S. Bopardikar, *Wavelet Transforms: Introduction to Theory and Applications*, Pearson Education, Indian Branch, Singapore, 3rd edition, 2002.

[28] A. Daamouche, L. Hamami, N. Alajlan, and F. Melgani, "A wavelet optimization approach for ECG signal classification," *Biomedical Signal Processing and Control*, vol. 7, no. 4, pp. 342–349, 2012.

[29] Md. A. Kabir and C. Shahnaz, "Denoising of ECG signals based on noise reduction algorithms in EMD and wavelet domains," *Biomedical Signal Processing and Control*, vol. 7, no. 5, pp. 481–489, 2012.

[30] V. S. Chourasia and A. K. Mittra, "Selection of mother wavelet and denoising algorithm for analysis of foetal phonocardiographic signals," *Journal of Medical Engineering and Technology*, vol. 33, no. 6, pp. 442–448, 2009.

[31] D. L. Donoho and I. M. Johnstone, "Adapting to unknown smoothness via wavelet shrinkage," *Journal of the American Statistical Association*, vol. 90, no. 432, pp. 1200–1224, 1995.

[32] G. Strang and T. Nguyen, *Wavelets and Filter Banks*, Wellesley, Cambridge, Mass, USA, 1997.

[33] I. Daubechies, "Wavelet transform, time-frequency localization and signal analysis," *IEEE Transactions on Information Theory*, vol. 36, no. 5, pp. 961–1005, 1990.

[34] P. P. Vaidyanathan, "Multirate digital filters, filter banks, polyphase networks, and applications: a tutorial," *Proceedings of the IEEE*, vol. 78, no. 1, pp. 56–93, 1990.

[35] T. Q. Nguyen and P. P. Vaidyanathan, "Two-channel perfect-reconstruction FIR QMF structures which yield linear-phase analysis and synthesis filters," *IEEE Transactions on Acoustics, Speech, and Signal Processing*, vol. 37, no. 5, pp. 676–690, 1989.

[36] M. Vetterli, "Wavelets and filter banks: theory and design," *IEEE Transactions on Signal Processing*, vol. 40, pp. 2207–2232, 1992.

[37] Simulink, "Simulation and model based design user's guide," Mathwork Incorporation, 2010, http://www.mathworks.com/.

[38] V. S. Chourasia and A. K. Tiwari, "Development of a signal simulation module for testing of phonocardiography based prenatal monitoring systems," in *Proceedings of the IEEE India Council Conference (INDICON '09)*, pp. 1–4, Ahmadabad, India, December 2009.

[39] V. S. Chourasia and A. K. Tiwari, "Wireless data acquisition for fetal phonographic signals using Bluetooth," *International Journal of Computers in Healthcare*, vol. 1, no. 3, pp. 240–253, 2012.

11

Sample Training Based Wildfire Segmentation by 2D Histogram θ-Division with Minimum Error

Jianhui Zhao,[1,2] Erqian Dong,[1] Mingui Sun,[3] Wenyan Jia,[3] Dengyi Zhang,[1] and Zhiyong Yuan[1]

[1] *School of Computer, Wuhan University, Wuhan, Hubei 430072, China*
[2] *Suzhou Institute of Wuhan University, Suzhou, Jiangsu 215123, China*
[3] *Department of Neurosurgery, University of Pittsburgh, Pittsburgh, PA 15213, USA*

Correspondence should be addressed to Jianhui Zhao; jianhuizhao@whu.edu.cn

Academic Editors: I. Korpeoglu, E. Pauwels, and S. Verstockt

A novel wildfire segmentation algorithm is proposed with the help of sample training based 2D histogram θ-division and minimum error. Based on minimum error principle and 2D color histogram, the θ-division methods were presented recently, but application of prior knowledge on them has not been explored. For the specific problem of wildfire segmentation, we collect sample images with manually labeled fire pixels. Then we define the probability function of error division to evaluate θ-division segmentations, and the optimal angle θ is determined by sample training. Performances in different color channels are compared, and the suitable channel is selected. To further improve the accuracy, the combination approach is presented with both θ-division and other segmentation methods such as GMM. Our approach is tested on real images, and the experiments prove its efficiency for wildfire segmentation.

1. Introduction

Image segmentation is an important research topic in the fields of image processing and computer vision, and its purpose is dividing one image into several different areas, while each region has certain characteristics and disjoints with the others. As the traditional segmentation approach, threshold value based methods [1, 2] have been widely used and studied for many years, mainly due to the fact of easy programming together with relatively low computational complexity.

The core issue of threshold methods is to construct the evaluation function to achieve the optimal segmentation threshold value. K-means clustering method [3] has two evaluation standards: the similarity among the elements in each subclass and the diversity among the elements in different subclasses. The main disadvantages of K-means method are that without prior knowledge the ideal number K cannot be determined before segmentation, and it is sensitive to noise and isolated targets, also sensitive to the initial clustering centers, thus is easy to obtain the local optimal solutions. Otsu method [4] is one adaptive thresholding approach. It divides the original image into two classes: object and background, according to the color histogram of image. The evaluation function is constructed based on the variance between two classes, and the best threshold value corresponds to the maximum variance. However, Otsu method ignores the spatial neighboring information of image and may fail when the size proportion of object and background is very small. Fuzzy C-means (FCM) [5] is also a clustering method, which promotes the target function of C-means with fuzzy mathematics [6]. Main advantage of FCM is using the degree of membership for classification, and thus it can conserve more information of the image and obtain better segmentation. The disadvantages of FCM are that the initial parameters are set arbitrarily, time and space complexities increase rapidly when number C becomes larger, and successful convergence in limited time is a problem [7]. Maximum entropy principle is used to help estimate the unknown probability distribution with limited conditions, while the main idea is to select the probability distribution with the maximum entropy value [8]. This principle is employed for

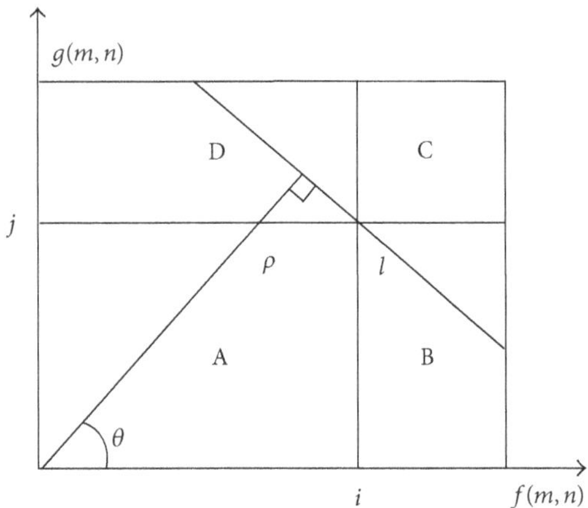

FIGURE 1: The 2D histogram with θ-division.

image segmentation assuming that the color value of each pixel in one image is a random event, and the normalized color histogram is taken as the probability distribution of pixels [9]. The minimum error thresholding is a kind of threshold image segmentation method based on Bayesian theory [10]. It assumes that the object and background obey the mixed normal distribution, and the classification is based on the minimum error principle. It can also be explained as the minimum relative entropy between actual distribution of color histogram and assumed mixed normal distribution [11]. Through quantitative comparison [12] of the typical threshold methods, it can be found that the minimum error thresholding derives good segmentations, even for the targets with very small sizes.

In threshold segmentation methods, color histogram is often used to describe the distribution of pixels. Since the information of spatial relationships among pixels is not considered in 1D histogram, it is very difficult to obtain satisfying segmentation results when the image has high complexity or significant noise. Therefore, 2D histogram has been employed using both color value of each pixel and its neighborhood averaged color [13–15]. As shown in Figure 1, threshold (i, j) divides the 2D histogram into 4 rectangular regions. Regions A and C represent the object and the background, respectively, while regions B and D represent edges, noise, and so on, where the probability distribution is assumed to be nearly zero. But in real images the assumption is hard to be satisfied, and thus the segmented results are less accurate. For this problem, the 2D linear division method with minimum error is introduced [16], and the line l perpendicular with the main diagonal line of histogram is used as threshold for segmentation. The method can divide more pixels into object or background, but it only considers the particular case; that is, angle θ is 45 degrees. Then a widely suitable thresholding method is proposed based on 2D histogram θ-division and maximum deviation criterion [17], and in this method the segmented results and running time related with different θ values are analyzed. More recently, new threshold method is presented based on 2D histogram θ-division with minimum error [18], and the influence of various θ values on both segmentation and computational expense is discussed with experiments.

Until now, the aforementioned threshold methods are automatic and unsupervised. Thus the segmented results cannot be affected by the prior knowledge and cannot be effectively evaluated either. Therefore, in our paper the sample training method is employed for 2D histogram θ-division with minimum error. Outline of our approach is to determine the suitable θ for specific application with sample training and to combine 2D θ-division with other kinds of segment techniques, for example, color ranges based methods [19, 20] such as GMM. To test our new approach, the combined algorithm is checked through segmentation of wildfire regions from some real images.

2. 2D Histogram and Sample Training

2.1. 2D Histogram θ-Division with Minimum Error. The 2D histogram θ-division with minimum error is illustrated in Figure 1, and θ is the angle between horizontal axis and the normal of division line l. When θ is set as 45 degrees, the weights between the pixel color and its neighboring average color are fixed at 1 : 1. When θ varies, the weights and the division line l also change, and the corresponding segmentation results are generated.

Given an image with the size of $M * N$, for any pixel (m, n), its color value is $f(m, n)$ and the total color level is L; there is $0 \le f(m, n) \le L - 1$. The neighboring average color value of pixel (m, n) is

$$g(m, n) = \frac{1}{w} \sum_{(x,y)\in D} f(x, y), \tag{1}$$

where w is the number of pixels in the neighborhood window area D of pixel (m, n).

Let $h(m, n)$ represent the number of pixels with the color value of $f(m, n)$ together with the neighboring average color value $g(m, n)$, and probability of $h(m, n)$ is computed as $p(m, n) = h(m, n)/M * N$.

The division line l in Figure 1 can be expressed by

$$f(m, n) \cos\theta + g(m, n) \sin\theta = \rho, \tag{2}$$

where ρ is the distance from the origin of coordinate to line l, and $0^\circ \le \theta \le 90^\circ$.

Based on definition of 2D color histogram, there are $0 \le f(m, n) \le L - 1$, $0 \le g(m, n) \le L - 1$; thus,

$$0 \le \rho \le (L - 1)(\cos\theta + \sin\theta). \tag{3}$$

To segment image, let $a = \cos\theta/(\cos\theta + \sin\theta)$ and $T = \rho/(\cos\theta + \sin\theta)$; line l can be expressed with the modified formula:

$$af(m, n) + (1 - a) g(m, n) = T. \tag{4}$$

Therefore, variable a can be taken as the weighting parameter of color value $f(m, n)$ and $0 \le a \le 1$, while T is the corresponding segmentation threshold value. For

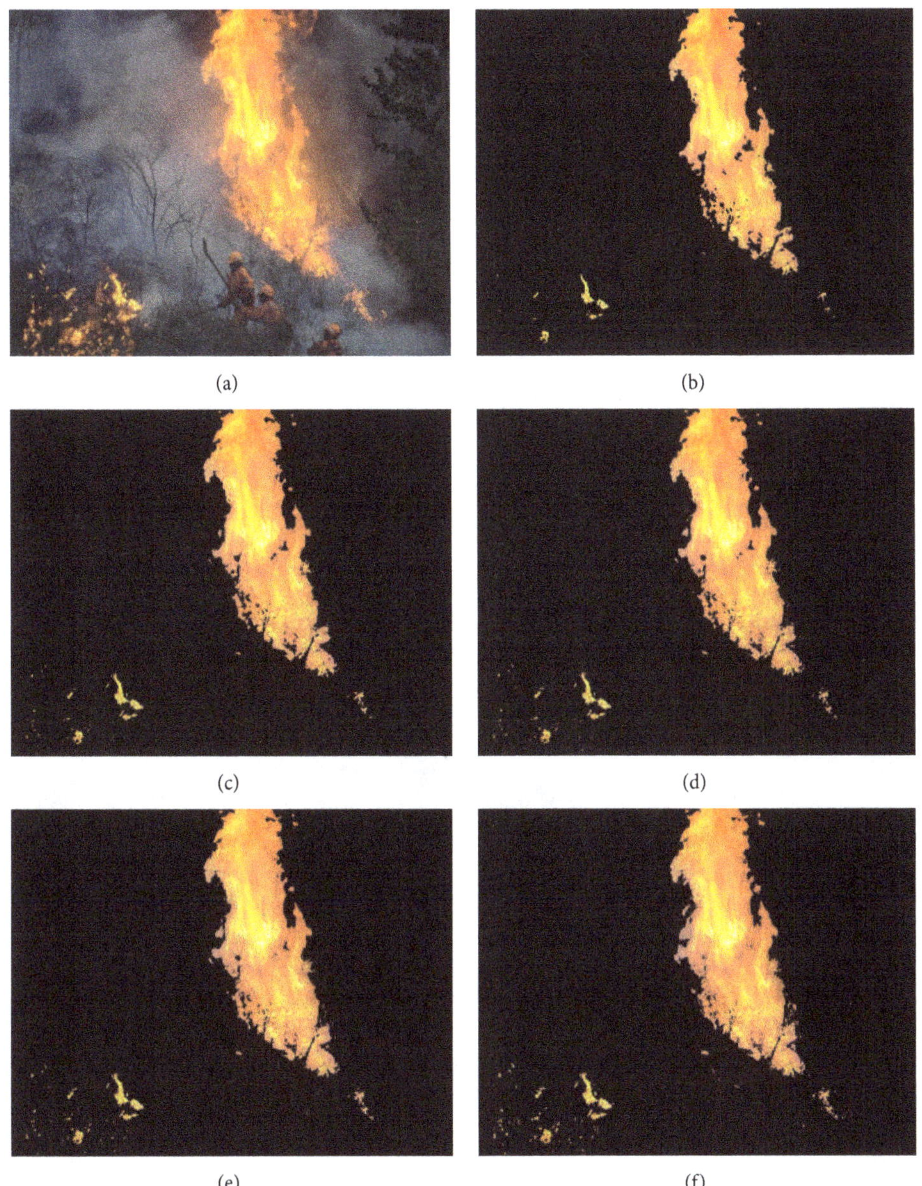

FIGURE 2: Fire segmentation with different angles θ: (a) input image; (b–f) results with $\theta = 0^{\circ}, 30^{\circ}, 45^{\circ}, 60^{\circ}, 90^{\circ}$.

pixel (m, n), its corresponding binarized value $b(m, n)$ in the segmented image is

$$b(m,n) = \begin{cases} 0, & af(m,n) + (1-a)\,g(m,n) \le T, \\ 1, & af(m,n) + (1-a)\,g(m,n) > T. \end{cases} \quad (5)$$

Obviously, weighting variable a is decided by angle θ of division line l, and the segmentation result varies with different value of a. When a is more than 0.5, the color value of one pixel has heavier weight, and the boundary of segmented region is more accurate, while when a is less than 0.5, the neighboring average color value of pixel has heavier weight; thus segmentation's insensitivity to noise is enhanced. Therefore, segmented results can be adjusted for different applications through modifying variable a or angle θ.

The division line l separates one image into object area C_o and background area C_b, and the corresponding probabilities are computed as $w_o(T) = \sum_{(i,j)\in C_o} p(i,j)$ and $w_b(T) = \sum_{(i,j)\in C_b} p(i,j)$.

Mean value vectors of object and background are

$$\begin{aligned} \mu_o(T) &= \left[\mu_{oi}(T), \mu_{oj}(T)\right]^T \\ &= \left[\frac{\sum_{(i,j)\in C_o} p(i,j)\,i}{w_o(T)}, \frac{\sum_{(i,j)\in C_o} p(i,j)\,j}{w_o(T)}\right]^T, \\ \mu_b(T) &= \left[\mu_{bi}(T), \mu_{bj}(T)\right]^T \\ &= \left[\frac{\sum_{(i,j)\in C_b} p(i,j)\,i}{w_b(T)}, \frac{\sum_{(i,j)\in C_b} p(i,j)\,j}{w_b(T)}\right]^T. \end{aligned} \quad (6)$$

FIGURE 3: Sample images and manually marked fires.

The global mean value vector is

$$\mu_T = [\mu_{Ti}, \mu_{Tj}]^T = \left[\sum_{i=0}^{M-1}\sum_{j=0}^{N-1} p(i,j)\, i, \sum_{i=0}^{M-1}\sum_{j=0}^{N-1} p(i,j)\, j\right]^T. \quad (7)$$

Variance vectors of object and background are

$$\sigma_o^2(T) = [\sigma_{oi}^2(T), \sigma_{oj}^2(T)]^T = \begin{bmatrix} \dfrac{\sum_{(i,j)\in C_o}[i-\mu_{oi}(T)]^2 p(i,j)}{w_o(T)} \\ \dfrac{\sum_{(i,j)\in C_o}[j-\mu_{oj}(T)]^2 p(i,j)}{w_o(T)} \end{bmatrix},$$

$$\sigma_b^2(T) = [\sigma_{bi}^2(T), \sigma_{bj}^2(T)]^T = \begin{bmatrix} \dfrac{\sum_{(i,j)\in C_b}[i-\mu_{bi}(T)]^2 p(i,j)}{w_b(T)} \\ \dfrac{\sum_{(i,j)\in C_b}[j-\mu_{bj}(T)]^2 p(i,j)}{w_b(T)} \end{bmatrix}. \quad (8)$$

Based on the previous probabilities and vectors, the evaluation function of 2D histogram thresholding with minimum error is defined as

$$\begin{aligned} J(T) = {} & 1 - w_o(T)\ln w_o(T) - w_b(T)\ln w_b(T) \\ & + w_o(T)\ln[\sigma_{oi}(T)\sigma_{oj}(T)] \\ & + w_b(T)\ln[\sigma_{bi}(T)\sigma_{bj}(T)]. \end{aligned} \quad (9)$$

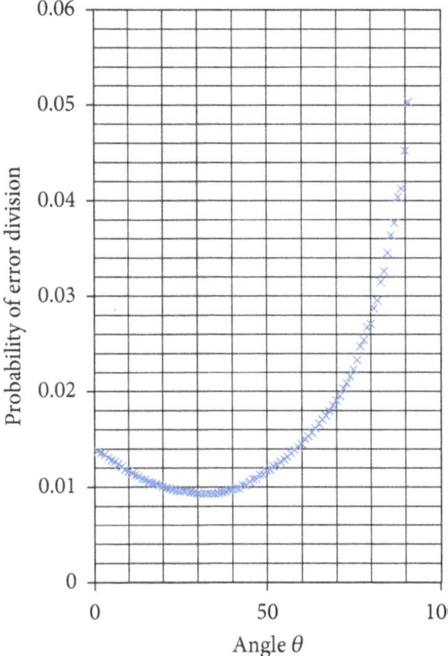

FIGURE 4: Relationship between $P_{err}(\theta)$ and angle θ.

When $J(T)$ obtains its minimum value, there is the following corresponding optimal threshold value T^* [16–18]:

$$T^* = \arg\min_{0 \le T \le L-1} J(T). \tag{10}$$

To illustrate the effects of angle θ on the segmented results, one image with wildfire is used. Taking the V channel of HSV color space as input for 2D histogram θ-division with minimum error, the segmentations with different angle θ are shown in Figure 2.

From the results of Figure 2, effects of angle θ on the segmentation can be found. When θ is small, most noises are deleted but some fire pixels are also discarded. When θ is large, most fire pixels are detected while some noises are falsely taken as flame too. Therefore, determining the optimal value of angle θ is very important.

2.2. Sample Training Based θ Determination. The 2D histogram θ-division with minimum error is an unsupervised threshold method, thus the application of prior knowledge helps to obtain better results. For the specific problem of fire segmentation, the optimal angle θ can be determined through sample learning. As shown in Figure 3, typical images with wildfire are collected, and the fire regions are marked manually, which are used to evaluate and train the segmentation method.

For any sample image (Figure 3(a)), we mark the background area with pure black color (value 0) and leave the object area unchanged, and the manual template image r is thus generated (Figure 3(b)). For the same sample image, 2D histogram division method with angle θ and minimum error is used for segmentation and the segmented image s is produced. For one pixel with coordinates (x, y), there are 4 possible cases as follows:

(1) if $r(x, y) = s_\theta(x, y) > 0$, the pixel has been correctly identified as the object point;

(2) if $r(x, y) = s_\theta(x, y) = 0$, the pixel has been correctly identified as the background point;

(3) if $r(x, y) = 0$ and $s_\theta(x, y) > 0$, the pixel has been wrongly identified as the object point;

(4) if $r(x, y) > 0$ and $s_\theta(x, y) = 0$, the pixel has been wrongly identified as the background point.

Based on comparison between image r and image s, we can obtain the total number of pixels, the number of object pixels, the number of background pixels, the number of pixels that are wrongly identified as object, and the number of pixels that are wrongly identified as background. Then the segmented result can be evaluated by the probability of error division as follows:

$$P_{err}(\theta) = P(O) \cdot P(B \mid O) + P(B) \cdot P(O \mid B), \tag{11}$$

where $P(O)$ and $P(B)$ are probabilities of object and background, respectively, $P(O \mid B)$ is probability that background is wrongly identified as object, $P(B \mid O)$ is probability that object is falsely identified as background, and the probabilities can be computed by aforementioned numbers of different kinds of pixels.

Obviously the probability of error division $P_{err}(\theta)$ is a function of angle θ, and the minimum function value is related with the optimal angle θ.

With a group of wildfire images having total 1767705 pixels and manually marked fire regions having 114320 pixels, relationship between $P_{err}(\theta)$ value and angle θ, varying in the range of $[0^\circ, 90^\circ]$ with the changing step of 1°, is shown in Figure 4.

The optimal angle θ corresponds with the minimum value of $P_{err}(\theta)$. Since distribution of discrete sampled points in Figure 4 approximates the shape of one unimodal function, the method of bisection search is used to find the minimum value of $P_{err}(\theta)$. For the example shown in Figure 4, the optimal angle θ is found as $\theta = 32.9375^\circ$ and the related minimum error probability of evaluation function $P_{err}(\theta)$ is 0.0093.

3. Application on Wildfire Segmentation

3.1. Selection of Color Channel. The 2D histogram θ-division with minimum error can deal with gray image or any channel of color image. To analyze its performance on different color channels, both RGB and HSV color models are used for testing, and all color channels together with gray images of three color fire images are processed.

The segmented results are shown in Figure 5; (a) shows input images, (b) shows results of R channel, (c) shows results of G channel, (d) shows results of B channel, (e) shows results of H channel, (f) shows results of S channel, (g) shows results of V channel, and (h) shows results of gray images which are converted from the input color images.

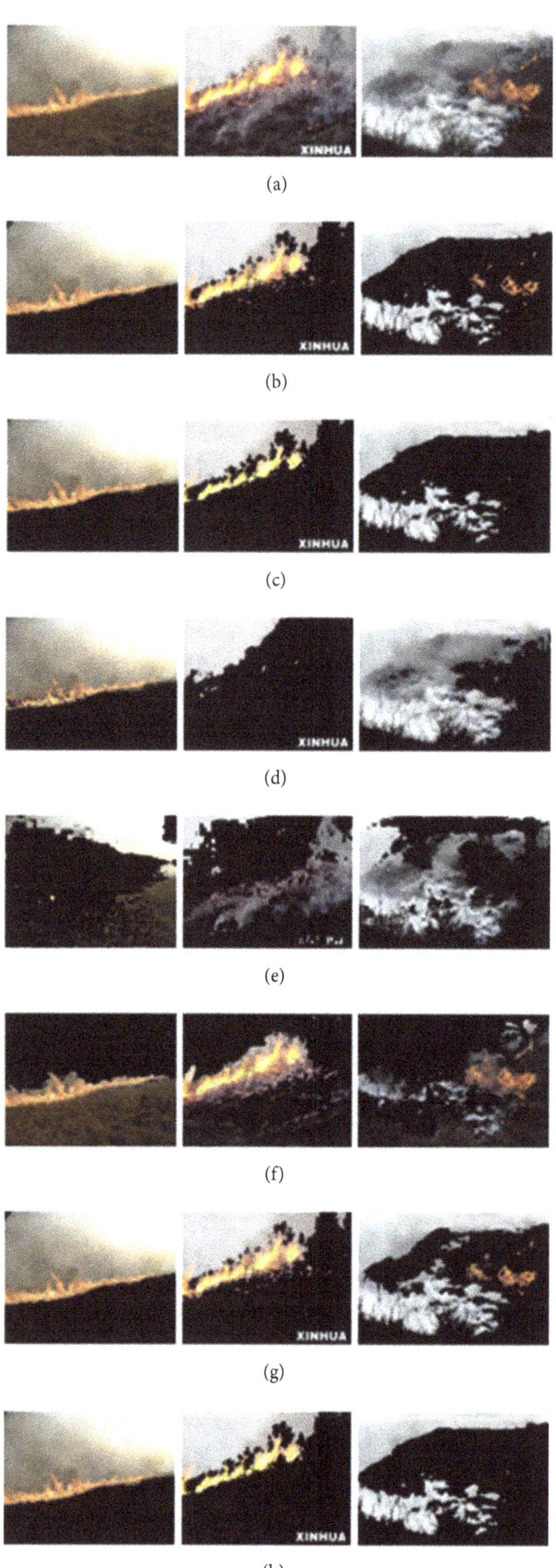

FIGURE 5: Results on color channels and gray images—the 1st row: input images; the 2nd row to 8th row: segmentations of R, G, B, H, S, V channels and gray images.

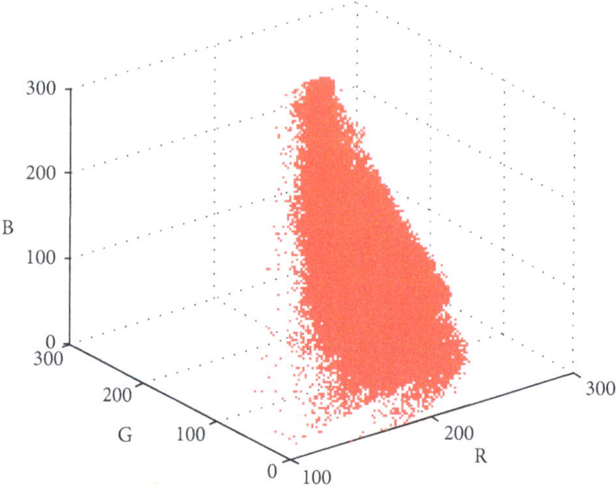

FIGURE 6: Sample pixels used for GMM training.

For wildfire images, the regions of interest are fire and smoke; thus the evaluation of color channels is based on whether there are successful segmentations of the regions from images. For G channel or B channel, the fire or the smoke regions cannot be identified correctly. R channel can obtain fire regions more accurately, but there are still some fire pixels wrongly classified as nonfire objects, while the segmented smoke regions are less accurate. H channel has worse results, since hue differences among real fire, smoke and background are not very obvious; for example, hue of dark fire and smoke is similar with background. S channel derives most fire regions but can also take some saturated pixels of background as fire. V channel has the ability to identify smoke and fire areas as high luminance objects at the same time. Comparatively, segmentations from gray images lose some fire regions.

It can be found that no channel successfully segments fire or smoke region separately, but V channel has the best performance to obtain both fire and smoke areas simultaneously. Therefore, V channel is selected for 2D histogram θ-division with minimum error. Of course, to obtain the individual objects of fire or smoke, segmented results need further process, for example, the popularly employed color range based segmentation methods.

3.2. Combination with Other Segmentation Methods. There are other kinds of methods for fire segmentation from images; among them the Gaussian mixture model (GMM) for sample distribution is an efficient one. In our previous work [20], 530,000 pixels manually segmented from the fire regions of 23 sample images are collected in RGB color space (shown in Figure 6), and then the 3D shape of fire sample pixels is represented by GMM. Parameters of GMM are trained with expectation maximization, and then the fire probability distribution in 3D color space is computed. Based on the calculated fire probability of one pixel under processing, we can decide whether it belongs to flame area or not.

(a) (b)

FIGURE 7: Segmented results from GMM with errors.

GMM parameters and models are represented as

$$g(x, \mu_i, \Sigma_i) = \frac{1}{\sqrt{(2\pi)^d |\Sigma_i|}} \exp\left(-\frac{1}{2}(x-\mu_i)^T \Sigma_i^{-1}(x-\mu_i)\right), \quad (12)$$

$$p(x) = \Sigma_i \alpha_i * g(x, \mu_i, \Sigma_i),$$

where the weighting value, kernel center, and covariance matrix of one single Gaussian model are α_i, μ_i, and Σ_i, respectively, while $p(x)$ illustrates how close of point x to the Gaussian mixture model. The suitable number of Gaussian models in GMM can be manually assigned or computed automatically.

However, performance of GMM is affected by the amount of sample pixels and the number of Gaussian models. If the Gaussian models or the sample pixels are not enough, GMM cannot precisely describe the object range in color space. In this case, application of GMM on the input image easily brings errors, for example, wrongly taking some nonfire objects as fires. As shown in Figure 7, figures of the 1st column are input images, while figures of the 2nd column are the segmented results. Having the similar color as fire, some nonfire pixels, such as smoke, wall, or words, are mistakenly detected as flame.

Therefore, combination of the 2D histogram θ-division with minimum error and other division methods such as GMM can further improve the segmentation. There are 2 possible ways for combination: (1) 2D θ-division first and then GMM segmentation; (2) GMM segmentation first and then 2D θ-division.

For (1), GMM deals with the already segmented fire and smoke areas from 2D θ-division; thus the pixels of other objects have no chance to be mistakenly chosen as flame. For (2), 2D θ-division method further divides the segmented result from GMM, which makes it possible to differentiate fire pixels and nonfire similar pixels again in images of the 2nd column of Figure 7.

Taking (1) combination as example, the basic steps of our algorithm using both 2D histogram θ-division with minimum error and GMM are the following.

Step 1. Determine the optimal angle θ based on sample training for wildfire detection.

Step 2. Segment the input image using 2D histogram θ-division with minimum error.

Step 3. Obtain wildfire regions with GMM processing from the segmented results of Step 2.

4. Results and Discussion

More real pictures with various wildfires are used for experiments, and the results are displayed in Figure 8. The 1st column shows input images; the 2nd column shows manually marked fire pixels used for comparison; the 3rd column shows segmented results from the 2D histogram θ-division ($\theta = 45^\circ$, untrained angle) with minimum error; the 4th column shows segmented results using only GMM; the 5th column shows final results of the combined algorithm, that is, first 2D histogram θ-division ($\theta = 32.9375^\circ$, the trained optimal angle) with minimum error and then GMM.

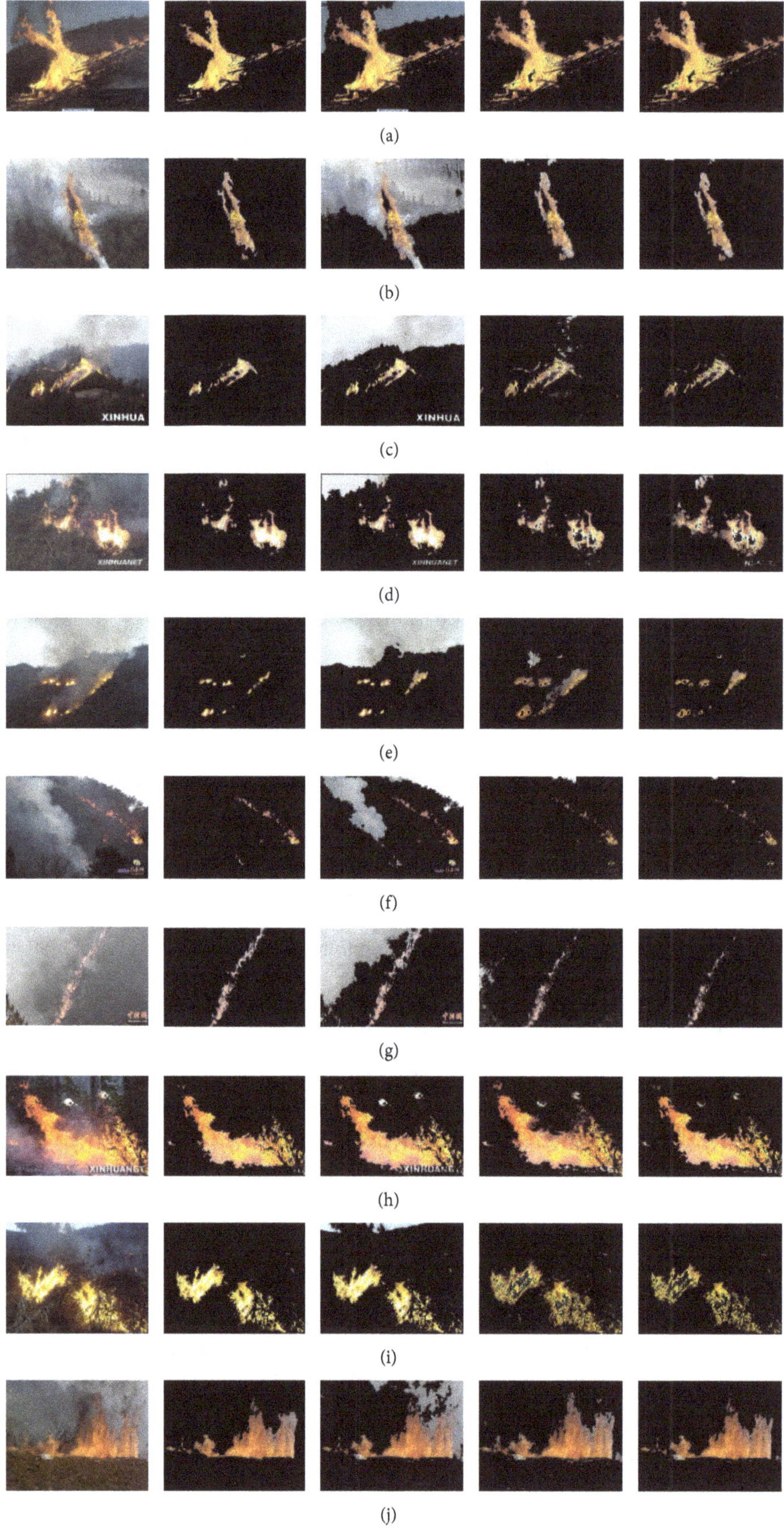

FIGURE 8: Experimental results of wildfire segmentation from images.

Table 1: Computing cost of the combined algorithm.

Image no.	Resolution	Time
Figure 8(a)	410 × 308	0.578
Figure 8(b)	519 × 389	0.400
Figure 8(c)	350 × 263	0.475
Figure 8(d)	480 × 318	0.438
Figure 8(e)	410 × 308	0.245
Figure 8(f)	410 × 285	0.487
Figure 8(g)	460 × 307	0.427
Figure 8(h)	450 × 305	0.839
Figure 8(i)	492 × 369	0.605
Figure 8(j)	300 × 225	0.269

From the experimental results, it can be found that the segmentation using only GMM still has the problem of wrongly taking non-fire pixels as fires. With the help of 2D histogram θ-division with minimum error and the sample trained optimal angle θ, most of the non-fire pixels are filtered; that is, our combined algorithm finally obtains better results of wildfire regions.

The computational expense for segmentation is also tested with a computer of Intel Core2 Duo CPU 2.93 GHz, 2 GB RAM. For the combined algorithm running on 10 pictures of Figure 8 with already trained angle θ and GMM parameters, their computational costs are listed in Table 1, where the unit of time is the second.

It can be found that the computational expense of our combined algorithm is not high, making it possible to be used in real applications with fire recognition. Having the ability of processing more than 1 frame per second, the method can work for video based approaches.

5. Conclusion

Based on 2D color histogram and the minimum error principle, θ-division methods have been studied in recent years. The segmented results are affected by angle θ, but its relationship with prior knowledge of certain problems still needs further study. Taking fire as a typical example, the sample training based 2D histogram θ-division with minimum error is implemented, and its application for segmentation is explored and analyzed.

Segmentation from angle θ is evaluated by the defined probability function of error division. Through collected samples and training, the optimal angle θ is found. Color channels are compared and the suitable one is chosen.

Then a new combined approach with 2D θ-division and GMM method is proposed to derive more accurate segmenting results. Compared with using only GMM, the combined method helps filter more errors. Our method has low computing cost and can be used for video based approaches with the help of temporal information.

The proposed techniques are verified with some real images, and the computing expenses are also tested. The combined algorithm obtains satisfying results, which has proved the ability of sample training base 2D histogram θ-division with minimum error. In the future, the presented methods will be further verified with the other kinds of segmentation techniques and will be tried for the other kinds of applications besides fire detection.

Acknowledgments

This work was supported by National Basic Research Program of China (973 Program, no. 2011CB707904), Science and Technology Bureau of Suzhou of China (no. SH201115), Natural Science Fund of Hubei Province of China, Natural Science Fund of China (no. 61272276), and Science and Technology Bureau of Wuhan of China (no. 201150124001).

References

[1] G. Csurka and F. Perronnin, "An efficient approach to semantic segmentation," *International Journal of Computer Vision*, vol. 95, no. 2, pp. 198–212, 2011.

[2] Y. Yang, S. Hallman, D. Ramanan, and C. C. Fowlkes, "Layered object models for image segmentation," *IEEE Transactions on Pattern Analysis and Machine Intelligence*, vol. 34, no. 9, pp. 1731–1743, 2012.

[3] J. B. MacQueen, "Some methods for classification and analysis of multivariate observations," in *Proceedings of the 5th Symposium on Math, Statistics and Probability*, pp. 281–297, Berkeley, Calif, USA, 1967.

[4] N. Otsu, "A threshold selection method from gray level histogram," *IEEE Transactions on Systems Man and Cybernetics*, vol. 9, no. 1, pp. 62–66, 1979.

[5] J. C. Bezdek, *Pattern Recognition with Fuzzy Objective Function Algorithms*, Plenum Press, New York, NY, USA, 1981.

[6] K.-S. Chuang, H.-L. Tzeng, S. Chen, J. Wu, and T.-J. Chen, "Fuzzy c-means clustering with spatial information for image segmentation," *Computerized Medical Imaging and Graphics*, vol. 30, no. 1, pp. 9–15, 2006.

[7] M. Filippone, F. Camastra, F. Masulli, and S. Rovetta, "A survey of kernel and spectral methods for clustering," *Pattern Recognition*, vol. 41, no. 1, pp. 176–190, 2008.

[8] E. T. Jaynes, "On the rationale of maximum entropy methods," *Proceedings of the IEEE*, vol. 70, no. 9, pp. 939–952, 1982.

[9] J. N. Kapur, P. K. Sahoo, and A. K. C. Wong, "A new method for gray-level picture thresholding using the entropy of the histogram," *Computer Vision, Graphics, & Image Processing*, vol. 29, no. 3, pp. 273–285, 1985.

[10] J. Kittler and J. Illingworth, "Minimum error thresholding," *Pattern Recognition*, vol. 19, no. 1, pp. 41–47, 1986.

[11] F. Morii, "A note on minimum error thresholding," *Pattern Recognition Letters*, vol. 12, no. 6, pp. 349–351, 1991.

[12] W. Liu and L. Wu, "Comparison of threshold selection for image segmentation," *Pattern Recognition and Artificial Intelligence*, vol. 10, no. 3, pp. 271–277, 1997.

[13] D. Zhou and J. Zong, "Minimum error thresholding based on two dimensional histogram," in *Proceedings of the WRI World Congress on Computer Science and Information Engineering (CSIE '09)*, vol. 7, pp. 169–175, April 2009.

[14] L. Zheng, G. Li, and Y. Bao, "Improvement of grayscale image 2D maximum entropy threshold segmentation method," in *Proceedings of the International Conference on Logistics Systems*

and Intelligent Management (ICLSIM '10), vol. 1, pp. 324–328, January 2010.

[15] P. Sthitpattanapongsa and T. Srinark, "A two-stage Otsu's thresholding based method on a 2D histogram," in *Proceedings of the 7th IEEE International Conference on Intelligent Computer Communication and Processing (ICCP '11)*, pp. 345–348, August 2011.

[16] J.-L. Fan and B. Lei, "Two-dimensional linear-type minimum error threshold segmentation method," *Journal of Electronics and Information Technology*, vol. 31, no. 8, pp. 1801–1806, 2009.

[17] Y.-Q. Wu and J.-K. Zhang, "Image thresholding based on two-dimensional histogram θ-division and maximum between-cluster deviation criterion," *Acta Automatica Sinica*, vol. 36, no. 5, pp. 634–643, 2010.

[18] Y. Q. Wu, X. J. Zhang, S. H. Wu, G. H. Zhang, S. W. Zhang, and S. F. Yu, "Image thresholding based on 2-D histogram θ-division and minimum error," *Journal of Shanghai Jiaotong University*, vol. 46, no. 6, pp. 892–899, 2012.

[19] B. U. Töreyin, Y. Dedeoğlu, U. Güdükbay, and A. E. Çetin, "Computer vision based method for real-time fire and flame detection," *Pattern Recognition Letters*, vol. 27, no. 1, pp. 49–58, 2006.

[20] J. Zhao, Z. Zhang, S. Han, C. Qu, Z. Yuan, and D. Zhang, "SVM based forest fire detection using static and dynamic features," *Computer Science and Information Systems*, vol. 8, no. 3, pp. 821–841, 2011.

12

Real-Time EEG-Based Happiness Detection System

Noppadon Jatupaiboon,[1] Setha Pan-ngum,[1] and Pasin Israsena[2]

[1] *Department of Computer Engineering, Faculty of Engineering, Chulalongkorn University, Bangkok 10330, Thailand*
[2] *National Electronics and Computer Technology Center, Pathumthani 12120, Thailand*

Correspondence should be addressed to Noppadon Jatupaiboon; noppadon_j@hotmail.com

Academic Editors: B.-W. Chen, S. Hsieh, and C.-H. Wu

We propose to use real-time EEG signal to classify happy and unhappy emotions elicited by pictures and classical music. We use PSD as a feature and SVM as a classifier. The average accuracies of subject-dependent model and subject-independent model are approximately 75.62% and 65.12%, respectively. Considering each pair of channels, temporal pair of channels (T7 and T8) gives a better result than the other area. Considering different frequency bands, high-frequency bands (Beta and Gamma) give a better result than low-frequency bands. Considering different time durations for emotion elicitation, that result from 30 seconds does not have significant difference compared with the result from 60 seconds. From all of these results, we implement real-time EEG-based happiness detection system using only one pair of channels. Furthermore, we develop games based on the happiness detection system to help user recognize and control the happiness.

1. Introduction

The aim of human computer interaction (HCI) is to improve the interactions between human and computers. Because most computers lack of understanding of user's emotions, sometimes they are unable to respond to the user's needs automatically and correctly [1]. One of the most interesting emotions is happiness. world happiness report reflects a new worldwide demand for more attention to happiness and absence of misery as criteria for government policy [2]. Being happy is related to many positive effects including confidence, optimism, self-efficacy, likability, activity, energy, physical well-being, flexibility, creativity, and the ability to cope with stress [3]. All of these benefits are the reasons why we should be happy.

In the past decades, most of emotion recognition researches have only focused on using facial expressions and speech. However, it is easy to fake facial expressions or change tone of speech and these signals are not continuously available, and they differ from using physiological signals, which occur continuously and are hard to conceal, such as Galvanic Skin Response (GSR), Electrocardiogram (ECG), Skin Temperature (ST), and, especially, Electroencephalogram (EEG). EEG is the signal from voltage fluctuations in the brain, that is, the center of emotions [1, 4]. Emotions are thought to be related with activity in brain areas that direct our attention, motivate our behavior, and determine the significance of what is going on around us. Emotion is related with a group of structures in the center of the brain called limbic system, which includes amygdala, thalamus, hypothalamus, and hippocampus [5, 6].

Electroencephalogram (EEG) is the recording of electrical activity on the scalp. EEG measures voltage changes resulting from ionic current flows within the neurons of the brain. There are five major brain waves distinguished by their different frequency bands (number of waves per second) as shown in Figure 1. These frequency bands from low to high frequencies, respectively, are called Delta (1–3 Hz), Theta (4–7 Hz), Alpha (8–13 Hz), Beta (14–30 Hz), and Gamma (31–50 Hz). Figure 2 shows the 10–20 system of electrode placement, that is, an internationally recognized method to describe and apply the location of scalp electrodes. Each site has a letter to identify the lobe and a number to identify the hemisphere location [7, 8].

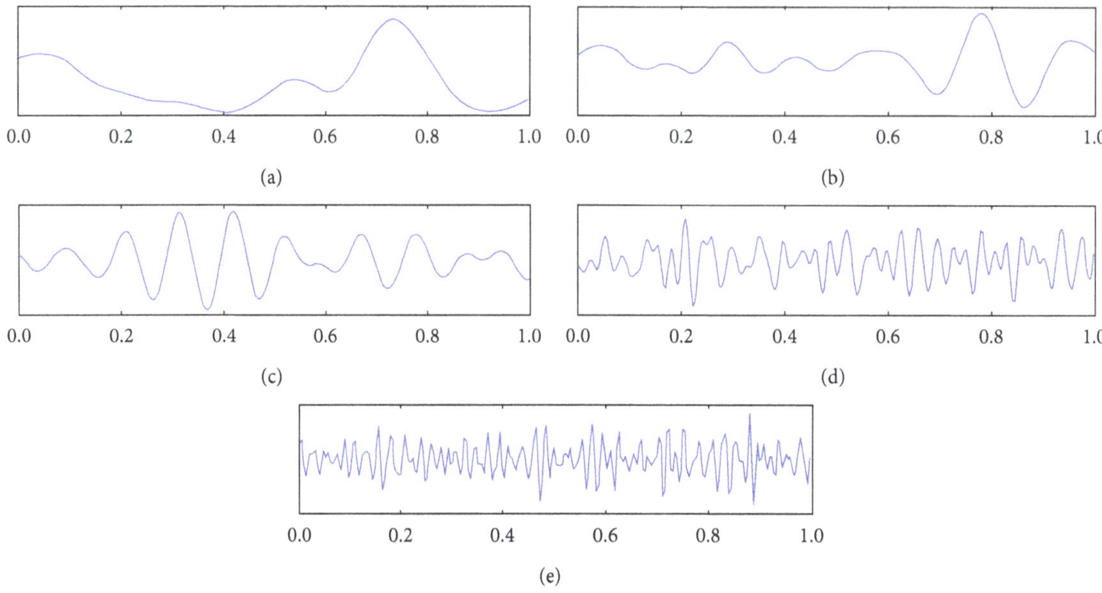

FIGURE 1: Brainwave: (a) Delta, (b) Theta, (c) Alpha, (d) Beta, and (e) Gamma [9].

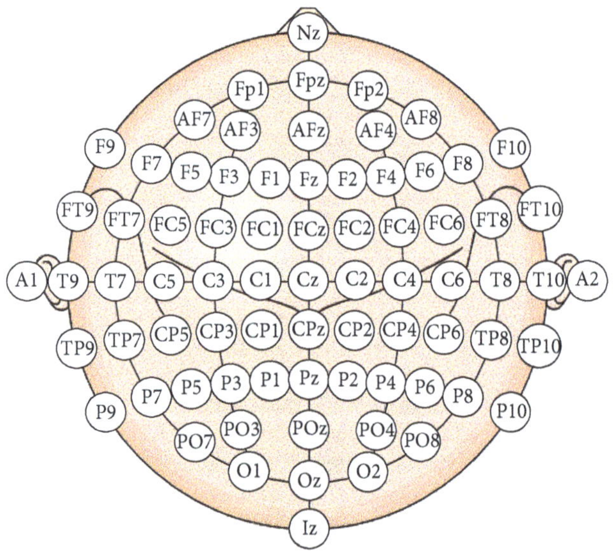

FIGURE 2: International 10–20 system of electrode placement [7].

2. The Literature Review

Nowadays, the EEG-based emotion recognition researches are highly active. The goal of these is to find suitable technique giving a good result that eventually can be implemented in real-time emotion recognition. The list of the EEG-based emotion recognition researches is shown in Table 1. It is difficult to compare results among them because there are a lot of factors that make different results from different researches including participant, model of emotion, stimulus, feature, temporal window, and classifier. The main six factors are described next to clarify the understanding.

2.1. Participant. The larger number of participants makes more reliable result. Moreover, we can divide the method for building emotion classification into subject-dependent and subject-independent models. The second model is harder than the first model due to interparticipants variability [10, 11]. The subject-dependent model avoids the problems related to interparticipant but a new classification model must be built for every new user. In this research, we build both subject-dependent and subject-independent models to compare the results.

2.2. Model of Emotion. The larger number of emotions makes emotion recognition harder, and some emotions may overlap. A good model of emotion should clearly separate these emotions. Several models have been proposed such as basic emotion and dimensional model. The most widely used basic emotions are the 6 basic emotions (i.e., anger, disgust, fear, joy, sadness, and surprise) that have been mostly used in facial expression recognition [12]. The common dimensional model is characterized by two main dimensions (i.e., valence and arousal). The valence emotion ranges from negative to positive, whereas the arousal emotion ranges from calm to excited [13]. This model is used in most researches because it is easier to express an emotion in terms of valence and arousal rather than basic emotions that can be confused by emotion names [14]. As shown in Figure 3, the emotions in any coordinates of the dimensional model are shown by facial expression. In this research, we use the dimensional models. The emotions used are happy and unhappy (sad). The happy emotion has positive valence and low arousal whereas the unhappy emotion has negative valence and low arousal.

2.3. Stimulus. There are various methods for emotion elicitation, which are self-eliciting, recalling, and using external stimulus such as picture, sound, and odor. The widely used

TABLE 1: EEG-based emotion recognition researches.

References	Year	Participant	Emotion	Stimulus	Feature	Temporal window	Classifier	Result	Real time
[10]	2006	4 subject-dependent	3 arousal classes	Picture	PSD	—	NB	58%	No
[11]	2008	26 subject-independent	4 classes (joy, anger, sadness, and pleasure)	Music	ASM	1 s	SVM	92.73%	No
[20]	2009	10 subject-dependent	2 valence classes	Picture	CSP	3 s	SVM	93.5%	No
[21]	2009	10 —	3 arousal classes	Recall	PSD	0.5 s	SVM	63%	No
[22]	2009	1 subject-dependent	3 classes (positively excited, negatively excited, and calm)	Picture	statistical features	—	QDA	66.66%	No
[23]	2009	3 subject-dependent	10 classes	Self-elicited	PSD	1 s	KNN	39.97–66.74%	No
[24]	2010	26 subject-independent	4 classes (joy, anger, sadness, and pleasure)	Music	ASM	1 s	SVM	82.29%	No
[25]	2010	6 subject-dependent	2 valence classes 2 arousal classes	Music video	PSD	—	SVM	58.8% (valence) 55.7% (arousal)	No
[26]	2010	26 subject-dependent	4 classes (calm, happy, sad, and fear)	Picture and music	SOM	2 s	KNN	84.5%	No
[28]	2010	15 —	2 classes (calm-neutral and negatively excited)	Picture	HOS	2 s	SVM	82%	No
[29]	2010	12 subject-dependent	2 valence classes 2 arousal classes	Sound	FD	—	threshold	—	Yes
[27]	2011	20 —	5 classes (happy, disgust, surprise, fear, and neutral)	video clip	Entropy	—	KNN	83.04%	No
[31]	2011	6 subject-dependent	2 valence classes	Movie clip	PSD	1 s	SVM	87.53%	No
[32]	2011	20 subject-independent	3 classes (boredom, engagement, and anxiety)	Game	PSD	—	LDA	56%	No
[33]	2011	5 subject-dependent	4 classes (joy, relax, sad, and fear)	Movie	PSD	1 s	SVM	66.51%	No
[34]	2011	11 —	3 valence classes	Picture	ASM	4 s	KNN	82%	No
[30]	2012	27 subject-independent	3 valence classes 3 arousal classes	Video	PSD and ASM	—	SVM	57.0% (valence) 52.4% (arousal)	No
[35]	2012	32 —	2 valence classes 2 arousal classes	Music video	PSD and ASM	—	NB	57.6% (valence) 62.0% (arousal)	No

Table 1: Continued.

References	Year	Participant	Emotion	Stimulus	Feature	Temporal window	Classifier	Result	Real time
[36]	2012	20 subject-dependent	5 classes (happy, angry, sad, relaxed, and neutral)	Picture	FD	—	SVM	70.5%	Yes
[37]	2012	5 subject-dependent	3 classes (positively excited, negatively excited, and calm)	Picture	HOC	—	KNN	90.77%	No
[38]	2012	4 —	2 valence classes 2 arousal classes	Video clip	ASP	—	—	66.05% (valence) 82.46% (arousal)	No
[39]	2012	32 —	2 classes (stress and calm)	Music video	PSD	—	KNN	70.1%	No
[40]	2012	36 —	3 classes	Music video	PSD	—	ANN	—	Yes
[41]	2013	11 subject-independent	2 valence classes	Picture	PSD	4 s	SVM	85.41%	No

*The feature, temporal window, and classifier shown in this table are the sets giving the best accuracy of each research.
Feature: Power Spectral Density (PSD), Spectral Power Asymmetry (ASM), Common Spatial Pattern (CSP), Higher Order Crossings (HOC), Self-Organizing Map (SOM), Higher Order Spectra (HOS), Fractal Dimension (FD), and Asymmetric Spatial Pattern (ASP).
Classifier: Support Vector Machine (SVM), Naïve Bayes (NB), Quadratic Discriminant Analysis (QDA), K-Nearest Neighbors (KNN), Linear Discriminant Analysis (LDA), Multilayer Perceptron (MLP), and Artificial Neural Network (ANN).

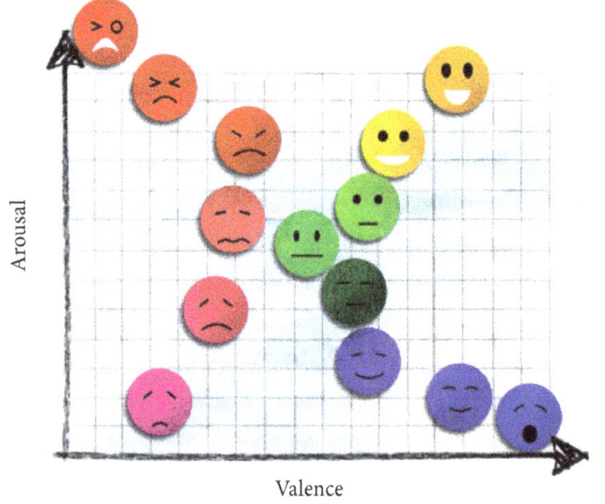

Figure 3: Dimensional model of emotion [14].

databases for emotion elicitation are International Affective Picture System (IAPS) [15] and International Digitized Sound System (IADS) [16]. These databases are generally accompanied by emotional evaluations from average judgments of several people. In this research, we choose pictures from Geneva Affective Picture Database (GAPED) [17] and sounds from classical emotion elicitation, because using visual-audio stimulus gives a better result than using either visual stimulus or audio stimulus [18].

2.4. Feature. Several signal characteristics of EEG have been used to be the features. The widely used feature is Power Spectral Density (PSD), the power of the EEG signal in focused frequency bands. In addition, others such as Spectral Power Asymmetry (ASM), Common Spatial Pattern (CSP), Higher Order Crossings (HOC), Self-Organizing Map (SOM), Higher Order Spectra (HOS), Fractal Dimension (FD), Asymmetric Spatial Pattern (ASP), and Entropy have been used as features and some give a good result. In this research, the feature we use is PSD since it gives a good performance in several researches as shown in Table 1, and it uses relatively little computation, which is suitable to implement in real-time emotion recognition.

2.5. Temporal Window. The appropriate length of temporal window depends on a type of emotion and physiological signal. Overall duration of emotions approximately falls between 0.5 and 4 seconds [42]. By using unsuitable window, the emotion may be misclassified because different emotions may be covered when too long or too short periods are measured. The existing literature does not provide suitable window size to be used to achieve optimal EEG-based emotion recognition [4]. In this research, we use temporal window 1 second.

2.6. Classifier. Several machine learning algorithms have been used as emotion classifiers such as Support Vector Machine (SVM), Naïve Bayes (NB), Quadratic Discriminant Analysis (QDA), K-Nearest Neighbors (KNN), Linear Discriminant Analysis (LDA), and Multilayer Perceptron (MLP). As shown in Table 1, SVM is implemented on many emotion classification researches because of many advantages. SVM is known to have good generalization properties and to be insensitive to overtraining and curse of dimensionality. The basic training principle of SVM is finding the optimal hyperplane where the expected classification error of test samples is minimized. The optimal hyperplane is the one that maximizes the margins. Maximizing the margins is known to increase the generalization capability. SVM uses regularization parameter (C) that enables accommodation to outliers and allows errors on the training set [43]. In this research, we use Gaussian SVM to be a classifier.

Beside the aforementioned factors, there is a factor that affects classification results from different researches. We found that some researches did not separate training set and test set completely although they did cross-validation (CV). Because simple cross-validation method randomly selects some data to be test set and the rest of data to be training set, some training data and test data may be in the same trial. Although the offline result is good, it does not guarantee the online result. In online emotion recognition, the training set is used to build the classification model, and the test set is a data from real-time EEG, so the training data and the test data are absolutely separated. For reliable result that can be guaranteed when using online emotion recognition, we should separate training set and test set completely. In this research, we use leave-one-trial-out cross-validation (LOTO-CV) and leave-one-subject-out cross-validation (LOSO-CV) for evaluating subject-dependent and subject-independent models, respectively.

As shown in Table 1, most of EEG-based emotion recognition researches are not for real-time implementation. There are a few researches that implement real-time emotion recognition such as [29, 40]. Wijeratne and Perera [40] proposed real-time emotion detection system using EEG and facial expression. However, the EEG signal acquisition part was still offline due to their time constraints, so they used pre-recorded EEG data instead of real-time EEG data. Liu et al. [29] proposed real-time emotion detection system using EEG. The user emotions are recognized and visualized in real time on his/her avatar. However, there is an issue in their approach that needs to be mentioned. In order to recognize an emotion, they did not use classifier and they only compared the Fractal Dimension (FD) values with predefined threshold, but they did not show how to define that threshold.

To fulfill these, we intend to implement EEG-based emotion detection system that can be truly implemented in real-time. Due to real-time processing, minimum computation time is required. We compare results among each pair of channels and different frequency bands in order to reduce insignificant channels and frequency bands. Furthermore, we develop games based on the happiness detection system to recognize and control happiness.

3. Methodology

The process of emotion classification consists of several steps as shown in Figure 4. First of all a stimulus such as picture, audio, and movie is needed. During experiment, the participant is exposed to the stimuli to elicit emotion, and EEG signal is recorded accordingly. Then artifacts that contaminate EEG signal are removed. These EEG data are analyzed and relevant features are extracted. Some parts of data are trained to build classification model and the rest of data, which are test data, are classified using this model.

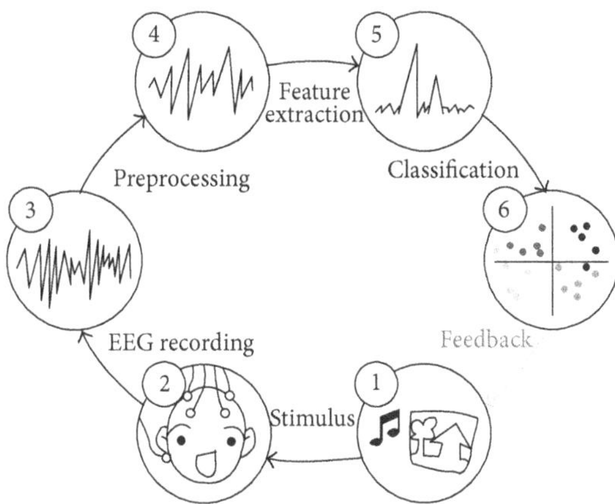

FIGURE 4: The process of emotion classification [19].

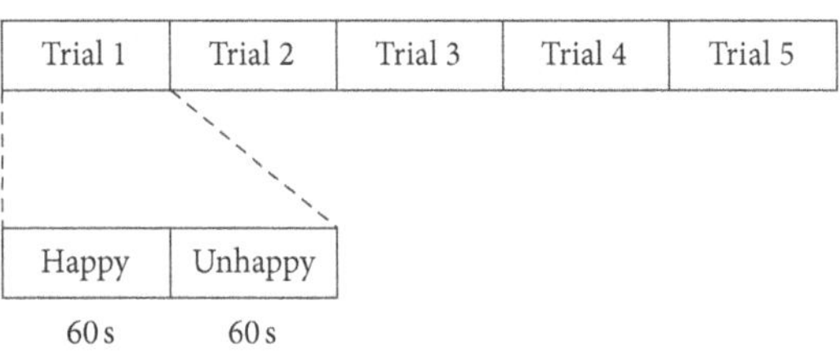

FIGURE 5: Procedure of experiment.

3.1. Stimulus. Both pictures and classical music were used to be the stimulus to elicit emotion. For pictures from GAPED [17], we selected the 50 highest valence scored pictures to be happy stimulus (i.e., pictures of human and animal babies as well as nature sceneries) and the 50 lowest valence scored pictures to be unhappy stimulus (i.e., pictures of human concerns and animal mistreatments). For classical music, we selected the highest and lowest valence scored pieces according to Vempala and Russo [44] to be happy and unhappy stimuli, respectively. The happy and unhappy pieces were Tritsch Tratsch Polka by Johann Strauss and Asas' Death by Edvard Grieg, respectively.

3.2. EEG Recording. We used 14-channels wireless EMOTIV [45] (i.e., AF3, AF4, F3, F4, F7, F8, FC5, FC6, P7, P8, T7, T8, O1, and O2). The sampling rate is 128 Hz. The resolution is 16 bits (14 bits effective). Before recording EEG, we put EMOTIV on the participant's head for a while to prevent undesired emotions that can arise from unfamiliar or uncomfortable feelings. Then we described the process of recording and advised the participant to stay as still as possible to prevent artifact that can occur from moving the body. When the participant was ready, we then recorded EEG and the experiment was started. As shown in Figure 5, there were 5 trials, where each trial consisted of one happy and one unhappy stimulus. Each stimulus was composed of 10 pictures and 1 piece of classical music that played along for 60 seconds. After that, a blank screen was shown for 12 seconds to adjust participant's

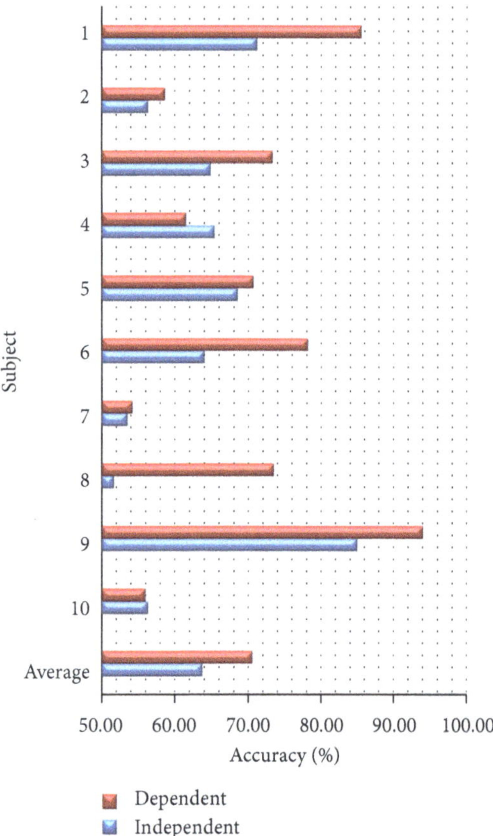

FIGURE 6: Accuracy from subject-dependent and subject-independent models.

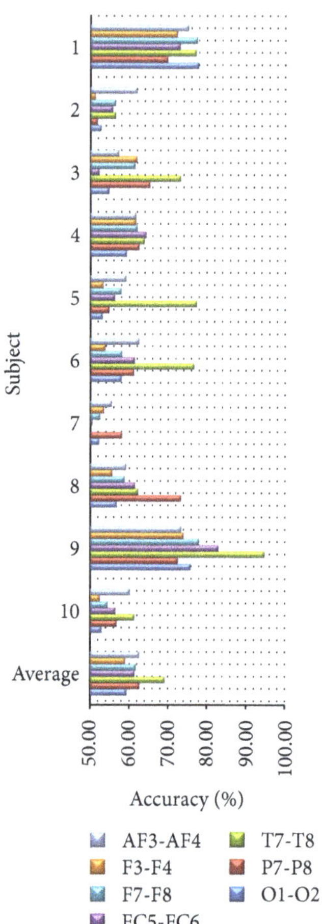

FIGURE 7: Accuracy from each pair of channels.

emotion to normal state and then the next stimulus was shown. When the 5 trials were completely shown, the process of recording ended. All these steps took approximately 15 minutes. There were 10 participants (i.e., 1 male and 9 females; average age is 34.60) taking part in this experiment.

3.3. Preprocessing. The EEG signal was filtered using a 5th-order sinc filter to notch out power line noise at 50 Hz and 60 Hz [45]. We removed baseline of the EEG signal for each channel so the values of the signal are distributed around 0.

3.4. Feature Extraction. The EEG signal with window 1 second was decomposed to 5 frequency bands that are Delta (0–4 Hz), Theta (4–8 Hz), Alpha (8–16 Hz), Beta (16–32 Hz), and Gamma (32–64 Hz) by Wavelet Transform as shown in Table 2. Then the PSD from each band was computed to be the feature. Since EMOTIV have 14 channels, the total features are 70. The features were normalized for each participant by scaling between 0 and 1 as shown in (1) to reduce inter-participant variability [11]:

$$\text{normalize}(X_i) = \frac{X_i - X_{\min}}{X_{\max} - X_{\min}}. \quad (1)$$

Since EEG signal from each trial has 120 seconds, there are 120 samples per trial. Due to 5 trials, there are 600 samples per participant. With 10 participants, the total samples are 6000. All samples were labeled whether happy or unhappy depending on the type of stimulus.

TABLE 2: EEG signal decomposition.

Frequency band	Frequency range (Hz)	Frequency bandwidth (Hz)	Decomposition level
Delta	0–4	4	A4
Theta	4–8	4	D4
Alpha	8–16	8	D3
Beta	16–32	16	D2
Gamma	32–64	32	D1

3.5. Classification. Gaussian SVM with leave-one-trail-out cross-validation (LOTO-CV) and leave-one-subject-out cross-validation (LOSO-CV) were used to compute accuracy for subject-dependent and subject-independent models, respectively. In the LOTO-CV method with 5 trials, one trial is set to be a test set and the rest to be a training set. Then the training set is built to be a classification model and the test set is classified using this model to evaluate accuracy. After that, we repeated the process using different trials as test sets, until all of the 5 trials had been test sets. The accuracy reported

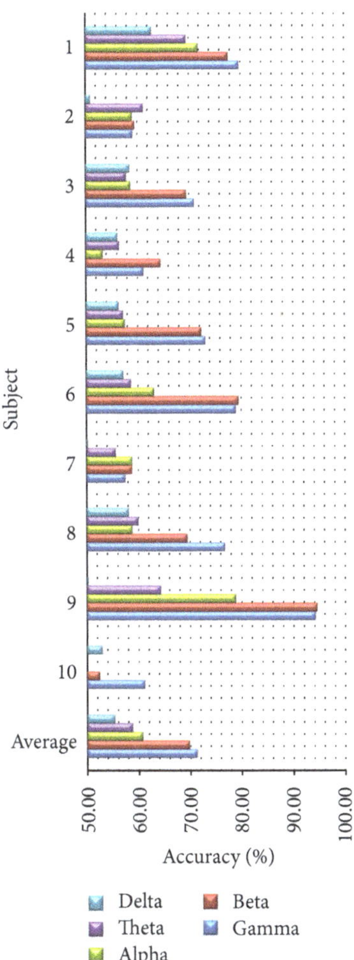

FIGURE 8: Accuracy from different frequency bands.

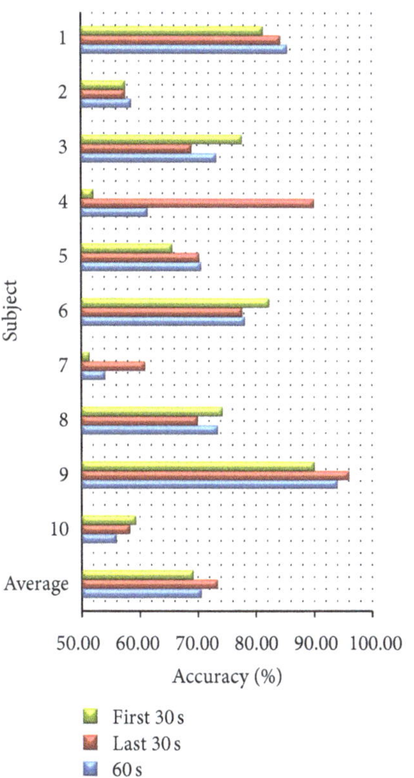

FIGURE 9: Accuracy from different time durations.

is the average accuracy of all 5 trials. The appropriate parameters are the set giving the best average of the 5 accuracies. In the LOSO-CV method with 10 subjects, one subject is set to be a test set and the rest to be a training set. Then the training set is built to be a classification model and the test set is classified using this model to evaluate accuracy. After that, we repeated the process using different subjects as test sets, until all of the 10 subjects had been test sets. The appropriate parameters are the set giving the best average of the 10 accuracies. The appropriate parameters C and γ of SVM were selected by grid search method. SVM implementation was done using LIBSVM [46].

4. Results and Discussion

4.1. Subject-Dependent and Subject-Independent Models. We compare subject-dependent and subject-independent accuracies using all features. As shown in Figure 6, we found that most of subject-independent accuracies are lower than subject-dependent accuracies. The average accuracies of subject-dependent model and subject-independent model are 70.55% and 63.67%, respectively. We can conclude that there are a lot of interparticipants. Different subjects may have different patterns of EEG when emotions are elicited. This conclusion is consistent with [24, 36]. As a result, we use only subject-dependent model to implement on real-time happiness detection system. Furthermore, we found that all of the older subjects (i.e., subject 2, 4, and 10; average age is 57.50) are giving low accuracies (accuracy of subject-dependent model lower than 65%). All of them confirm that they were elicited well by stimulus. We suppose as Levenson et al. [47] found that the magnitude of change in physiological signal was smaller in older than in younger subjects during emotion elicitation. So the accuracies of older subjects are low. When we exclude these older subjects, the average accuracies of subject-dependent model and subject-independent model are up to 75.62% and 65.12%, respectively.

4.2. Varying Pairs of Channels. We compare subject-dependent accuracy among each pair of channels (i.e., AF3-AF4, F3-F4, F7-F8, FC5-FC6, P7-P8, T7-T8, and O1-O2) using all frequency bands. As shown in Figure 7, we found that the highest average accuracy at 69.20% given by the pair of T7-T8 is very close to the average accuracy given by all channels. When we exclude older subjects, the average accuracy of T7-T8 is still highest at 72.90%. With PSD feature, we can conclude that temporal lobe is more effective for classifying happy and unhappy emotions than the others. This conclusion is consistent with [35, 48]. As a result, we can use this pair of channels instead of fourteen channels to reduce the number of channels and save computation time.

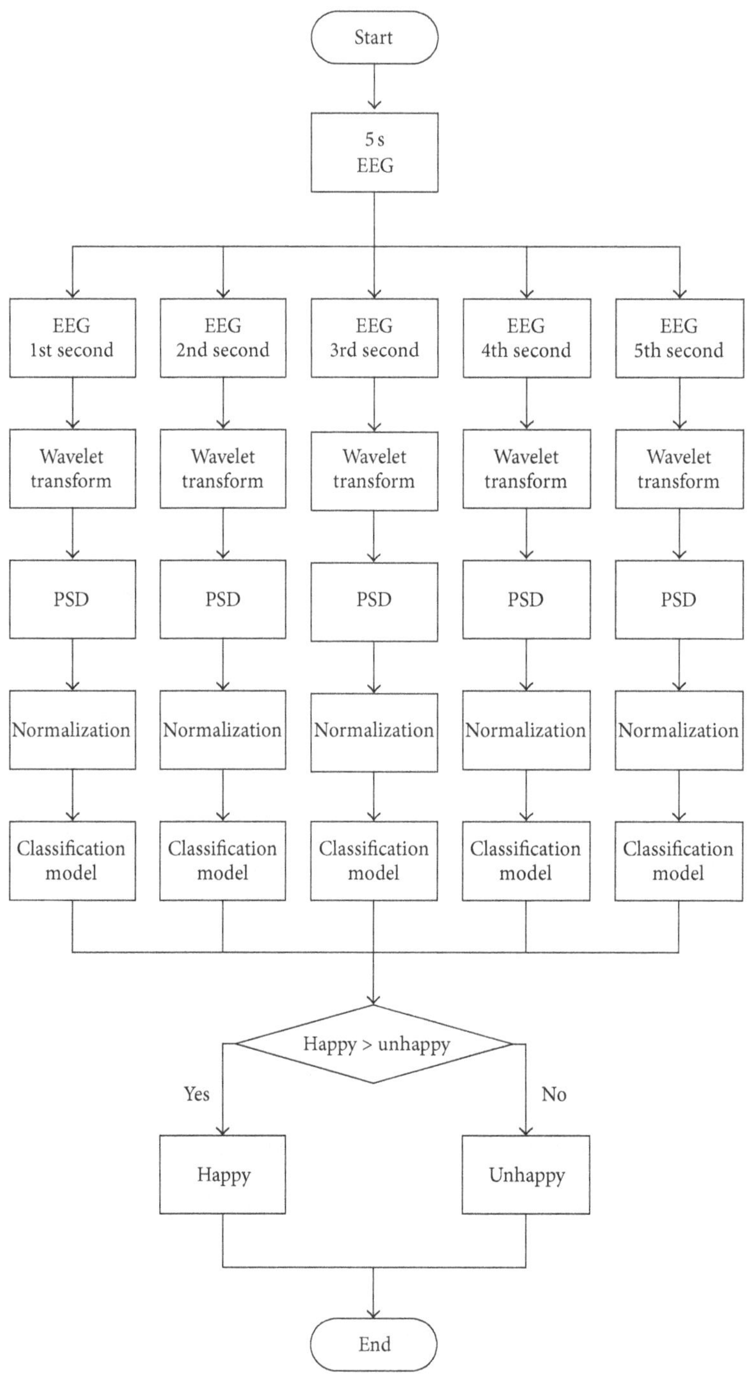

FIGURE 10: Flowchart of real-time happiness detection system.

4.3. Varying Frequency Bands. We compare subject-dependent accuracy among different frequency bands (i.e., Delta, Theta, Alpha, Beta, and Gamma) using all channels. As shown in Figure 8, we found that the average accuracies of Beta and Gamma are 69.83% and 71.28%, respectively, which are clearly higher than these of the other bands. When we exclude older subjects, the average accuracies of Beta and Gamma are still clearly higher than these of the other bands at 74.55% and 75.90%, respectively. With PSD feature, we can conclude that high frequency bands are more effective for classifying happy and unhappy emotions than low frequency bands. This conclusion is consistent with [20, 31, 48]. As a result, we can omit low-frequency bands such as Delta and Theta in order to save computation time.

4.4. Varying Time Durations. We compare subject-dependent accuracy from different time durations for emotion elicitation using all features. We consider accuracy from the first 30 seconds and the last 30 seconds of each stimulus. As shown in Figure 9, we found that the average accuracies

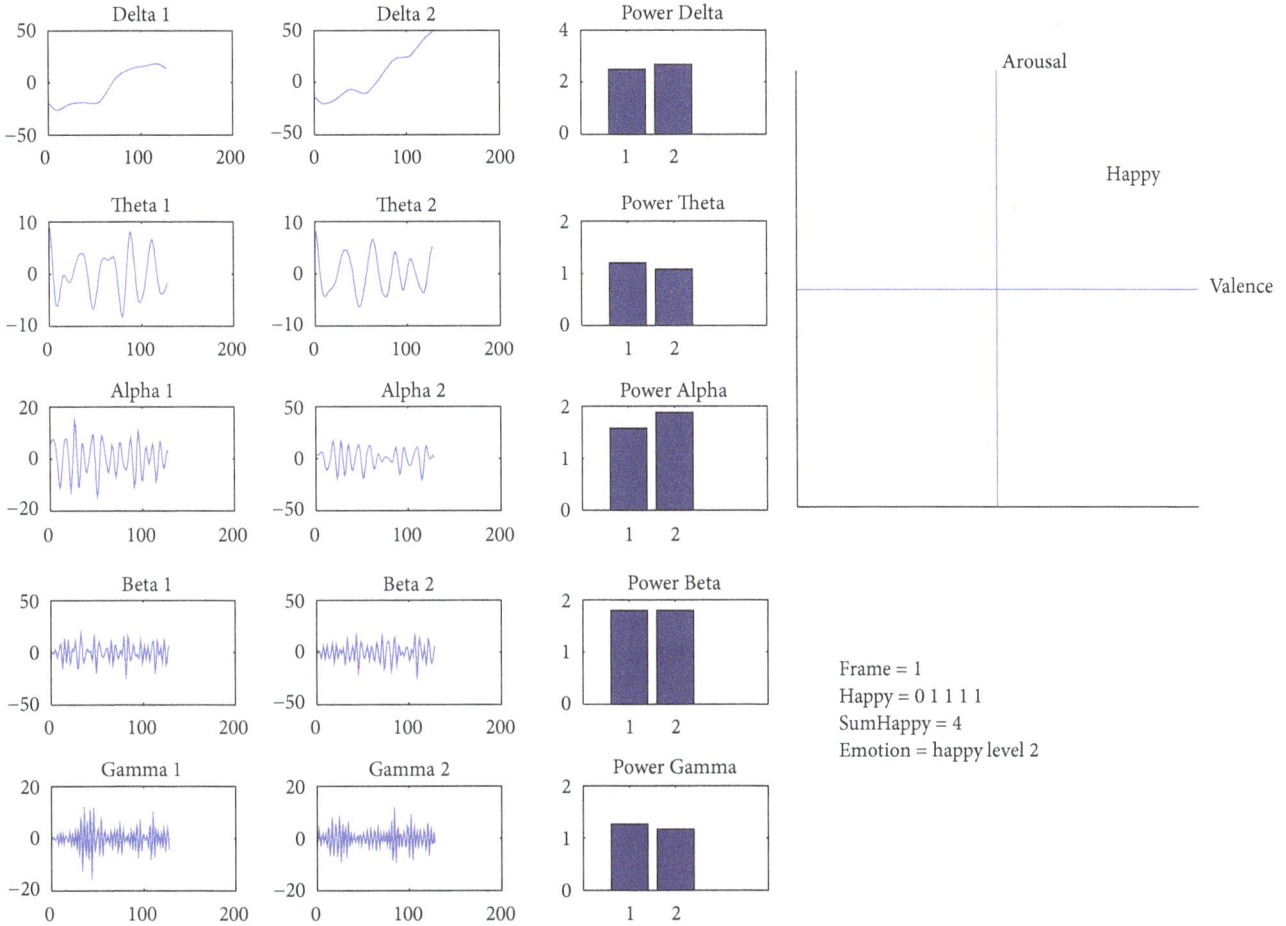

FIGURE 11: Screenshot of real-time happiness detection system.

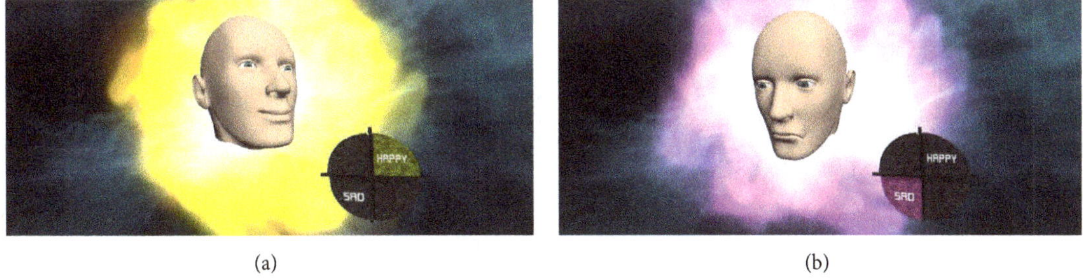

(a) (b)

FIGURE 12: Screenshot of AVATAR game: (a) happy and (b) unhappy.

of the first 30 seconds and the last 30 seconds are 69.17% and 73.43%, respectively. When we exclude older subjects, the average accuracies of the first 30 seconds and the last 30 seconds are up to 74.67% and 75.48%, respectively. Some subjects have higher accuracy in the first 30 seconds than the last 30 seconds and some subjects have higher accuracy in the last 30 seconds than the first 30 seconds. It shows that the time duration to elicit emotion is different depending on subjects. Considering statistical significance, we found that result from the first 30 seconds does not have significant difference from the result from the last 30 seconds (P value > 0.05). Furthermore, result from the first 30 seconds does not have significant difference from the result from 60 seconds (P-value > 0.05). As a result, we may reduce time to elicit emotion from 60 to 30 seconds to save time duration for emotion elicitation.

5. Real-Time Happiness Detection System

From the results of the tests in Section 4, we implement real-time EEG-based happiness detection system using only one pair of channels. Figure 10 shows the flowchart of the happiness detection system that can be described as follows. The EEG signals with window 1 second are decomposed into

FIGURE 13: Screenshot of RUNNING game.

TABLE 3: Level of happiness.

Happy	Unhappy	Emotion
0	5	Unhappy level 3
1	4	Unhappy level 2
2	3	Unhappy level 1
3	2	Happy level 1
4	1	Happy level 2
5	0	Happy level 3

5 frequency bands (i.e., Delta, Theta, Alpha, Beta, and Gamma) by Wavelet Transform. Then we compute PSD of each band as features. With 2 channels, there are 10 features. After that, each feature is normalized by scaling between 0 and 1. Then the normalized features are inserted to classification model, built from previous experiment, to classify emotion. The selected appropriate parameters are derived from LOTO-CV method from previous experiment. The system detects the happy emotion every 5 seconds. Since emotion is classified every second, there are 5 classifications. Majority vote among classifications is used for system detection output. If the number of classifications during consecutive 5 seconds is happy more than unhappy, the detected emotion is happy. Otherwise, the detected emotion is unhappy. We divide the level of emotion from happy to unhappy depending on the number of happy classifications as shown in Table 3. The real-time happiness detection system is implemented using BCI2000 [49] and Matlab as shown in Figure 11. It is run on ASUS K45A with Intel Core i3-3110 M (2.4 GHz, 3 MB L3 Cache).

Furthermore, we develop games for recognizing and controlling happiness that consist of AVATAR and RUNNING. Both games are implemented using UINITY3D based on the real-time happiness detection system that was presented.

AVATAR. We develop AVATAR game to demonstrate real-time facial expression depending on user's emotion. When the user is happy, the program shows happy face with happy music. Conversely, when the user is unhappy, the program shows unhappy face with unhappy music as shown in Figure 12. This is the game that can help user recognize the happiness.

RUNNING. We develop RUNNING game. The aim of this game is to control the character to run as far as possible within time constraint as shown in Figure 13. The speed of character depends on how happy the user is at the moment. The happier the user is, the more speed the character has. The speed is divided into 6 levels depending on the level of happiness. If the user can sustain their happiness, the character can cover long distance. This is the game that can help user control the happiness.

6. Conclusions and Future Work

In this research we propose to use real-time EEG signal to classify happy and unhappy emotions elicited by pictures and classical music. Considering each pair of channels and different frequency bands, temporal pair of channels gives a better result than the other area does, and high frequency bands give a better result than low frequency bands do. All of these are beneficial to the development of emotion classification system using minimal EEG channels in real time. From these results, we implement real-time happiness detection system using only one pair of channels. Furthermore, we develop games to help users recognize and control the happy emotion to be what they want. In the future, we will use other physiological signals such as Galvanic Skin Response (GSR), Electrocardiogram (ECG), and Skin Temperature (ST) combined with EEG to enhance the performance of emotion recognition in the aspect of accuracy and number of emotions.

Acknowledgments

The authors would like to thank all participants for the valuable time to be a part of this research, and they would also like to thank GAPED for the effective pictures. This research has been granted by the National Science and Technology Development Agency University-Industry Research Collaboration (NUI-RC).

References

[1] A. Luneski, E. Konstantinidis, and P. D. Bamidis, "Affective medicine: a review of affective computing efforts in medical informatics," *Methods of Information in Medicine*, vol. 49, no. 3, pp. 207–218, 2010.

[2] J. Helliwell, R. layard, and J. Sachs, "World Happiness Report," http://www.earth.columbia.edu/sitefiles/file/Sachs%20Writing/2012/World%20Happiness%20Report.pdf.

[3] S. Lyubomirsky, L. King, and E. Diener, "The benefits of frequent positive affect: does happiness lead to success?" *Psychological Bulletin*, vol. 131, no. 6, pp. 803–855, 2005.

[4] H. Gunes and M. Pantic, "Automatic, dimensional and continuous emotion recognition," *International Journal of Synthetic Emotions*, vol. 1, pp. 68–99, 2010.

[5] J. W. Papez, "A proposed mechanism of emotion," *Archives of Neurology and Psychiatry*, vol. 38, no. 4, pp. 725–743, 1937.

[6] P. D. MacLean, "Some psychiatric implications of physiological studies on frontotemporal portion of limbic system (Visceral brain)," *Electroencephalography and Clinical Neurophysiology*, vol. 4, no. 4, pp. 407–418, 1952.

[7] F. Sharbrough, G. E. Chatrian, R. P. Lesser, H. Luders, M. Nuwer, and T. W. Picton, "American electroencephalographic society guidelines for standard electrode position nomenclature," *Journal of Clinical Neurophysiology*, vol. 8, no. 2, pp. 200–202, 1991.

[8] E. Niedermeyer and F. L. da Silva, *Electroencephalography: Basic Principles, Clinical Applications, and Related Fields*, 2004.

[9] Wikipedia, "Electroencephalography," http://en.wikipedia.org/wiki/Electroencephalography.

[10] G. Chanel, J. Kronegg, D. Grandjean, and T. Pun, "Emotion assessment: arousal evaluation using EEG's and peripheral physiological signals," in *Multimedia Content Representation, Classification and Security*, B. Gunsel, A. Jain, A. M. Tekalp, and B. Sankur, Eds., vol. 4105, pp. 530–537, Springer, Berlin, Germany, 2006.

[11] Y. P. Lin, C. H. Wang, T. L. Wu, S. K. Jeng, and J. H. Chen, "Support vector machine for EEG signal classification during listening to emotional music," in *Proceedings of the 10th IEEE Workshop on Multimedia Signal Processing (MMSP '08)*, pp. 127–130, Cairns, Australia, October 2008.

[12] P. Ekman and W. Friesen, "Measuring facial movement with the facial action coding system," in *Emotion in the Human Face*, Cambridge University Press, New York, NY, USA, 2nd edition, 1982.

[13] J. A. Russell, "A circumplex model of affect," *Journal of Personality and Social Psychology*, vol. 39, no. 6, pp. 1161–1178, 1980.

[14] R. Horlings, *Emotion Recognition Using Brain Activity*, Department of Mediamatics, Delft University of Technology, 2008.

[15] M. M. Bradley, P. J. Lang, and B. N. Cuthbert, *International Affective Picture System (IAPS): Digitized Photographs, Instruction Manual and Affective Ratings*, University of Florida, Gainesville, Fla, USA, 2005.

[16] M. M. Bradley and P. J. Lang, *The International Affective Digitized Sounds (IADS-2): Affective Ratings of Sounds and Instruction Manual*, University of Florida, Gainesville, Fla, USA, 2nd edition, 2007.

[17] E. S. Dan-Glauser and K. R. Scherer, "The Geneva affective picture database (GAPED): a new 730-picture database focusing on valence and normative significance," *Behavior Research Methods*, vol. 43, no. 2, pp. 468–477, 2011.

[18] T. Baumgartner, M. Esslen, and L. Jäncke, "From emotion perception to emotion experience: emotions evoked by pictures and classical music," *International Journal of Psychophysiology*, vol. 60, no. 1, pp. 34–43, 2006.

[19] D. Bos, "EEG-based emotion recognition," http://hmi.ewi.utwente.nl/verslagen/capita-selecta/CS-Oude_Bos-Danny.pdf.

[20] M. Li and B. L. Lu, "Emotion classification based on gamma-band EEG," in *Proceedings of the Annual International Conference of the IEEE Engineering in Medicine and Biology Society (EMBC '09)*, pp. 1223–1226, Minneapolis, Minn, USA, September 2009.

[21] G. Chanel, J. J. M. Kierkels, M. Soleymani, and T. Pun, "Short-term emotion assessment in a recall paradigm," *International Journal of Human Computer Studies*, vol. 67, no. 8, pp. 607–627, 2009.

[22] Z. Khalili and M. H. Moradi, "Emotion recognition system using brain and peripheral signals: using correlation dimension to improve the results of EEG," in *Proceedings of the International Joint Conference on Neural Networks (IJCNN '09)*, pp. 1571–1575, Atlanta, Ga, USA, June 2009.

[23] O. AlZoubi, R. A. Calvo, and R. H. Stevens, "Classification of EEG for affect recognition: an adaptive approach," in *AI 2009: Advances in Artificial Intelligence*, A. Nicholson and X. Li, Eds., vol. 5866 of *Lecture Notes in Computer Science*, pp. 52–61, Springer , Berlin, Germany, 2009.

[24] Y. P. Lin, C. H. Wang, T. P. Jung et al., "EEG-based emotion recognition in music listening," *IEEE Transactions on Biomedical Engineering*, vol. 57, no. 7, pp. 1798–1806, 2010.

[25] S. Koelstra, A. Yazdani, M. Soleymani et al., "Single trial classification of EEG and peripheral physiological signals for recognition of emotions induced by music videos," in *Proceeding of the International Conference on Brain Informatics (BI '10)*, pp. 89–100, Toronto, Canada, 2010.

[26] R. Khosrowabadi, H. C. Quek, A. Wahab, and K. K. Ang, "EEG-based emotion recognition using self-organizing map for boundary detection," in *Proceedings of the 20th International Conference on Pattern Recognition (ICPR '10)*, pp. 4242–4245, Istanbul, Turkey, August 2010.

[27] M. Murugappan, R. Nagarajan, and S. Yaacob, "Combining spatial filtering and wavelet transform for classifying human emotions using EEG Signals," *Journal of Medical and Biological Engineering*, vol. 31, no. 1, pp. 45–51, 2011.

[28] S. A. Hosseini, M. A. Khalilzadeh, M. B. Naghibi-Sistani, and V. Niazmand, "Higher order spectra analysis of EEG signals in emotional stress states," in *Proceedings of the 2nd International Conference on Information Technology and Computer Science (ITCS '10)*, pp. 60–63, ukr, July 2010.

[29] Y. Liu, O. Sourina, and M. K. Nguyen, "Real-time EEG-based human emotion recognition and visualization," in *Proceedings of the International Conference on Cyberworlds (CW '10)*, pp. 262–269, Singapore, October 2010.

[30] M. Soleymani, J. Lichtenauer, T. Pun, and M. Pantic, "A multimodal database for affect recognition and implicit tagging," *IEEE Transactions on Affective Computing*, vol. 3, no. 1, pp. 42–55, 2012.

[31] D. Nie, X. W. Wang, L. C. Shi, and B. L. Lu, "EEG-based emotion recognition during watching movies," in *Proceedings of the 5th International IEEE/EMBS Conference on Neural Engineering (NER '11)*, pp. 667–670, Cancun, Mexico, May 2011.

[32] G. Chanel, C. Rebetez, M. Bétrancourt, and T. Pun, "Emotion assessment from physiological signals for adaptation of game difficulty," *IEEE Transactions on Systems, Man, and Cybernetics A*, vol. 41, no. 6, pp. 1052–1063, 2011.

[33] X. W. Wang, D. Nie, and B. L. Lu, "EEG-based emotion recognition using frequency domain features and support vector machines," in *Neural Information Processing*, B. L. Lu, L. Zhang, and J. Kwok, Eds., vol. 7062 of *Lecture Notes in Computer Science*, pp. 734–743, Springer, Berlin, Germany, 2011.

[34] L. Brown, B. Grundlehner, and J. Penders, "Towards wireless emotional valence detection from EEG," in *Proceedings of the Annual International Conference of the IEEE Engineering in Medicine and Biology Society (EMBC '11)*, pp. 2188–2191, Boston, Mass, USA, September 2011.

[35] S. Koelstra, C. Mühl, M. Soleymani et al., "DEAP: a database for emotion analysis; using physiological signals," *IEEE Transactions on Affective Computing*, vol. 3, no. 1, pp. 18–31, 2012.

[36] V. H. Anh, M. N. Van, B. B. Ha, and T. H. Quyet, "A real-time model based support vector machine for emotion recognition through EEG," in *Proceedings of the International Conference on Control, Automation and Information Sciences (ICCAIS '12)*, pp. 191–196, Ho Chi Minh City, Vietnam, November 2012.

[37] H. Xu and K. N. Plataniotis, "Affect recognition using EEG signal," in *Proceedings of the 14th IEEE International Workshop on Multimedia Signal Processing (MMSP '12)*, pp. 299–304, Banff, Canada, September 2012.

[38] D. Huang, C. Guan, K. K. Ang, H. Zhang, and Y. Pan, "Asymmetric spatial pattern for EEG-based emotion detection," in

Proceedings of the International Joint Conference on Neural Networks (IJCNN '12), pp. 1–7, Brisbane, Australia, June 2012.

[39] T. F. Bastos-Filho, A. Ferreira, A. C. Atencio, S. Arjunan, and D. Kumar, "Evaluation of feature extraction techniques in emotional state recognition," in *Proceedings of the 4th International Conference on Intelligent Human Computer Interaction (IHCI '12)*, pp. 1–6, Kharagpur, India, December 2012.

[40] U. Wijeratne and U. Perera, "Intelligent emotion recognition system using electroencephalography and active shape models," in *Proceedings of the IEEE EMBS Conference on Biomedical Engineering and Sciences (IECBES '12)*, pp. 636–641, Langkawi, Malaysia, December 2012.

[41] N. Jatupaiboon, S. Pan-ngum, and P. Israsena, "Emotion classification using minimal EEG channels and frequency bands," in *Proceedings of the 10th International Joint Conference on Computer Science and Software Engineering (JCSSE '13)*, pp. 21–24, 2013.

[42] R. W. Levenson, "Emotion and the autonomic nervous system: a prospectus for research on autonomic specificity," in *Social Psychophysiology and Emotion: Theory and Clinical Applications*, H. L. Wagner, Ed., pp. 17–42, John Wiley & Sons, New York, NY, USA, 1988.

[43] F. Lotte, M. Congedo, A. Lécuyer, F. Lamarche, and B. Arnaldi, "A review of classification algorithms for EEG-based brain-computer interfaces," *Journal of Neural Engineering*, vol. 4, no. 2, pp. R1–R13, 2007.

[44] N. N. Vempala and F. A. Russo, "Predicting emotion from music audio features using neural networks," in *Proceedings of the 9th International Symposium on Computer Music Modeling and Retrieval (CMMR '12)*, 2012.

[45] "Emotiv EEG Neuroheadset," http://emotiv.com/upload/manual/EEGSpecifications.pdf.

[46] C. C. Chang and C. J. Lin, "LIBSVM: a library for support vector machines," *ACM Transactions on Intelligent Systems and Technology*, vol. 2, no. 3, article 27, 2011.

[47] R. W. Levenson, L. L. Carstensen, W. V. Friesen, and P. Ekman, "Emotion, physiology, and expression in old age," *Psychology and Aging*, vol. 6, no. 1, pp. 28–35, 1991.

[48] J. A. Onton and S. Makeig, "High-frequency broadband modulations of electroencephalographic spectra," *Frontiers in Human Neuroscience*, vol. 3, article 61, 2009.

[49] G. Schalk, D. J. McFarland, T. Hinterberger, N. Birbaumer, and J. R. Wolpaw, "BCI2000: a general-purpose brain-computer interface (BCI) system," *IEEE Transactions on Biomedical Engineering*, vol. 51, no. 6, pp. 1034–1043, 2004.

13

A Spectrum Sensing Scheme for Partially Polarized Waves over α-μ Generalized Gamma Fading Channels

Mohamed A. Hankal, Islam A. Eshrah, and Hazim Tawfik

Electronics and Electrical Communications Engineering Department, Faculty of Engineering, Cairo University, Giza 12613, Egypt

Correspondence should be addressed to Mohamed A. Hankal; mohamedhankal@gmail.com

Academic Editors: N. Bouguila and C.-M. Kuo

Schemes for spectrum holes sensing for cognitive radio based on the estimation of the Stokes parameters of monochromatic and quasimonochromatic polarized electromagnetic waves are developed. Statistical information that includes the variations of the polarization state in both cases (present and absent) of Primary User (PU) is accounted for. A detector based on the fluctuation of the Stokes parameters is analyzed, and its performance is compared with that of energy detectors, which use only the scalar amplitude information to sense the PU signal. The cooperative spectrum sensing based on the polarization in which the reporting channels are noisy will be investigated. The cluster technique is proposed to reduce the bit error probability due to channel impairment. A closed-form expression for the polarization detection is derived using α-μ generalized fading model, which provides directly an expression for the special cases of Nakagami-m and Weibull models as well as their derivatives. These expressions are verified using simulation. The results show that the polarization spectrum sensing gives superior performance for a wide range of SNR over the conventional energy detection method.

1. Introduction

Cognitive radio (CR) technology has witnessed a growing interest over the past decade, as it promises more efficient use of the available spectrum [1, 2]. A key stage in CR is spectrum sensing, in which the Secondary User (SU) must detect the presence of a Primary User (PU) in a certain channel, and thus, deems this part of the spectrum unused, and make the decision to share it. This entails a sequence of functions that the CR system should perform, such as power control [3] and spectrum management [4]. Several techniques were proposed to improve spectrum sensing such as energy detection [5], cyclostationary feature detection [6], sensing based on smart antennas [7, 8], and wideband spectrum sensing [9, 10]. These techniques primarily make use of the amplitude, frequency, and phase information of the PU signal.

It is possible, however, to improve the spectrum sensing process by exploiting the polarization state of the signal. In radar systems [11], the polarization state was used to improve the detection capability of the system. The new polarization-dependent detection statistics, which use the power and relative phase of the two orthogonal polarization components was proposed to enhance radar detection in homogenous channels [12]. The radar detection performance was enhanced based on the polarization difference between the clutter and the target [11]. The sine of the relative phase between two orthogonally polarized received signals has been proposed and tested as detection statistic in radar systems [13]. New statistics use two orthogonal polarization component powers and their relative phase to enhance target detection [12]. They are thus fundamentally different from the well-known Marcum-Swerling envelope detector [14], which operates on only a single polarization component of the received power, and the pseudocoherent detector [13], which operates on only the relative phase of the polarization components.

Recent research on polarization detection in CR was concerned with completely polarized waves. A Virtual Polarization Detection (VPD) method based on the vector signal processing was presented for effective spectrum sensing of cognitive radios [15]. Polarization spectrum hole sensing was proposed for cognitive radio to optimize the received polarization at the SU in order to protect the PU from interference caused by the SU and to reduce the interference from PU to SU [16]. Optimal Polarization Reception (OPR) was proposed for CR to improve the SINR [17]. A new blind spectrum sensing method based on the polarization characteristic of the received signal, which is completely represented by the orientation of a polarization vector, was proposed [18], and a closed-form expression for the probability of false alarm and probability of detection under Additive White Gaussian Noise (AWGN) and Rayleigh-fading channels was derived. The fading and noisy nature of a wireless communication channel places a major challenge for spectrum sensing. Since sensing decisions based on a single SU measurements may be unreliable, the idea of collaborative spectrum sensing has attracted a lot of research interest [19]. However, the polarization state is changed with spectrum sensing time which was not considered in this work.

In this paper, the above constraint is addressed by proposing a polarization-based spectrum sensing scheme where a statistical model for the PU signal polarization parameters is adopted. This model takes into account the channel backscatter and the partial polarization nature of the PU signal, with the assumption that the channel experiences slow fading. Cooperative polarization spectrum sensing is proposed to mitigate the effects of fading and shadowing, which can seriously degrade the sensing performance. A cluster-based cooperation scheme is proposed to decrease bit error probability. All SUs are grouped into few clusters and one cluster head is set for each cluster to collect the sensing results, make cluster decisions, and forward measurements to the central unit. Thus the bit error probability will be reduced greatly because most of SUs will be closer to the cluster heads than to the central unit. Analytical results show that significant improvement can be achieved with our proposed method.

The rest of this paper is organized as follows. In Section 2, the method of characterizing the signal polarization is provided. The spectrum sensing models are developed in Section 3. Section 4 describes the cooperative spectrum sensing mechanism and the cluster technique of the cooperative sensing is shown in Section 5. Simulation results are illustrated in Section 6, followed by the conclusion in Section 7.

2. The Method of Characterizing the Signal Polarization

The polarization of a monochromatic plane wave is completely specified by constant amplitude and relative phase of the two orthogonal electric-field components.

In a cognitive radio system, the SU uses orthogonally dual polarized antennas to detect the PU signal $\vec{E}$, which can be completely described in vector form as

$$\vec{E} = \begin{bmatrix} E_h \\ E_v \end{bmatrix} = \begin{bmatrix} E_h^i + jE_h^q \\ E_v^i + jE_v^q \end{bmatrix}, \tag{1}$$

where i and q indicate the in-phase and quadrate phase components, respectively. Typically, the polarization of a monochromatic wave is determined by the Jones vector $\vec{J}$ of the signal $\vec{E}$ defined as [20]

$$\vec{J} = \begin{bmatrix} \cos\zeta \\ e^{j\theta}\sin\zeta \end{bmatrix}, \tag{2}$$

where $\zeta = \tan^{-1}(|E_v|/|E_h|)$ and $\theta = \tan^{-1}(|E_v^q|/|E_v^i|) - \tan^{-1}(|E_h^q|/|E_h^i|)$.

Alternatively, the polarization is determined by the geometrical parameters, namely the ellipticity angle τ, and orientation angle ϕ, which can be uniquely represented by a point on the Poincare sphere. The field component parameters ζ and θ, and the geometrical parameters τ and ϕ, are described in the following set of equations [21]

$$\begin{aligned} \sin(2\zeta)\cos(\theta) &= \sin(2\phi)\cos(2\tau), \\ \sin(2\zeta)\sin(\theta) &= \sin(2\tau), \\ \cos(2\zeta) &= \cos(2\phi)\cos(2\tau). \end{aligned} \tag{3}$$

However, if the amplitudes and phases encounter slow fluctuations with time, which is the case if the polarization of primary signal suffers from either noise or fading, the components of the Jones vector are said to be quasimonochromatic or narrowband. The polarization of a quasimonochromatic wave is quantified using an average polarization state vector, which may be defined in terms of four measurable components; namely, $\vec{X} = [x_0\, x_1\, x_2\, x_3]^T$. These components are known as the Stokes parameters (SPs) and are given by [12]

$$\begin{aligned} x_0 &= \frac{1}{N}\sum_{n=1}^{N}\Big[\left(E_h^i(nT_s)\right)^2 + \left(E_h^q(nT_s)\right)^2 \\ &\qquad + \left(E_v^i(nT_s)\right)^2 + \left(E_v^q(nT_s)\right)^2\Big], \\ x_1 &= \frac{1}{N}\sum_{n=1}^{N}\Big[\left(E_h^i(nT_s)\right)^2 + \left(E_h^q(nT_s)\right)^2 \\ &\qquad - \left(E_v^i(nT_s)\right)^2 - \left(E_v^q(nT_s)\right)^2\Big], \\ x_2 &= \frac{2}{N}\sum_{n=1}^{N}\left[E_h^i(nT_s)E_v^i(nT_s) + E_h^q(nT_s)E_v^q(nT_s)\right], \\ x_3 &= \frac{2}{N}\sum_{n=1}^{N}\left[E_h^i(nT_s)E_v^q(nT_s) - E_v^i(nT_s)E_h^q(nT_s)\right], \end{aligned} \tag{4}$$

where T_s is the sampling time, and N is the number of samples which is restricted by the bandwidth of a narrowband filter. x_0, x_1, x_2, and x_3 are physically recognizable quantities as follows.

(1) x_0 is the sum of the power in the h and v electric-field components and thus represents the total power of the received signal.

(2) x_1 is the difference between the power in the h and v electric-field components.

(3) x_2 is the difference between the power in the two orthogonal electric-field components whose axes are rotated 45° relative to the h and v axes.

(4) x_3 is the difference between the right-hand and the left-hand circularly polarized power.

The component x_0 satisfies the relation

$$x_0 \geq x_1 + x_2 + x_3. \tag{5}$$

Alternatively, the average polarization state can be obtained from the coherence matrix $\mathbf{C}$ of the received field, such that

$$\mathbf{C} = \left\langle \begin{bmatrix} E_h E_h^* & E_h E_v^* \\ E_v E_h^* & E_v E_v^* \end{bmatrix} \right\rangle, \tag{6}$$

where $\langle\cdot\rangle$ denotes averaging over NT_s and $*$ represents the complex conjugate. The coherence matrix $\mathbf{C}$ is a linear combination of the SPs such that:

$$\mathbf{C} = x_0\mathbf{F}_0 + x_1\mathbf{F}_1 + x_2\mathbf{F}_2 + x_3\mathbf{F}_3, \tag{7}$$

where $\mathbf{F}_0 = 0.5\left[\begin{smallmatrix} 1 & 0 \\ 0 & 1 \end{smallmatrix}\right]$, $\mathbf{F}_1 = 0.5\left[\begin{smallmatrix} 1 & 0 \\ 0 & -1 \end{smallmatrix}\right]$, $\mathbf{F}_2 = 0.5\left[\begin{smallmatrix} 0 & 1 \\ 1 & 0 \end{smallmatrix}\right]$, and $\mathbf{F}_3 = 0.5\left[\begin{smallmatrix} 0 & 1 \\ -1 & 0 \end{smallmatrix}\right]$.

The relationship of the SPs to the geometrical parameters (ϕ, τ) is shown in the following set of equations:

$$\begin{aligned} x_1 &= \cos(2\phi)\cos(2\tau), \\ x_2 &= \sin(2\phi)\cos(2\tau), \\ x_3 &= \sin(2\tau). \end{aligned} \tag{8}$$

Thus far, the orientation of a polarization vector on the unit Poincare sphere can completely represent the polarization state of the received signal.

3. Spectrum Sensing Based on the Polarization

Spectrum sensing is essentially a binary hypothesis testing problem, which indicates the PU's absence or presence, respectively, such that

$$Y(t) = \begin{Bmatrix} hE(t) + W(t) & H_1 \\ W(t) & H_0 \end{Bmatrix}, \tag{9}$$

where $Y(t)$ is the observed signal at the CR, $E(t)$ is the PU signal, $W(t)$ is the zero mean Gaussian random process with identical autocorrelation and power spectral density N_0 Watts/Hz, and h is the amplitude gain of the channel having mean-square value $\Omega = \overline{h}^2$ and Probability Density Function (PDF) $f_h(h)$. The received instantaneous signal power is modulated by h^2 and consequently the instantaneous Signal-to-Noise Ratio (SNR) can be expressed as $\gamma = h^2(Es/N_0)$ with an average $\overline{\gamma} = \Omega(Es/N_0)$ where E_s is the signal energy accumulated over the observation period.

In this paper, we propose two schemes which use the degree of polarization and the axial ratio to detect the PU signal.

3.1. Spectrum Sensing Based on the Degree of Polarization.

The degree of polarization (D) is a quantity used to describe the polarized portion of an electromagnetic wave. A perfectly polarized wave has D equal to 1, whereas an unpolarized wave has D equal to 0. A wave, which is partially polarized, can therefore be represented by superposition of a polarized and unpolarized component. This implies a D somewhere in between 0 and 1. Alternatively, the degree of polarization D is defined as the ratio of the polarized power to the total power in the wave; that is,

$$D = \frac{\sqrt{x_1^2 + x_2^2 + x_3^2}}{x_0}. \tag{10}$$

The estimate of the ratio of the polarized power to the total power in the received signal is used as detection statistics in radar systems [12]. Figure 1 depicts a block diagram of the proposed spectrum sensing system where $D_S = D$. There are two orthogonal antennas, which detect the horizontal and vertical components E_h and E_v of the signal E, respectively. The Stokes vector estimation block delivers the Stokes vector $\vec{X}$ to the polarization degree estimator, which, in turn, produces the detection statistic D. This statistic serves as the input to the threshold detector to decide whether the PU is present or not.

The distribution that governs the statistic D can be used to estimate the probability of detection and the probability of false alarm. Here, the in-phase and quadrature components of the quasimonochromatic wave are assumed to be zero-mean Gaussian random processes. The estimation of the SP and the elements of the sample correlation random process is equivalent, which is called Wishart distribution [22]. Thus, the probability density function of the detection statistic D is given by [12]

$$\begin{aligned} P(x) = {} & \frac{2\Gamma(N-1/2)}{\sqrt{\pi}\Gamma(N-1)} \frac{\left[1-D_\infty^2\right]^N\left[1-x^2\right]^{N-2}}{\left[1-D_\infty^2 x^2\right]^{2N-1}} \\ & \times \sum_{k=1}^{N} \frac{\Gamma(2N) D_s^{2k-2} x^{2k}}{\Gamma(2k)\Gamma(2N-2k+1)}, \end{aligned} \tag{11}$$

where D_∞ and $\{s_0, s_1, s_2, s_3\}$ are the actual degree of polarization, and the actual Stokes vector components, respectively. They are obtained in the limit as the number of independent samples approaches infinity. For a fixed threshold λ, the conditional probability of false alarm P_f and detection P_d can be expressed as [12]

$$\begin{aligned} P_f &= \frac{1}{2} I_{1-\lambda^2}\left(N-1, \frac{3}{2}\right), \\ P_d &= 1 - A\left[1 - \left(\frac{\gamma}{1+\gamma}\right)^2\right]^N \\ & \quad \times \sum_{k=1}^{N} B_k \left(\frac{\gamma}{1+\gamma}\right)^{2k-2} \sum_{n=0}^{\infty} C_{k,n} \left(\frac{\gamma}{1+\gamma}\right)^{2n}, \end{aligned} \tag{12}$$

where γ is the Signal-to-Noise Ratio (SNR), $I_{1-\lambda^2}(n+1/2, N)$ is the incomplete beta function, $A = 2\Gamma(N+1/2)/\sqrt{\pi}$,

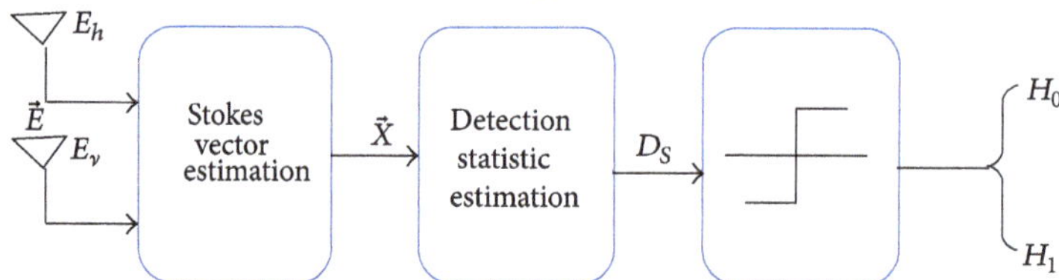

FIGURE 1: The proposed spectrum sensing system model.

$B_k = 1/\Gamma(2k)\Gamma(2N-2k+1)$, and $C_{k,n} = \Gamma(2N+n-1)\Gamma(n+k-1/2)/(\Gamma(n+1)\Gamma(N+n+k-1/2))I_{\chi^2}(k+n+1/2, N-1)$. The SNR γ can be affected by fading that in turn affects P_d and hence in order to incorporate its influence, P_d must be averaged over all possible values of γ according to

$$\overline{P_d} = \int_0^{\infty} P_d(\gamma) f_\gamma(\gamma) d\gamma, \tag{13}$$

where $f(\cdot)$ represents the PDF of the channel. For a fading signal with envelope h, an arbitrary parameter $\alpha > 0$, and a α-root mean value $\overline{h} = \sqrt[\alpha]{E(h^\alpha)}$, the α-μ PDF $f_h(h)$ is given by [23]

$$f_h(h) = \frac{\alpha \mu^\mu h^{\alpha\mu-1} e^{-\mu(h/\overline{h})^\alpha}}{\Gamma(\mu) \overline{h}^{\alpha\mu}}, \tag{14}$$

where $\mu = \mathscr{E}^2(h^\alpha)/\mathscr{V}(h^\alpha)$ and $\mathscr{E}(\cdot)$ and $\mathscr{V}(\cdot)$ are the expectation and variance operators, respectively [23]. The PDF of the SNR is obtained by a change of variables as shown in [24]

$$f_\gamma(\gamma) = \frac{(\alpha/2)\mu^\mu \gamma^{\alpha\mu/2-1}}{\Gamma(\mu)\overline{\gamma}^{\alpha\mu/2}} e^{-\mu(\gamma/\overline{\gamma})^{\alpha/2}}. \tag{15}$$

By substituting (15) in (13), $\overline{P_d}$ can be written as

$$\overline{P}_d = 1 - A' \sum_{k=1}^{N} B_k \sum_{n=0}^{\infty} C_{k,n} \int_0^{\infty} g_n(\gamma) d\gamma, \tag{16}$$

where

$$\begin{aligned} g_n(\gamma) &= \left[1 - \left(\frac{\gamma}{1+\gamma}\right)^2\right]^N \\ &\quad \times \left(\frac{\gamma}{1+\gamma}\right)^{2n+2k-2} \gamma^{\alpha/2\mu-1} e^{-\mu(\gamma/\overline{\gamma})^{\alpha/2}}, \\ A' &= \frac{\alpha/2\mu^\mu}{\Gamma(\mu)\overline{\gamma}^{-\alpha\mu/2}} A. \end{aligned} \tag{17}$$

Using [25], one can get the average probability of detection $\overline{P}_d$ as follows:

$$\overline{P}_d = 1 - \sum_{k=1}^{N} B_k \sum_{n=0}^{\infty} C_{k,n} \sum_{r=0}^{N} \binom{N}{r} (-1)^r \times \begin{Bmatrix} A(\nu,\mu,z) & \alpha = 1 \\ \Gamma(\nu+\mu) U(\nu, 1-\mu, z) & \alpha = 2 \\ \dfrac{(z/k)^\nu l^{\mu+1/2}}{\sqrt{2\pi}} G_{k+l,k}^{k,k+l} \times \left[\dfrac{2}{z^k} \,\middle|\, \begin{matrix} \Delta(l,-\mu), \Delta(k,\nu+1) \\ \Delta(k,0) \end{matrix}\right] & \alpha \neq 1,2 \end{Bmatrix}, \tag{18}$$

where

$$\begin{aligned} A(\nu,\mu,z) &= \frac{1}{\Gamma(\nu)} \\ &\quad \times \left[{}_1F_2\left(\nu; \frac{1}{2} - \frac{\mu}{2}, 1 - \frac{\mu}{2}; -\frac{z}{4}\right) \Gamma(\mu)\Gamma(\nu) \right] \\ &\quad \times {}_1F_2\left(\nu + \frac{\mu}{2}; \frac{1}{2}, \frac{\mu}{2} + 1; -\frac{z}{4}\right) \frac{z^{\mu/2}}{2} \\ &\quad \times \Gamma\left(-\frac{\mu}{2}\right) \Gamma\left(\nu + \frac{\mu}{2}\right) \\ &\quad - {}_1F_2\left(\nu + \frac{\mu}{2} + \frac{1}{2}; \frac{3}{2}, \frac{\mu}{2} + \frac{3}{2}; -\frac{z}{4}\right), \\ z &= \frac{\mu}{\overline{\gamma}}, \\ \nu &= 2(N+n+k-1), \\ &{}_1F_2\left(\nu; \frac{1}{2} - \frac{\mu}{2}, 1 - \frac{\mu}{2}; -\frac{z}{4}\right) \end{aligned} \tag{19}$$

is the Hypergeometric function, $G(\cdot,:)$ is the Meijer G function, and U is the Hypergeometric U function [26].

3.2. Spectrum Sensing Based on the Axial Ratio of a Polarization Ellipse. The polarized portion of the PU signal represents a net polarization ellipse traced by the electric field vector as a function of time. The ellipse has a magnitude (R) such that

$$R = \left|\frac{x_3}{x_0}\right|. \tag{20}$$

The ellipticity is the ratio of the minor to the major axis of the corresponding electric field polarization ellipse and varies from 0 for linearly polarized wave to 1 for circularly polarized wave. The polarization ellipse is alternatively described by its eccentricity, which is zero for a circularly polarized wave, and increases as the ellipse becomes thinner. It then becomes one for a linearly polarized wave. Alternatively R is defined as the ratio of the polarized power to the total power in the wave.

Figure 1 depicts a block diagram of the proposed spectrum sensing system, where $D_S = R$. This statistic serves as the input to the threshold detector to decide whether the PU is present or not. The distribution that governs the statistic R is to be determined to estimate the probability of detection and the probability of false alarm [12], which is given by

$$\begin{aligned} P_R(x) &= \frac{\Gamma(N+1/2)}{\sqrt{\pi}\Gamma(N)}\left[1-D_{\infty}^2\right]^2\left[1-x^2\right]^{N-1} \\ &\times \left\{\frac{\left[1-(s_3x/s_0)\right]}{\left[(1-(s_3x/s_0))^2-(s_1^2+s_2^2/s_0^2)(1-x^2)^2\right]^{N+1/2}}\right\} \\ &+ \left\{\frac{\left[1+s_3x/s_0\right]}{\left[(1+s_3x/s_0)^2+((s_1^2+s_2^2)/s_0^2)(1-x^2)^2\right]^{N+1/2}}\right\}. \end{aligned} \tag{21}$$

For a fixed threshold λ, the conditional probability of false alarm P_f can be expressed as

$$P_f = \frac{1}{2}I_{1-\lambda^2}\left(N, \frac{1}{2}\right), \tag{22}$$

and the probability of detection P_d can be given by

$$\begin{aligned} P_d &= 1-\left[1-\left(\frac{\gamma}{1+\gamma}\right)^2\right]^N \\ &\times \sum_{n=0}^{\infty}\frac{\Gamma(N+n)}{\Gamma(n+1)\Gamma(N)}I_{\lambda^2}\left(n+\frac{1}{2},N\right)\left(\frac{\gamma}{1+\gamma}\right)^{2n}. \end{aligned} \tag{23}$$

P_d must be averaged over all possible values of γ as follows:

$$\overline{P_d} = \int_0^{\infty} P_d(\gamma) f_{\gamma}(\gamma)\,d\gamma, \tag{24}$$

where $f_\gamma(\gamma)$ represents the α-μ probability density function. Following a similar approach to that of Section 3.1 and by substituting (15) in (24), $\overline{P_d}$ can be written as

$$\overline{P}_d = 1-k\sum_{n=0}^{\infty}\frac{\Gamma(N+n)}{\Gamma(n+1)\Gamma(N)}I_{\lambda^2}\left(n+\frac{1}{2},N\right)\int_0^{\infty} g_n(\gamma)\,d\gamma, \tag{25}$$

where

$$\begin{aligned} g_n(\gamma) &= \left[1-\left(\frac{\gamma}{1+\gamma}\right)^2\right]^N\left(\frac{\gamma}{1+\gamma}\right)^{2n}\gamma^{\alpha/2\mu-1}e^{-\mu(\gamma/\overline{\gamma})^{\alpha/2}}, \\ k &= \frac{\alpha\mu^{\mu}}{2\Gamma(\mu)\overline{\gamma}^{\alpha\mu/2}}. \end{aligned} \tag{26}$$

Using binomial theorem and [25], $\overline{P}_d$ will be given by

$$\begin{aligned} \overline{P}_d &= 1-\sum_{n=0}^{\infty}\frac{\Gamma(N+n)}{\Gamma(n+1)\Gamma(N)} \\ &\times I_{Y_{\text{th}}}\left(n+\frac{1}{2},N\right)\sum_{r=0}^{N}\binom{N}{r}(-1)^r \\ &\times \left\{\begin{array}{ll} A(v,\mu,z) & \alpha=1 \\ \Gamma(v+\mu)U(v,1-\mu,z) & \alpha=2 \\ \dfrac{(z/k)^v l^{\mu+(1/2)}}{\sqrt{2\pi}}G_{k+l,k}^{k,k+l} & \\ \times\left[\dfrac{2}{z^k}\,\middle|\,\begin{array}{c}\Delta(l,-\mu),\Delta(k,v+1)\\ \Delta(k,0)\end{array}\right] & \alpha\neq 1,2 \end{array}\right\}. \end{aligned} \tag{27}$$

It is worth mentioning that the restriction for this scheme is that the polarization information of the primary signal must be known a priori. If the PU signal is linearly polarized, then R will be close to 0 for high SNR, and if the PU signal is circularly polarized, then R will be close to 1 for high SNR.

4. Spectrum Sensing Based on Cooperative Polarization Detection

The fading and noisy nature of a wireless communication channel places a major challenge on the accuracy of spectrum sensing. Sensing decisions that is based on measurements of a single SU may be unreliable. Cooperative spectrum sensing is one possible solution to overcome this unreliability. In Figure 2, a number of SUs, which are distributed in different locations independently, can detect the PU and make the decision whether the signal exists or not. According to the information received from various SUs, the central unit makes the final decision based on some rules. A general fusion rule is when a final decision of 1 is taken when *m-out-of-N* SUs report 1. When $m = 1$, the *m-out-of-N* rule is equivalent to the OR rule. When $m = N$, the decision rule becomes the AND rule. By selecting different values of m, different detection performances are obtained.

Two cases are considered, namely, cooperative spectrum sensing with perfect and imperfect reporting channels.

4.1. Perfect Reporting Channels. If the channels between each SU and the central unit are noise free, then the overall probability of false alarm Q_F and the overall probability of detection Q_D of the cooperative spectrum sensing for *m-out-of-N* rule of the cooperative spectrum sensing are given by [27]

$$\begin{aligned} Q_F &= \sum_{i=m}^{N}\binom{N}{i}(P_{f,i})^i(1-P_{f,i})^{N-i}, \\ Q_D &= \sum_{i=m}^{N}\binom{N}{i}(P_{d,i})^i(1-P_{d,i})^{N-i}, \end{aligned} \tag{28}$$

where the $P_{f,i}$ and $P_{d,i}$ are the probabilities of false alarm and detection of the degree of polarization and axial ratio as derived in (14), (18), (22), and (27), respectively.

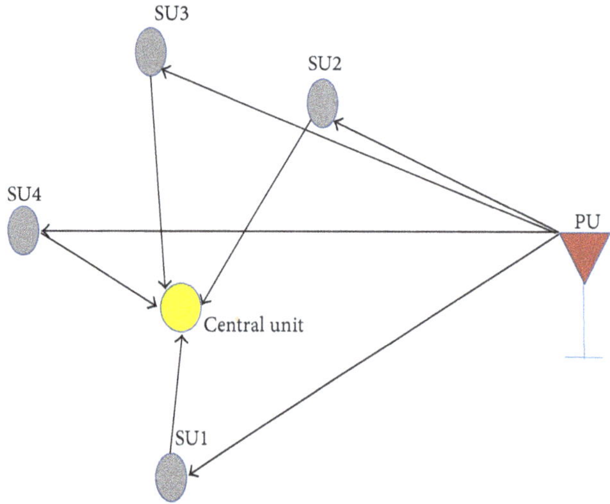

FIGURE 2: System model of the cooperative network.

4.2. Imperfect Reporting Channels. In practical systems, the reporting channels between the SUs and the central unit will experience fading. This, in turn, will degrade transmission reliability of the sensing results that are reported from the SUs to the central unit. Let $P_{E,i}$ denote the probability of receiving H_1 at the central unit when the ith Secondary User sends H_0 and the probability of receiving H_0 at the central unit when the ith Secondary User sends H_1 which can be calculated over α-μ fading channel as follows:

$$P_{E,i} = \int_0^{\infty} \frac{1}{2} \operatorname{erfc}\left(\sqrt{\gamma}\right) \frac{(\alpha/2)\,\mu^{\mu}\gamma^{(\alpha\mu/2)-1}}{\Gamma(\mu)\,\overline{\gamma}^{\alpha\mu/2}} e^{-\mu(\gamma/\overline{\gamma})^{\alpha/2}} d\gamma, \quad (29)$$

where $(1/2)\operatorname{erfc}(\sqrt{\gamma})$ is the probability of error for BPSK in an *AWGN* channel. The shown integral can be put in the form of Laplace transform and hence using [25] (3.7.2–18) and substituting $v = \mu - 1$, $a = \sqrt{\overline{\gamma}(1/\mu)^{2/\alpha}}$, $l/2k = 1/\alpha$, the probability of error can be calculated as

$$\overline{P}_{E,i} = \frac{\alpha}{4\Gamma(\mu)} \times \left[\frac{\sqrt{2}l^{\mu-1/2}}{(2\pi)^{(k+l-1)/2}} G^{k+1,l}_{l+1,k+1}\left(\left(\frac{\overline{\gamma}(1/\mu)^{2/\alpha}}{k}\right)^{k} l^{l} \,\middle|\, \begin{matrix} \Delta(l,-\mu+1),1 \\ \Delta\left(k,\frac{1}{2}\right),0 \end{matrix}\right)\right], \quad (30)$$

where $\Delta(k, a) = a/k, (a+1)/k, \ldots, (a+k-1)/k$.

Consequently, the overall probability of false alarm Q_F is obtained as

$$Q_F = \sum_{i=m}^{N} \binom{N}{i} \operatorname{Prob}\left[\left\{\frac{H_1^{\mathrm{CU}}}{H_0^{\mathrm{PU}}}\right\}\right]^{i} \times \operatorname{Prob}\left[\left\{\frac{H_0^{\mathrm{CU}}}{H_0^{\mathrm{PU}}}\right\}\right]^{N-i}, \quad (31)$$

$$Q_F = \sum_{i=m}^{N} \binom{N}{i} \operatorname{Prob}\left[\left\{\frac{H_1^{\mathrm{CU}}}{H_0^{\mathrm{PU}}}\right\}\right]^{i} \times \operatorname{Prob}\left[1 - \left\{\frac{H_1^{\mathrm{CU}}}{H_0^{\mathrm{PU}}}\right\}\right]^{N-i}, \quad (32)$$

$$\operatorname{Prob}\left\{\frac{H_1^{\mathrm{CU}}}{H_0^{\mathrm{PU}}}\right\} = \left[\operatorname{Prob}\left\{\frac{H_1^{\mathrm{CU}}}{H_1^{\mathrm{SU}}}\right\} \times \operatorname{Prob}\left\{\frac{H_1^{\mathrm{SU}}}{H_0^{\mathrm{PU}}}\right\}\right] + \operatorname{Prob}\left\{\frac{H_1^{\mathrm{CU}}}{H_0^{\mathrm{SU}}}\right\} \cdot \operatorname{Prob}\left\{\frac{H_0^{\mathrm{SU}}}{H_0^{\mathrm{PU}}}\right\}, \quad (33)$$

$$Q_F = \sum_{i=m}^{N} \binom{N}{i} \left[\left(1 - P_{E,i}\right) P_{F,i} + P_{E,i}\left(1 - P_{F,i}\right)\right]^{i} \times \left[\left(1 - P_{E,i}\right)\left(1 - P_{F,i}\right) + P_{E,i}P_{F,i}\right]^{N-i}, \quad (34)$$

where H_1^{CU}, H_1^{PU}, and H_1^{SU} are the hypothesis in the central unit, PU, and SU, respectively.

Substituting (12) and (30) into (34) yields

$$Q_F = \sum_{i=m}^{N} \binom{N}{i} \left\{ \begin{array}{l} \left(1 - \frac{\alpha}{4\Gamma(\mu)} J(k,l,\alpha,\mu)\right) \frac{\Gamma(m,\lambda/2)}{\Gamma(m)} \\ + \frac{\alpha}{4\Gamma(\mu)} J(k,l,\alpha,\mu) \left(1 - \frac{\Gamma(m,\lambda/2)}{\Gamma(m)}\right) \end{array} \right\}^{i} \times \left\{ \begin{array}{l} \left(1 - \frac{\alpha}{4\Gamma(\mu)} J(k,l,\alpha,\mu)\right)\left(1 - \frac{\Gamma(m,\lambda/2)}{\Gamma(m)}\right) \\ + \frac{\alpha}{4\Gamma(\mu)} J(k,l,\alpha,\mu) \frac{\Gamma(m,\lambda/2)}{\Gamma(m)} \end{array} \right\}^{N-i}, \quad (35)$$

where

$$J(k,l,\alpha,\mu) = \frac{\sqrt{2} l^{\mu-1/2}}{(2\pi)^{(k+l-1)/2}} \times G_{l+1,k+1}^{k+1,l} \left(\left(\frac{\overline{\gamma}(1/\mu)^{2/\alpha}}{k} \right)^{k} l^{l} \middle| \begin{array}{c} \Delta(l,-\mu+1), 1 \\ \Delta\left(k,\frac{1}{2}\right), 0 \end{array} \right). \quad (36)$$

Similarly, the detection probability Q_D can be computed as

$$Q_D = \sum_{i=m}^{N} \binom{N}{i} \left\{ \begin{array}{l} \left(1 - \frac{\alpha}{4\Gamma(\mu)} J(k,l,\alpha,\mu)\right) A \sum_{n=0}^{\infty} a_n G_{l,k}^{k,l}\left(S; \begin{array}{c} \Delta(l,-v) \\ \Delta(k,0) \end{array}\right) \\ + \frac{\alpha}{4\Gamma(\mu)} J(k,l,\alpha,\mu) \left(1 - A \sum_{n=0}^{\infty} a_n G_{l,k}^{k,l}\left(S; \begin{array}{c} \Delta(l,-v) \\ \Delta(k,0) \end{array}\right)\right) \end{array} \right\}^{i} \times \left\{ \begin{array}{l} \left(1 - \frac{\alpha}{4\Gamma(\mu)} J(k,l,\alpha,\mu)\right) \left(1 - A \sum_{n=0}^{\infty} a_n G_{l,k}^{k,l}\left(S; \begin{array}{c} \Delta(l,-v) \\ \Delta(k,0) \end{array}\right)\right) \\ + \frac{\alpha}{8\Gamma(\mu)} J(k,l,\alpha,\mu) A \sum_{n=0}^{\infty} a_n G_{l,k}^{k,l}\left(S; \begin{array}{c} \Delta(l,-v) \\ \Delta(k,0) \end{array}\right) \end{array} \right\}^{N-i}, \quad (37)$$

where $P_{D,i}$ is the probability of detection of the polarization obtained in (18), (38) for degree detection and axial ratio detection methods, respectively.

Therefore, for AND rule, the overall probability of false alarm and detection are given by

$$Q_F = \prod_{i=0}^{N} \left\{ \begin{array}{l} \left(1 - \frac{\alpha}{4\Gamma(\mu)} J(k,l,\alpha,\mu)\right) \frac{\Gamma(m,\lambda/2)}{\Gamma(m)} \\ + \frac{\alpha}{4\Gamma(\mu)} J(k,l,\alpha,\mu) \left(1 - \frac{\Gamma(m,\lambda/2)}{\Gamma(m)}\right) \end{array} \right\}^{i}, \quad (38)$$

$$Q_D = \prod_{i=0}^{N} \left\{ \begin{array}{l} \left(1 - \frac{\alpha}{4\Gamma(\mu)} J(k,l,\alpha,\mu)\right) A \sum_{n=0}^{\infty} a_n G_{l,k}^{k,l}\left(S; \begin{array}{c} \Delta(l,-v) \\ \Delta(k,0) \end{array}\right) \\ + \frac{\alpha}{4\Gamma(\mu)} J(k,l,\alpha,\mu) \left(1 - A \sum_{n=0}^{\infty} a_n G_{l,k}^{k,l}\left(S; \begin{array}{c} \Delta(l,-v) \\ \Delta(k,0) \end{array}\right)\right) \end{array} \right\}^{i}. \quad (39)$$

5. Cluster-Based Cooperative Spectrum Sensing

The clustering method is proposed in the cooperative spectrum sensing scheme in order to improve the sensing performance by decreasing the reporting channel error, which is proved to exploit a selection diversity gain [28]. In this paper, we assume that the instantaneous channel state information of the channel is available for the Secondary Users, and the channel between any two Secondary Users in the same cluster is perfect because they are in the vicinity of each other. Figure 3 shows the system model of cluster-based cooperative spectrum sensing. We assume that there are K SUs, who are divided into L clusters. The ith cluster has an integer number N_i of SUs, which satisfies

$$\sum_{i=l}^{L} N_i = K. \quad (40)$$

Firstly, all SUs are assumed to belong to few clusters by the distributed clustering algorithms [29]. Secondly, the Secondary User who has the largest instantaneous reporting channel gain will be selected as the cluster head. Then, the cooperative sensing is carried out through the following steps. Firstly, all SUs perform the local spectrum sensing. Every SU sends a decision to the cluster head. Secondly, the cluster head receives those local decisions from the SUs in the same cluster and then makes the decision according to certain fusion rule. All cluster heads send their decisions to the central unit. In the end, the central unit makes the final decision according to the fusion rule. The cluster head and central unit make the decision according to the OR-rule in order to limit the interference from the SUs to the PU. The *m-out-of-N* fusion rule is adopted in the cluster head and the OR fusion rule is adopted in the central unit [30]. If each cluster has K/L SUs, then the global false alarm probability and detection probability will be given by

$$Q_F = 1 - \left(\begin{array}{c} 1 - \sum_{m=l}^{K/L} \left(\begin{array}{c} \left(\frac{K}{L}\right) \\ m \end{array} \right) \\ \times \left(\frac{\Gamma(m,\lambda/2)}{\Gamma(m)}\right)^{K} \left(1 - \frac{\Gamma(m,\lambda/2)}{\Gamma(m)}\right)^{K/L-K} \end{array} \right)^{L}, \quad (41)$$

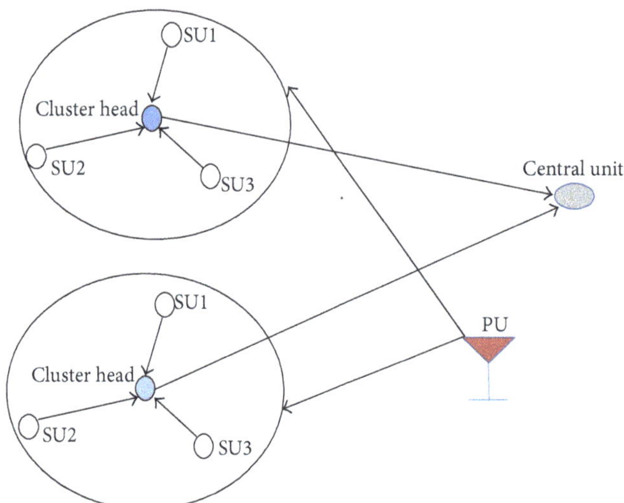

FIGURE 3: Cluster-based spectrum sensing mode.

$$Q_D = 1 - \left(1 - \sum_{m=l}^{K/L} \left(\binom{K}{L} \atop m\right) (P_d)^K (1 - P_d)^{K/L-K}\right)^L. \quad (42)$$

Substituting P_d for the degree detection in (18) into (42) yields

$$Q_D = 1 - \left(\begin{array}{c} 1 - \sum_{m=l}^{K/L} \left(\binom{K}{L} \atop m\right) \\ \times \left(A \sum_{n=0}^{\infty} a_n G_{l,k}^{k,l} \left(S; \begin{array}{c} \Delta(l,-v) \\ \Delta(k,0) \end{array} \right) \right)^K \\ \left(1 - A \sum_{n=0}^{\infty} a_n G_{l,k}^{k,l} \left(S; \begin{array}{c} \Delta(l,-v) \\ \Delta(k,0) \end{array} \right) \right)^{K/L-K} \end{array} \right)^L, \quad (43)$$

where

$$U(v,\mu,z) = \left\{ \begin{array}{ll} A(v,\mu,z) & \alpha = 1 \\ \Gamma(v+\mu) U(v, 1-\mu, z) & \alpha = 2 \\ \frac{(z/k)^v l^{\mu+1/2}}{\sqrt{2\pi}} G_{k+l,k}^{k,k+l} & \\ \times \left[\frac{2}{z^k} \mid \begin{array}{c} \Delta(l,-\mu), \Delta(k, v+1) \\ \Delta(k,0) \end{array} \right] & \alpha \neq 1, 2. \end{array} \right\}, \quad (44)$$

and substituting P_d for the axial ratio detection in (27) into (42) yields

$$Q_D = 1 - \left(\begin{array}{c} 1 - \sum_{m=l}^{K/L} \left(\binom{K}{L} \atop m\right) \\ \times \left(\begin{array}{c} 1 - \sum_{n=0}^{\infty} \frac{\Gamma(N+n)}{\Gamma(n+1)\Gamma(N)} I_{Y_{\text{th}}} \left(n + \frac{1}{2}, N\right) \\ \times \sum_{r=0}^{N} \binom{N}{r} (-1)^r U(v,\mu,z) \end{array} \right)^K \\ \left(\begin{array}{c} \sum_{n=0}^{\infty} \frac{\Gamma(N+n)}{\Gamma(n+1)\Gamma(N)} I_{Y_{\text{th}}} \left(n + \frac{1}{2}, N\right) \\ \times \sum_{r=0}^{N} \binom{N}{r} (-1)^r U(v,\mu,z) \end{array} \right)^{K/L-K} \end{array} \right)^L. \quad (45)$$

6. Results and Discussion

Results obtained using the proposed methods are presented. Monte Carlo simulation consisting of 100000 independent trials was performed. The degree of freedom is set to 3.5 to make fair comparison between the proposed methods and the energy detection method [31]. The equal and normalized polarization states of the PUs are randomly generated with $\zeta \in (0, \pi/2)$, $\phi \in (-\pi/2, \pi/2)$. It is assumed that SUs use dual polarized antennas, which is necessary to detect the horizontal and vertical components of the wave.

6.1. The Degree of Polarization (DoP) Method. Figure 4 compares the Receiver Operating Characteristic (ROC) of the DoP method with the directional method [18]. It can be noticed from this figure that the DoP method improves the P_d at low SNR since the signal is polarized and the noise is unpolarized wave. This makes the signal detection, in particular at low SNR, more efficient, which is the range of interest for CR systems. Therefore, it cannot be used in high SNR but it offers a baseline for comparison.

Figure 5 shows the ROC under AWGN compared with the energy detection (ED) method. It is clear from this figure that the proposed method improves the probability of detection over a wide range of SNR (from −5 dB to 5 dB). For example, in the case of SNR = 0 dB, P_d improved from 22% to 88%. Moreover, it is observed that P_d at SNR = 0 dB for the DoP method is better than P_d for the ED method at SNR = 5 dB, meaning the improvement of more than +5 dB for the DoP method relative to the ED method. It is also noticed that at SNR = 10 dB, the DoP method offers 100% detection probability for all values of P_f. This improvement is due to the more accurate detection of the polarized portion.

The case of Rayleigh fading channel is considered in Figure 6, which illustrates that fading has a stronger impact on the ED than on the proposed DoP method. This is attributed to the fact that the DoP is based on a second order statistic; that is, the improvement in the probability of

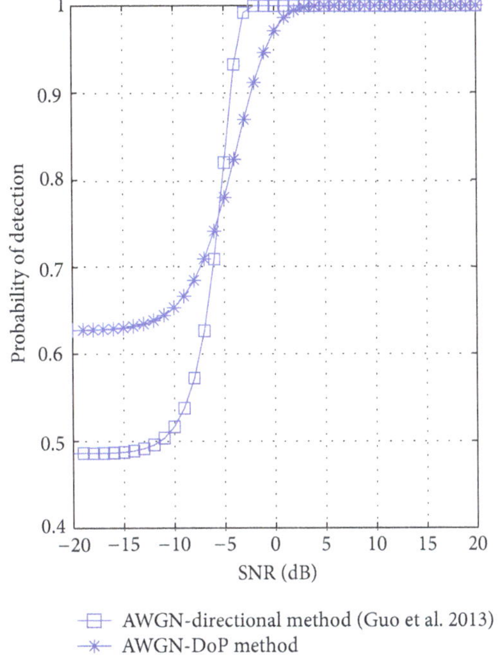

FIGURE 4: Probability of detection versus SNR for directional and DoP methods.

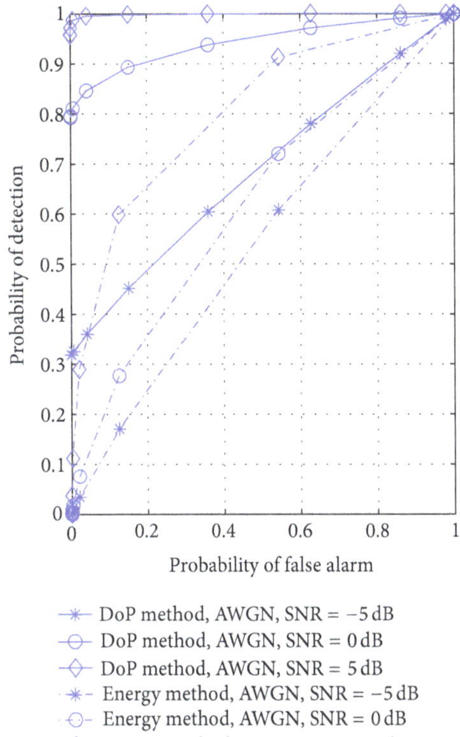

FIGURE 5: ROC of the DoP method in AWGN.

detection for SNR = 5 dB is approximately 30% at $P_f = 0.2$. A better performance for the DoP is reported for Nakagami fading as shown in Figure 7, where, at SNR = 0 dB, the improvement is approximately 20% when the $P_f = 0.2$. Moreover, the P_d at SNR = 5 dB for the proposed method is greater than the P_d for the ED method at SNR = 10 dB, this means that the improvement is nearly +7 dB.

It can be noticed that at fading conditions, the depolarization effect increases, which means higher dispersion of polarization states. Therefore, the detection performance decreases. Thus, the smaller the depolarization effect on the primary signal, the more constant the polarization state and thus the better the detection performance.

6.2. The Axial Ratio (AR) Method. To demonstrate the performance of the axial ratio (AR) method, Figure 8 shows the characteristics for different values of α and μ. For $\alpha = 2$ and $\mu = 2$, the probability of detection is 91% at $P_f = 0.2$, which is a superior value relative to the ED method. This can be easily explained since at strong fading less energy is coupled between the cross-polarized channels, which results in small dispersion of the polarization state leading to better detection performance.

Figure 9 shows the superior performance of the AR method relative to the ED method over a wide range of SNR under Rayleigh fading channel. In Figure 10, where Nakagami fading is considered, for SNR = −5 dB, an improvement of approximately 60% in the probability of detection is achieved for $P_f = 0.2$ and 70% for SNR = 5 dB relative to the ED method. Hence, the AR method is the most robust and applicable detection method in the case of fading condition and/or presence of noise power uncertainty.

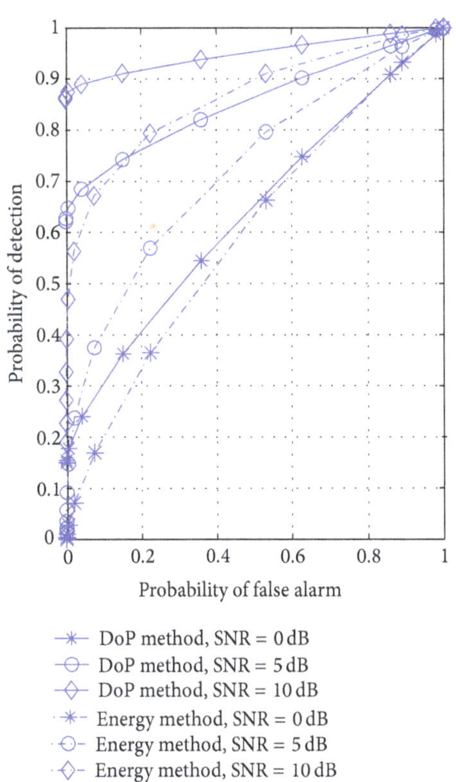

FIGURE 6: ROC of the DoP method under Rayleigh fading channel.

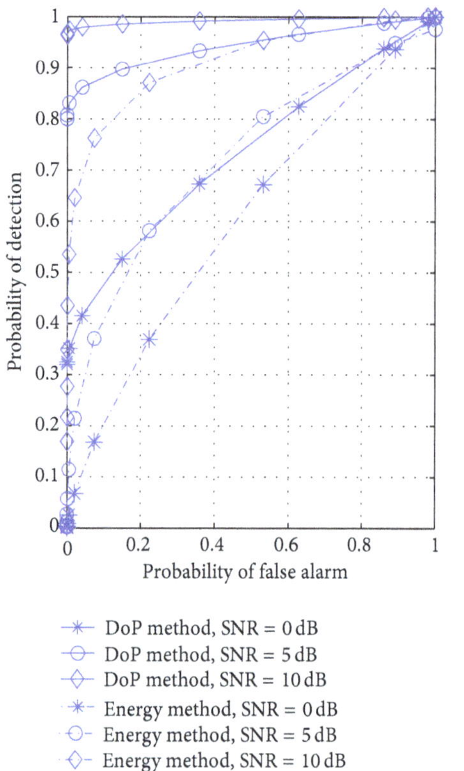

FIGURE 7: ROC of the DoP method under Nakagami fading channel.

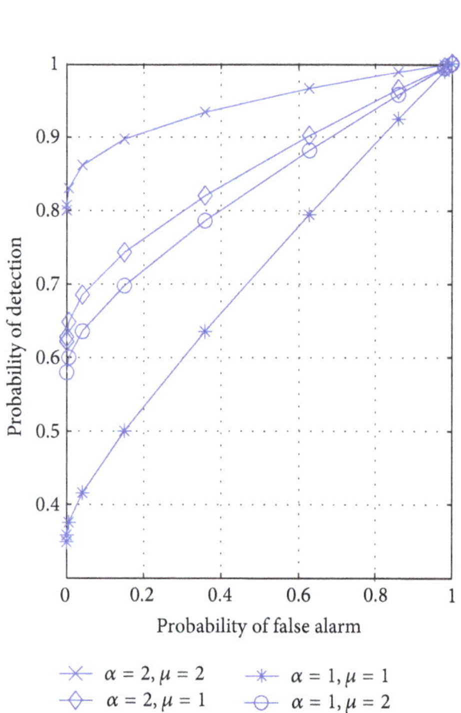

FIGURE 8: ROC of the DoP method under different α, μ at SNR = 5 dB.

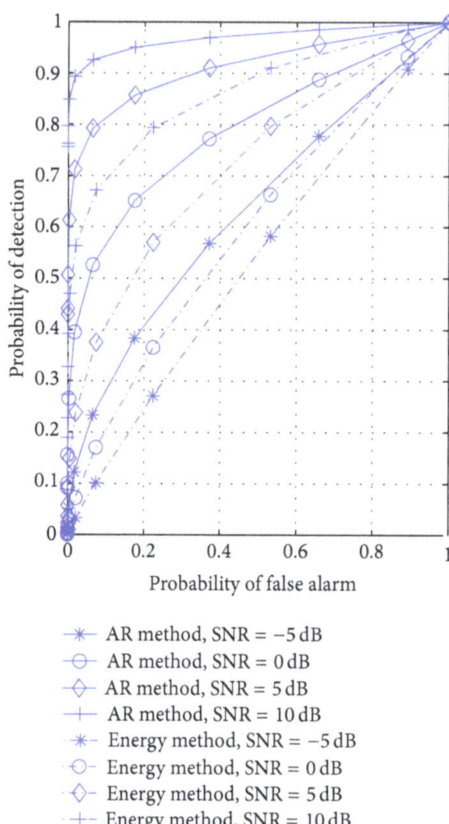

FIGURE 9: ROC of the AR method under Rayleigh fading channel.

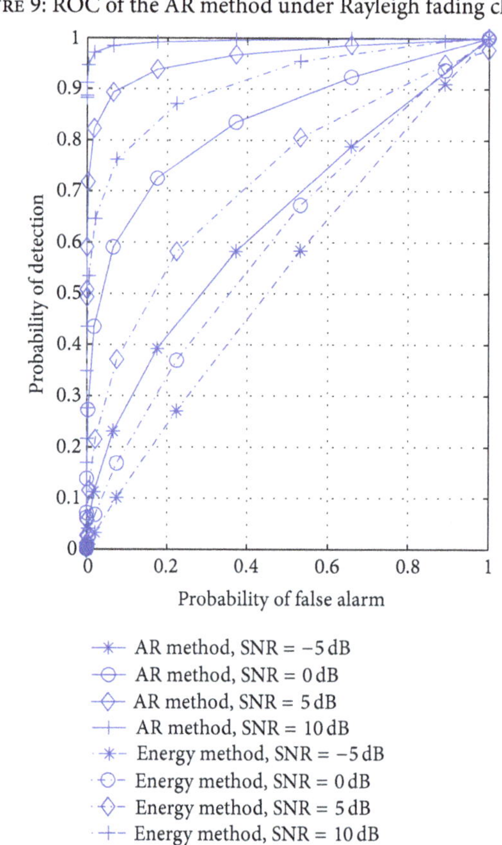

FIGURE 10: ROC of the AR method under Nakagami fading channel.

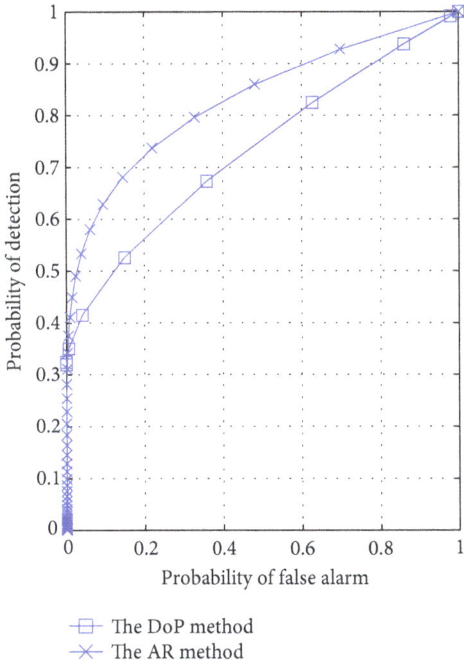

FIGURE 11: ROC of the two methods (AR, DoP) at SNR=0 dB under Nakagami fading channel.

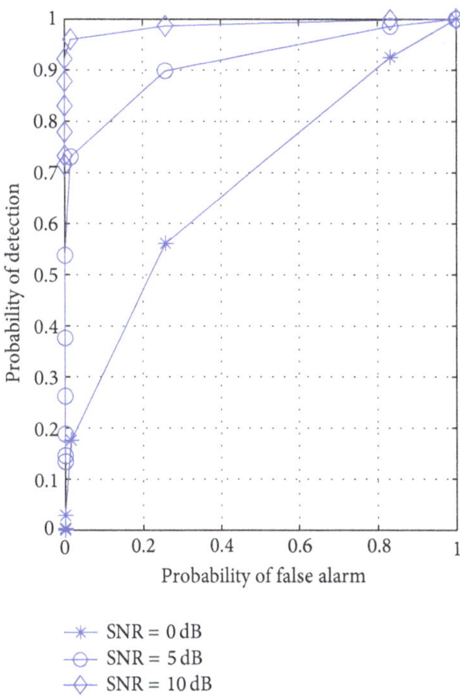

FIGURE 12: ROC of the cooperative scheme of the DoP method under imperfect Nakagami fading channel.

A comparison between both approaches is drawn in Figure 11 showing the superiority of the AR method relative to the DoP, where for $P_f = 0.2$, the AR method gives $P_d = 0.73$ compared to $P_d = 0.58$ for the DoP. It can be seen that the sensing performance of the AR is better than the DoP method due to the fact that the AR method resists the depolarization effect on the primary signal.

6.3. The Cooperative Scheme. Figure 12 shows the ROC of the cooperative scheme for the DoP method over a wide range of SNR under imperfect Nakagami fading channel. It is clear that the P_d is very high at low P_f especially at SNR = 5 dB and 10 dB. It can be observed that the P_d increase rapidly with the increase of the SNR. When the SNR = 10 dB, the P_d is almost 100% for all values of P_f.

7. Conclusions

Spectrum sensing based on Stokes parameters was thoroughly analyzed using new detection statistics, namely, the degree of polarization and the axial ratio. The proposed approaches were studied under different fading scenarios and the obtained results demonstrated superior performance relative to the conventional energy detection method. An extensive study is reported on the two methods and their performance under α-μ fading channels. The results demonstrated that the proposed algorithms are particularly applicable for the case of unknown primary polarization and/or presence of noise power uncertainty. In general, the proposed methods are better than energy detection in different ranges of the SNR and under different fading conditions. They are, however, more complex in terms of implementation due to the use of two orthogonally polarized antennas. It is also worth noting that the axial ratio method is generally better than the degree of polarization method.

Cooperative spectrum sensing was then considered and shown to be a powerful method for dealing with the hidden terminal problem. Simulation results show that the cluster rule gives superior performance for a wide range of SNR.

Conflict of Interests

The authors declare that there is no conflict of interests regarding the publication of this paper.

References

[1] S. Haykin, "Cognitive radio: brain-empowered wireless communications," *IEEE Journal on Selected Areas in Communications*, vol. 23, no. 2, pp. 201–220, 2005.

[2] A. Sahai and D. Cabric, "Spectrum sensing: fundamental limits and practical challenges," in *Proceedings of the IEEE International Symposium on New Frontiers in Dynamic Spectrum Access Network (DySPAN '05)*, Baltimore, Md, USA, November 2005.

[3] W. Wei, P. Tao, and W. Wenbo, "Optimal power control under interference temperature constraints in cognitive radio network," in *Proceedings of the IEEE Wireless Communications and Networking Conference (WCNC '07)*, pp. 116–120, March 2007.

[4] I. F. Akyildiz, W.-Y. Lee, M. C. Vuran, and S. Mohanty, "A survey on spectrum management in cognitive radio networks," *IEEE Communications Magazine*, vol. 46, no. 4, pp. 40–48, 2008.

[5] F. F. Digham, M.-S. Alouini, and M. K. Simon, "On the energy detection of unknown signals over fading channels," *IEEE Transactions on Communications*, vol. 55, no. 1, pp. 21–24, 2007.

[6] A. V. Dandawate and G. B. Giannakis, "Statistical tests for presence of cyclostationarity," *IEEE Transactions on Signal Processing*, vol. 42, no. 9, pp. 2355–2369, 1994.

[7] L. Zhang, Y.-C. Liang, Y. Xin, and H. V. Poor, "Robust cognitive beamforming with partial channel state information," *IEEE Transactions on Wireless Communications*, vol. 8, no. 8, pp. 4143–4153, 2009.

[8] H. Kim, J. Kim, S. Yang, M. Hong, and Y. Shin, "An effective MIMO-OFDM system for IEEE 802.22 WRAN channels," *IEEE Transactions on Circuits and Systems II*, vol. 55, no. 8, pp. 821–825, 2008.

[9] Z. Quan, S. Cui, A. H. Sayed, and H. V. Poor, "Wideband spectrum sensing in cognitive radio networks," in *Proceedings of the IEEE International Conference on Communications (ICC '08)*, pp. 901–906, May 2008.

[10] Z. Tian, "Compressed wideband sensing in cooperative cognitive radio networks," in *Proceedings of the IEEE Global Telecommunications Conference (GLOBECOM '08)*, pp. 3756–3760, December 2008.

[11] A. J. Poelman, "On using orthogonally polarized noncoherent receiving channels to detect target echoes in gaussian noise," *IEEE Transactions on Aerospace and Electronic Systems*, vol. 11, no. 4, pp. 660–663, 1975.

[12] G. M. Vachula and R. M. Barnes, "Polarization detection of a fluctuating radar target," *IEEE Transactions on Aerospace and Electronic Systems*, vol. 19, no. 2, pp. 250–257, 1983.

[13] R. E. Stovall, "A gaussian noise analysis of the pseudo-coherent discriminant," Tech. Rep. Note 1978-46, M.I.T. Lincoln Laboratory, 1978.

[14] D. P. Meyer and H. A. Mayer, *Radar Target Detection: Handbook of Theory and Practice*, Academic Press, New York, NY, USA, 1973.

[15] F. Liu, C. Feng, C. Guo, Y. Wang, and D. Wei, "Polarization spectrum sensing scheme for cognitive radios," in *Proceedings of the 5th International Conference on Wireless Communications, Networking and Mobile Computing (WiCOM '09)*, September 2009.

[16] D. Wei, C. Guo, F. Liu, and Z. Zeng, "A SINR improving scheme based on optimal polarization receiving for the cognitive radios," in *Proceedings of the IEEE International Conference on Network Infrastructure and Digital Content (IC-NIDC '09)*, pp. 100–104, November 2009.

[17] F. Liu, C. Feng, C. Guo, Y. Wang, and D. Wei, "Virtual polarization detection: a vector signal sensing method for cognitive radios," in *Proceedings of the 71st IEEE Vehicular Technology Conference (VTC-Spring '10)*, pp. 1–5, Taipei, Taiwan, 2010.

[18] C. Guo, X. Wu, C. Feng, and Z. Zeng, "Spectrum sensing for cognitive radios based on directional statistics of polarization vectors," *IEEE Journal on Selected Areas in Communications*, vol. 31, no. 3, pp. 379–393, 2013.

[19] K. B. Letaief and W. Zhang, "Cooperative communications for cognitive radio networks," *Proceedings of the IEEE*, vol. 97, no. 5, pp. 878–893, 2009.

[20] L. P. Murza, "The non coherent polarimetry of noise like radiation," *Radio Engineering and Electronic Physics*, vol. 23, no. 7, pp. 57–63, 1978.

[21] G. Deschamps, "Techniques for handling elliptically polarized waves with special reference to antennas: part II—geometrical representation of the polarization of a plane electromagnetic wave," *Proceedings of the IRE*, vol. 39, no. 5, pp. 540–544, 1951.

[22] N. Goodman, "Statistical analysis based on a certain multivariate complex gaussian distribution (an introduction)," *Annals of Mathematical Statistics*, vol. 34, no. 1, pp. 152–177, 1963.

[23] M. D. Yacoub, "The α-μ distribution: a physical fading model for the Stacy distribution," *IEEE Transactions on Vehicular Technology*, vol. 56, no. 1, pp. 27–34, 2007.

[24] M. K. Simon and M. S. Alouini, *Digital Communication over Fading Channels*, vol. 86, Wiley-IEEE Press, 2004.

[25] A. Prudnikov, Y. A. Brychkov, and O. Marichev, *Integrals and Series*, vol. 4, Gordon and Breach Science, 1986.

[26] F. W. Olver, D. W. Lozier, R. F. Boisvert, and C. W. Clark, *NIST Handbook of Mathematical Functions*, Cambridge University Press, 2010.

[27] A. Ghasemi and E. S. Sousa, "Collaborative spectrum sensing for opportunistic access in fading environments," in *Proceedings of the 1st IEEE International Symposium on New Frontiers in Dynamic Spectrum Access Networks (DySPAN '05)*, pp. 131–136, November 2005.

[28] C. Sun, W. Zhang, and K. B. Letaief, "Cluster-based cooperative spectrum sensing in cognitive radio systems," in *Proceedings of the IEEE International Conference on Communications (ICC '07)*, pp. 2511–2515, June 2007.

[29] O. Younis and S. Fahmy, "Distributed clustering in ad-hoc sensor networks: a hybrid, energy-efficient approach," in *Proceedings of the 23rd Annual Joint Conference of the IEEE Computer and Communications Societies (INFOCOM '04)*, vol. 1, pp. 629–640, March 2004.

[30] A. Ghasemi and E. S. Sousa, "Spectrum sensing in cognitive radio networks: the cooperation-processing tradeoff," *Wireless Communications and Mobile Computing*, vol. 7, no. 9, pp. 1049–1060, 2007.

[31] J. Ma, G. Zhao, and Y. Li, "Soft combination and detection for cooperative spectrum sensing in cognitive radio networks," *IEEE Transactions on Wireless Communications*, vol. 7, no. 11, pp. 4502–4507, 2008.

14

Complex Cepstrum Based Voice Conversion Using Radial Basis Function

Jagannath Nirmal,[1] Suprava Patnaik,[1] Mukesh Zaveri,[2] and Pramod Kachare[1]

[1] *Department of Electronics Engineering, SVNIT, Surat, India*
[2] *Department of Computer Engineering, SVNIT, Surat, India*

Correspondence should be addressed to Jagannath Nirmal; jhnirmal1975@gmail.com

Academic Editors: K. S. Chuang and A. Maier

The complex cepstrum vocoder is used to modify the speaker specific characteristics of the source speaker speech to that of the target speaker speech. The low time and high time liftering are used to split the calculated cepstrum into the vocal tract and the source excitation parameters. The obtained mixed phase vocal tract and source excitation parameters with finite impulse response preserve the phase properties of the resynthesized speech frame. The radial basis function is explored to capture the nonlinear mapping function for modifying the complex cepstrum based real and imaginary components of the vocal tract and source excitation of the speech signal. The state-of-the-art Mel cepstrum envelope and the fundamental frequency (F_0) are considered to represent the vocal tract and the source excitation of the speech frame, respectively. Radial basis function is used to capture and formulate the nonlinear relations between the Mel cepstrum envelope of the source and target speakers. Mean and standard deviation approach is employed to modify the fundamental frequency (F_0). The Mel log spectral approximation filter is used to reconstruct the speech signal from the modified Mel cepstrum envelope and fundamental frequency. A comparison of the proposed complex cepstrum based model has been made with the state-of-the-art Mel Cepstrum Envelope based voice conversion model with objective and subjective evaluations. The evaluation measures reveal that the proposed complex cepstrum based voice conversion system approximate the converted speech signal with better accuracy than the model based on the Mel cepstrum envelope based voice conversion.

1. Introduction

The voice conversion (VC) system extracts the features of the source and the target speaker sound's and formulates the mapping function to modify the features of the source speaker sound's such that the resynthesized speech sound's as if spoken by a target speaker [1]. Application of VC includes the personification of text to speech, design of multispeaker based speech synthesis system, audio dubbing, karaoke applications, security related system, the design of speaking aids for the speech impaired patient, broadcasting, and multimedia applications [2–4]. The VC involves the transformation of speaker specific characteristics such as vocal tract parameters, source excitation, and long term prosodic parameters with that of desired speaker parameters [5]. The vocal tract parameters are relatively more prominent for identifying the speaker uniqueness than the source excitation [5].

Several methods have been reported in the literature to characterize the spectrum of the speech frame, namely, Formant Frequency (FF), Formant Bandwidth (FBW) [1], Linear Predictive Coefficients (LPC) [6], Reflection Coefficients (RC) [7], Log Area Ratio (LAR) [8], Cepstrum Coefficients [9], Mel cepstrum envelope (MCEP) [10], Wavelet Transform (WT) [11], and Mel generated spectra [12]. Line Spectral Frequency (LSF) [13, 14] is a direct mathematical transformation of LPC, which has a special attraction in representing the vocal tract as it smoothly traces the shape of formants and antiformants and overcomes the interpolation, quantization, and stability issues of the LPC. However, LP related features does not assume nonstationary characteristics of the speech signal within a frame and therefore fail to analyze the local speech events accurately [15]. Further, a very accurate approach STRAIGHT [16] has also been proposed. It needs enormous computations and therefore, it is inappropriate for real time applications. Another approach using Mel Frequency Cepstrum Coefficients (MFCC) have been proposed [17], which properly model both spectral peaks and valleys. However, the main toil of MFCC synthesis is to loose pitch and phase related information [17].

The conventional parametric speech production model like LPC, real cepstrum [18–20], and Liljencrants-Fant (LF) [21] models is based on minimum phase model with infinite impulse response [22]. In fact, a completely different category of glottal flow estimation relies on the mixed-phase model of speech [22, 23]. According to this estimation, the speech signal is composed of both maximum (i.e., anticausal) and minimum phase (i.e., causal) components. The return phase of the glottal pulse components and vocal tract impulse response is part of minimum phase signals, whereas the open phase of the glottal flow is considered as maximum phase of the signal [24]. It has been shown in the literature that the mixed phase models are appropriate for representing the voiced speech [25]. The real cepstrum with minimum phase discards the glottal flow information of speech. However, the complex cepstrum incorporates phase as glottal pulse information during speech synthesis [25]. The complex cepstrum representation of the speech signal allows noncasual modeling of short time speech frame, which is actually observed in natural speech [22–24]. Complex cepstrum perform well in speech synthesis and speech modeling [25, 26].

For the development of appropriate transformation model, various mapping functions have been proposed in the literature such as Vector Quantization (VQ) based codebook mapping [6] and Gaussian Mixture Model (GMM) based transformation models [3, 9, 10]. Fuzzy vector quantization [27] and a Speaker Transformation Algorithm using Segmental Code-book (STASC) have been proposed to overcome limitations of VQ based model [14]. In addition Dynamic Frequency Warping (DFW) [28] have also been used for transformation of the spectral envelope. The GMM oversmoothing issue is resolved via maximum likelihood estimators and hybrid methods [29]. The dynamic kernel partial least square regression technique has also implied [12] for spectral transformation. In fact, the relation between the shapes of the vocal tracts of the different speakers are highly nonlinear, to capture this nonlinearity between the vocal tracts artificial neural network has been explored in the literature [10, 11, 14, 18, 30].

In addition to vocal tract, the source excitation contains vital speaker-specific characteristics [1, 3], so it is necessary to properly modify the excitation signal to accurately synthesize the target speaker's voice [4]. Very few methods have been discussed in the literature for excitation signal transformation such as residual copying, but the converted sound seems to be a third speaker's voice [31], another method is residual prediction [3]. However, it has the problem of over smoothening. In order to alleviate the over smoothening problem of residual prediction, residual selection method, unit selection method [31], and combination of residual selection and unit selection have been also explored in the literature [32]. The Artificial Neural Network model has also applied to modify the residual signal but time domain residual transformation loses the correlation in the speech production model which leads to distortion in speech signal [12].

In this paper, the prominent complex cepstrum vocoder is employed to model the vocal tract and source excitation of the speech. The low time and high time lifters are designed to separate the complex cepstrum into vocal tract and source excitation parameters with real and imaginary components. The reasons behind the use of radial basis function (RBF) based the transformation model are its fast training ability, desirable computational efficiency, and interpolation property. The RBF based mapping function are trained separately to capture the nonlinear relations for modifying the real and the imaginary components of cepstrum based vocal tract and source excitation of the source speaker to that of the target speaker utterance's. Similarly, the MCEP parameters of source speaker's utterances are also modified according to the target speaker's utterances using RBF. The fundamental frequency between source and target speaker's utterances is modified using mean and standard deviation approach [10]. Mel log spectral approximation (MLSA) filter [33] is used to reconstruct the speech signal from modified MCEP and fundamental (F_0).

Finally, the performance of the proposed complex cepstrum based VC approach is compared with MCEP [34] based VC approach. This is done using various objective measures such as a performance index (P_{LSF}) [3], formant deviation [14, 30], and spectral distortion [14]. The commonly used subjective measures such as Mean Opinion Score (MOS) and ABX verify the quality and speaker identity of the converted speech signal.

This paper is organized as follows. Section 2 describes the complex cepstrum analysis with low time and high time lifters which are used to extract the cepstrum based features of the vocal tract and excitation based signals. Section 3 explains the proposed VC system based on complex cepstrum and the state-of-the-art MCEP based VC system. Radial basis based spectral mapping is described in Section 4. The experimental environment, database, and objective measures, such as performance index, formant deviation, spectrograph, and the perceptual tests, namely, Mean Opinion Score (MOS) and ABX, conducted with different human listeners are presented in Section 5. The last Section gives the overall conclusions of the paper.

2. Complex Cepstrum Analysis

According to the source-filter model of the human speech production system, the source signal excites the vocal tract and it generates the speech signal. The human speech is two-sided real and asymmetrical in nature. Hence, a mixed phase Finite Impulse Response (FIR) system may be realized which preserves the phase related information to give more accurate synthesized speech. From the signal processing point of view, the short time speech signal $s(n)$ can be considered as linear convolution of the source excitation $g(n)$ with the impulse function of the vocal tract $v(n)$. It can be defined as follows:

$$s(n) = v(n) * g(n). \quad (1)$$

By applying DTFT to the speech signal we obtain

$$S(\omega) = \sum_{n=-M}^{M} s(n) e^{-j\omega n}, \quad (2)$$

where M is the order of cepstrum, that is, number of one sided frequencies. The time domain convolution can be modeled as spectral multiplication of the vocal tract filter response $V(\omega)$ and source excitation response $G(\omega)$ giving the short time speech spectrum $S(\omega)$ as shown,

$$S(\omega) = V(\omega) G(\omega). \quad (3)$$

Cepstral analysis includes transforming the multiplied source excitation and vocal tract responses in the frequency domain into linear combination of the two components in the cepstral domain. The analysis of the speech signal needs to separate two components $V(\omega)$ and $G(\omega)$. In frequency domain logarithmic representation is used to linearly combine the components $V(\omega)$ and $G(\omega)$. The complex spectrum $S(\omega)$ can be rewritten by performing logarithmic compression

$$\widehat{S}(\omega) = \log S(\omega). \quad (4)$$

Therefore the log spectrum is further separated into two parts

$$\log S(\omega) = \log V(\omega) + \log G(\omega). \quad (5)$$

Thus, the log spectrum can be decomposed as addition of magnitude and phase components

$$\widehat{S}(\omega) = \log |S(\omega)| + j \arg S(\omega). \quad (6)$$

The imaginary part of the logarithmic spectrum is the unwrapped phase sequence [23]. Thus, phase information is no more ignored giving rise to a complex cepstrum. Hence comprising of a mixed phase system, with a finite impulse response (FIR) type, which is stable. The cepstrum is defined as

$$c(n) = \frac{1}{2\pi} \int_{-\pi}^{\pi} \widehat{S}(\omega) e^{j\omega n} d\omega, \quad (7)$$

where $c(n)$ can be given as

$$c(n) = \frac{1}{2\pi} \int_{-\pi}^{\pi} \log |S(\omega)| e^{j\omega n} d\omega + \frac{1}{2\pi} \int_{-\pi}^{\pi} e^{j(\omega + \arg(S(\omega)))n} d\omega. \quad (8)$$

The log spectral components that vary rapidly with frequency ω are denoted as a high time component $\log G(\omega)$ and the log spectral components that slowly with frequency ω are designated as a low time component $\log V(\omega)$ [20]. Here, $c(n)$ is time aliased version, therefore, $M > N$ condition avoids aliasing effect; N is total number of cepstrum samples.

Consider

$$\begin{aligned} l_l(n) &= \begin{cases} 1, & 0 \le n < L_c, \\ 0, & L_c \ge n \le N, \end{cases} \\ c_v(n) &= l_l(n) c(n), \\ l_h(n) &= \begin{cases} 1, & L_c \le n \le N, \\ 0, & \text{elsewhere}, \end{cases} \\ c_e(n) &= l_h(n) c(n), \end{aligned} \quad (9)$$

where the $c(n)$ represents complex cepstrum of speech frame, $l_l(n)$ is low time lifter, $l_h(n)$ is high time lifter. In the deconvolution stage an appropriate value of lifter index L_c is chosen to separate the two components, namely, the fast changing excitation parameter $c_e(n)$ and the slowly changing parameters, that is, vocal tract parameter $c_v(n)$. The windowed signal, the complex cepstrum with magnitude, and phase spectra are shown in Figure 1. The coefficient, $c(0)$ is the speech signal energy and the coefficients $c(n)$ for $n \geq 1$ signifies the magnitude and phase at the quefrency n in the spectrum. The vocal tract cepstrum $c_v(n)$ has coefficients with significant magnitudes at lower values of n and source excitation cepstrum; $c_e(n)$ has relatively lower magnitude coefficients for higher values of n. Theoretically, the complex cepstrum being a mixed phase results in a more accurate model of the speech signal, when compared to the minimum phase synthesis filter approach which discard the glottal flow information content in the cepstrum [18]. The cepstrum values lower than zero represents the maximum phase (i.e., anticausal) response, whereas the values above zero can be considered as the minimum phase (i.e., causal) response are shown in Figure 2. Mathematically, it can be modeled as

$$\begin{aligned} \text{Minimum Phase} &= c_{\min}(n) \\ &= \begin{cases} 0, & n = -M, \ldots, -2, -1, \\ c(n), & n = 0, \\ c(n) + c(-n) & n = 1, 2, \ldots, M, \end{cases} \\ \text{Maximum Phase} &= c_{\max}(n) = c(n) - c_{\min}(n), \\ c_{\max}(n) &= \begin{cases} c(n), & n = -M, \ldots, -2, -1, \\ 0, & n = 0, \\ -c(-n) & n = 1, 2, \ldots, M. \end{cases} \end{aligned} \quad (10)$$

The anticausal and casual cepstrum parts with the corresponding magnitude and phase spectrum are shown in Figure 3. It has been observed that the logarithmic compression involved in the cepstrum analysis helps in obtaining the mixed phase response for both voiced as well as unvoiced signals.

3. Voice Conversion Framework

In this section, the complex cepstrum based VC algorithm is proposed. The MCEP-MLSA based VC algorithm is also

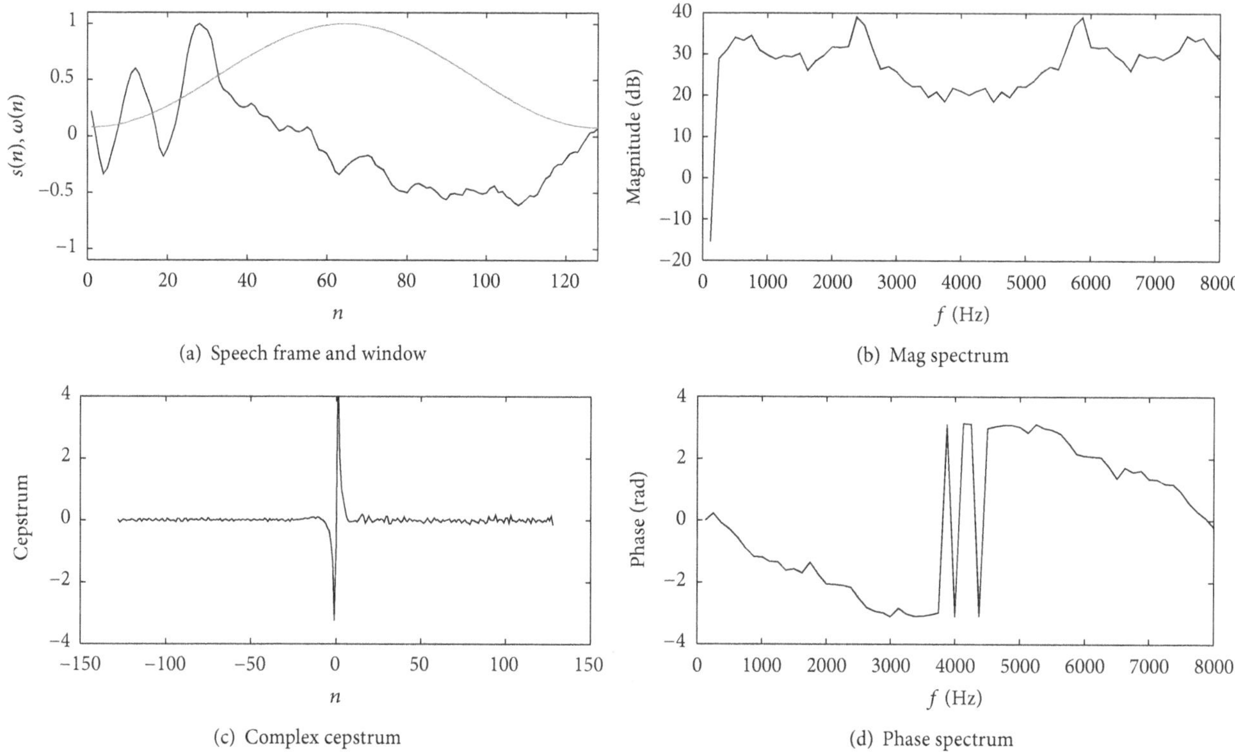

(a) Speech frame and window

(b) Mag spectrum

(c) Complex cepstrum

(d) Phase spectrum

FIGURE 1: (a) Speech frame and window, (b) magnitude spectra, (c) complex cepstrum, and (d) phase spectra.

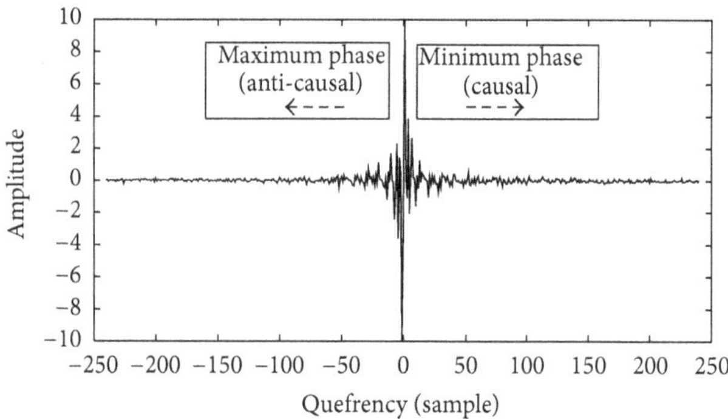

FIGURE 2: Complex cepstrum decomposition into maximum and minimum phase speech components.

developed for comparing the performance with the proposed algorithm.

3.1. Proposed Complex Cepstrum Vocoder Based VC. The proposed algorithm is implemented in two distinct phases: (i) training and (ii) transformation phase, as depicted in Figure 4. In the training phase, the input speech signal of the source and target speakers are normalized and silence frames are removed. The normalized speech frame is represented using homomorphic decomposition. It takes the advantages of the logarithmic scaling and the theory of convolution. The low time portion of the complex cepstrum can be approximated as a vocal tract impulse response (VT), where as high time portion of the complex cepstrum is considered as source excitation (GE) of the speech frame. The length of the rectangular lifter is chosen with regard to the accuracy of the vocal tract model and sampling frequency. Thus, the cepstrum frame is split into vocal tract impulse response and source excitation of the speech using low time and high time liftering, respectively. Even if the source and the target speaker utter the same sentence, the length of their feature vectors may be different so dynamic time warping is used to align these feature vectors. The separate RBF based mapping functions are developed for modifying the cepstrum based real and imaginary components of the vocal tract and source excitation of the source speaker according to the target speaker.

In the transformation phase followed by training phase, the parallel utterance of the test speaker speech is preprocessed to derive vocal tract and source excitation feature set based on cepstral analysis. The test feature vectors are projected to the trained RBF, in order to obtain the transformed feature vectors. The time domain features are computed by inverse transforming complex cepstrum based parameters. The modified speech frame is reconstructed by convolving the transformed vocal tract and source excitation. The similar process is adapted for all remaining frames. The overlap and add method is used to resynthesize speech from modified speech frames. Finally, the speech quality is enhanced through the postfiltering, applied to the modified speech. Figure 4 depicts the training and testing phase details of the proposed approach. The resynthesized speech from the complex cepstrum has higher perceptual quality than the speech signal constructed from the real cepstrum.

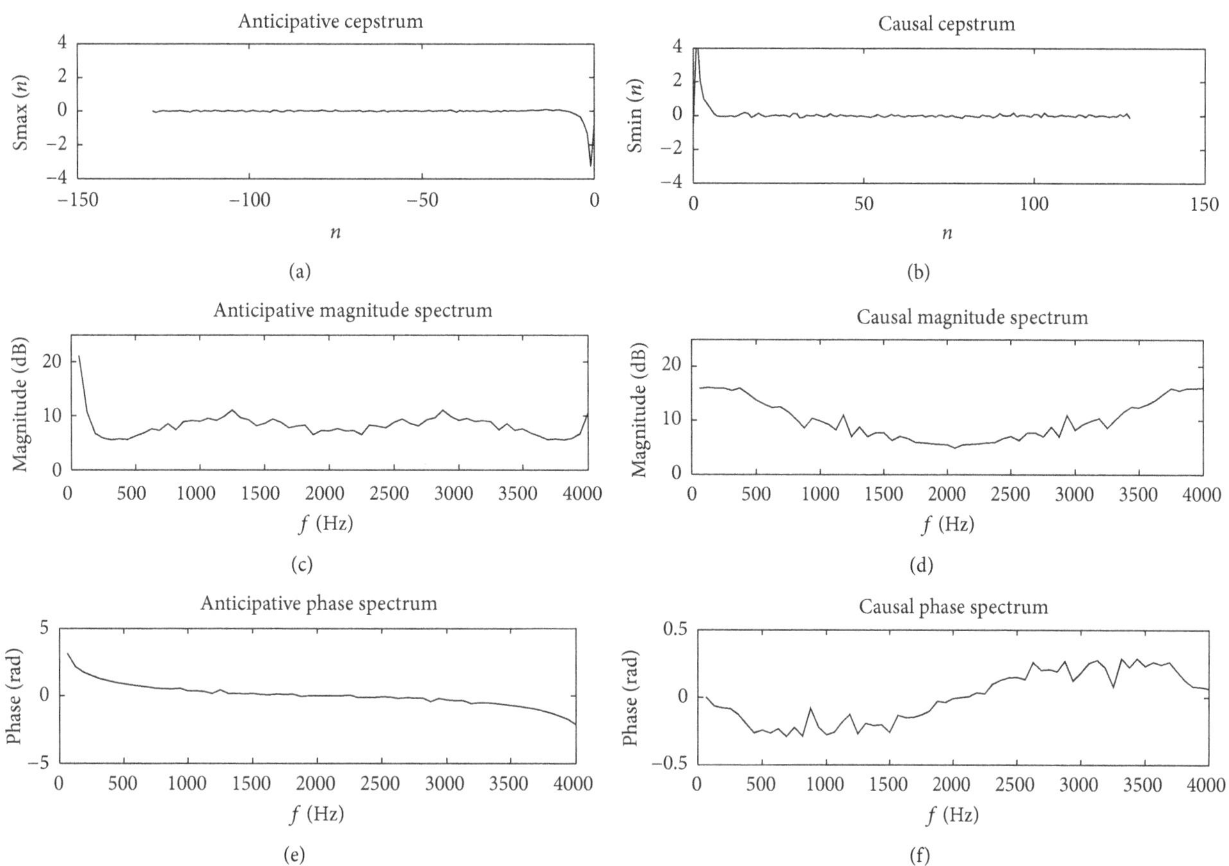

FIGURE 3: Anticausal and causal cepstrum with corresponding magnitude and phase spectrum.

3.2. Baseline Mel Cepstral Envelope Vocoder Based VC. Figure 5 depicts a block diagram of a VC system using baseline features. During the analysis step, the MCEPs are derived as spectral parameters and the fundamental frequency (F_0) is derived as excitation parameter for every 5 msec [10]. As discussed in the earlier section the feature sets obtained from the source and target speakers usually differ in time duration. Therefore, the source and target speaker's utterances are aligned using DTW. The feature set captures the joint distribution of source and target speaker using RBF to carry out VC. The excitation features (F_0) use the cepstrum method to calculate the pitch period for the frame size of 25 msec resulting into 25 MCEP features. Mean and standard deviation statistics are obtained from $\log(F_0)$ and used as feature set. In the testing phase, the parallel utterances of test speaker are used to obtain the feature vector with the procedure similar to that of the training set feature vector. In order to produce transformed feature vector, the test speaker feature vector is projected through the trained RBF model. In the synthesis stage, the transformed MCEP and F_0 are passed through the MLSA [10, 33, 35] filter. The postfiltering applied to the transformed speech signal ensures its high quality.

4. Radial Basis Function Based VC

The RBF is used to model the nonlinearity between the source and the target speaker feature vectors [11]. It is a special case of feed forward network which nonlinearly maps input space to hidden space followed by a linear mapping from a hidden space to the output space. The network represents a map from M_0 dimensional input space to N_0 dimensional output space written as $S : R_0^M \rightarrow R_0^N$. When a training dataset of input output pairs $[x_k, d_k]$; $k = 1, 2, \ldots, M_0$ is applied to the RBF model; the mapping function F is computed as

$$F_k(x) = w_{j0} + \sum_{j=1}^{m} w_{jk} \Phi\left(\left\| x - d_j \right\|\right), \tag{11}$$

where $\|\cdot\|$ is a norm usually Euclidian and computes the distance between applied input x and training data point d_j and $\Phi(\|x - d_j\|) \mid j = 1, 2 \ldots, m$ is the set of m arbitrary functions known as radial basis functions. The commonly considered form of Φ is Gaussian function defined as

$$\Phi(x) = \exp\left(\frac{\|x - \mu\|^2}{2\sigma^2}\right). \tag{12}$$

RBF neural network learning process includes training and generalized phase. The training phase constitutes the optimization of basis function parameters using input dataset to evaluate k-means algorithm in an unsupervised manner [11]. In the second phase, hidden-output neurons weight matrix is

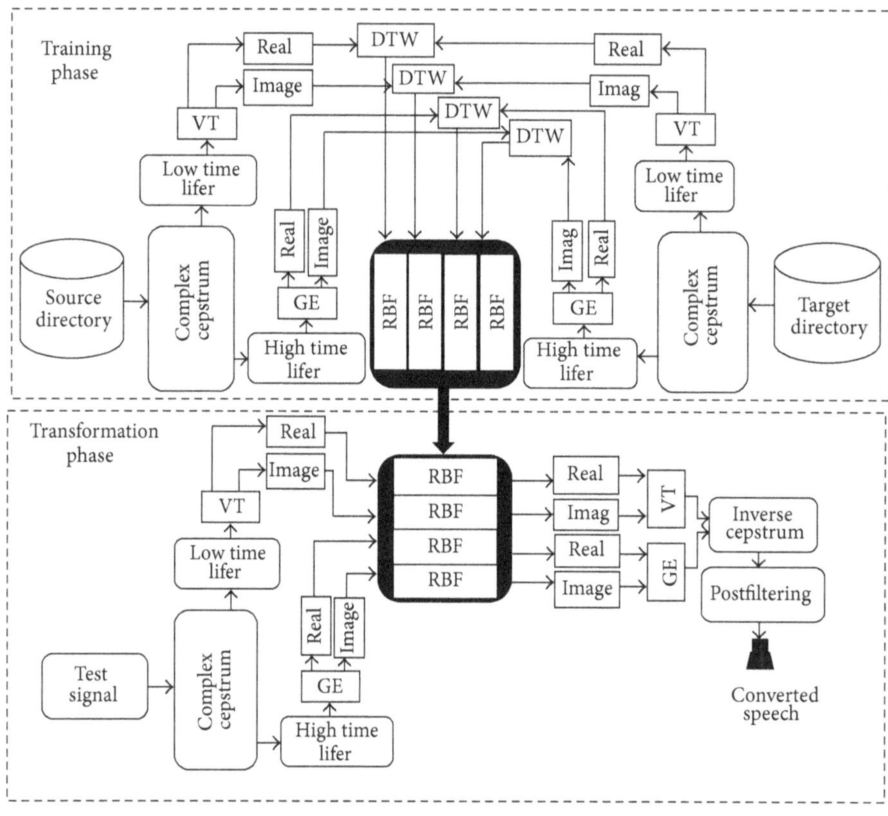

FIGURE 4: Functional block diagram of the complex cepstrum based VC.

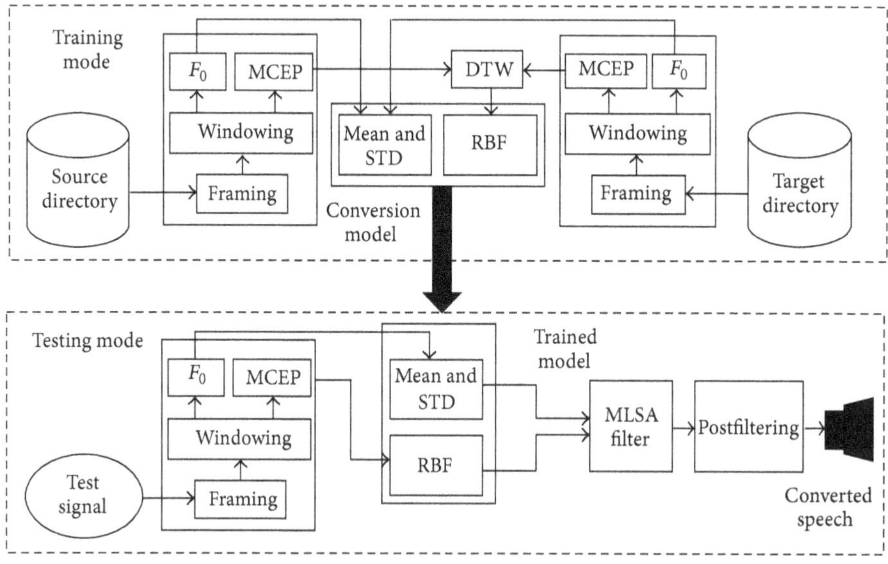

FIGURE 5: Block diagram of analysis and Synthesis of Mel cepstrum based VC system.

optimized by the least square sense to minimize the squared error function using the equation

$$E = \frac{1}{2}\sum_{n}\sum_{k}\left[f_k\left(\left(x^n\right) - \left(d_k\right)^n\right)\right]^2, \tag{13}$$

where $(d_k)^n$ is desired value for kth output unit when input to the network is x^n. The weight vector is determined as

$$W = \Phi^T D, \tag{14}$$

where Φ: matrix of size $(n \times j)$, D: matrix of size $(n \times k)$, and Φ^T: transpose of matrix Φ:

$$\left(\Phi^T\Phi\right)W = \Phi^T D,$$
$$W = \left(\Phi^T\Phi\right)^{-1}\Phi^T D, \tag{15}$$

where $(\Phi^T\Phi)^{-1}\Phi^T$ represents the pseudoinverse of matrix Φ and D denotes the target matrix for d_k^n. The weight matrix W can be calculated by linear inverse matrix technique and used for mapping between the source and target acoustic feature vector. The exact interpolation of RBF is acquainted with two serious problems, namely, (i) poor performance for noisy data and (ii) increased computational complexity. These problems can be addressed by modifying two RBF parameters. The first one is the spread factor which is calculated as

$$\sigma_j = 2 \times \text{avg}\left\{\left\|x - \mu_j\right\|\right\}. \tag{16}$$

The selected spread factor confirms that the individual RBFs are neither wide nor narrow. The second one is an extra bias unit which is introduced into the linear sum of activations at the desired output layer to compensate for the difference between the mean over the data set of the basis function activations and the corresponding mean of the targets. Hence, we achieve the RBF network for mapping as

$$F_k(x) = \sum_{j=0}^{m} w_{jk}\Phi\left(\left\|x - d_j\right\|\right). \tag{17}$$

In this work RBF neural networks are initialized and best networks are developed to obtain the mapping between the cepstral based acoustic parameters of the source and the target speakers. The trained networks are used to predict real and imaginary components of the vocal tract and source excitation of the target speaker's speech signal. In the baseline approach, the MCEP based feature matrices of the source and target utterances with the order of 25 are formed. Radial basis function is trained to obtain best mapping function. The best mapping function is obtained using RBF network and used to predict the MCEP parameters of the target speaker's speech signal.

5. Experimental Results

In this paper, the RBF based mapping functions are developed using CMU-ARCTIC corpus. The corpus consists of different sets of 1132 phonetically balanced parallel utterances of each speaker, sampled at 16 kHz. The corpus includes two female, that is, CLB (US Female) and SLT (US Female), and five different male such as AWB (Scottish Male), BDL (US Male), JMK (Canadian Male), RMS (US Male), and KSP (Indian Male) [36]. In this work, we have made use of the parallel utterances of the AWB (M1), CLB (F1), BDL (M2), and SLT (F2) with different speaker combinations like M1-F1, F2-M2, M1-M2 and F1-F2. For each of the speaker pairs 50 parallel sentences of source and target speakers are used for VC system training and system evaluations are made using a separate set of 25 source speaker sentences. The performance of homomorphic vocoder based VC system is compared with the state-of-the-art MCEP based VC system using different objective and subjective measures.

5.1. Objective Evaluation. The objective measures provide the mathematical analysis for determining the similarity index and quality inspection score between desired (target) and transformed speech signal. In this work, performance index, spectral distortion and formant deviation are considered as objective measures.

The performance index (P_{LSF}) is computed for investigating the requirement of normalized error for different pairs. The spectral distortion between desired and transformed utterances, $D_{\text{LSF}}(d(n), \hat{d}(n))$ and the interspeaker spectral distortion, $D_{\text{LSF}}(d(n), s(n))$ are used for computing the P_{LSF} measure. In general, the speaker spectral distortion between signals u and v, $D_{\text{LSF}}(u, v)$ is defined as

$$D_{\text{LSF}}(u, v) = \left[\frac{1}{N}\sum_{i=1}^{N}\sqrt{\frac{1}{P}\sum_{j=1}^{P}\left(\text{LSF}_u^{i,j} - \text{LSF}_v^{i,j}\right)^2}\right], \tag{18}$$

where N represents the number of frames, P refers to a LSF order, and $\text{LSF}_u^{i,j}$ is the jth LSF component in the frame i. The P_{LSF} measure is given as

$$P_{\text{LSF}} = \left[1 - \frac{D_{\text{LSF}}\left(d(n), \hat{d}(n)\right)}{D_{\text{LSF}}\left(d(n), s(n)\right)}\right]. \tag{19}$$

The performance index $P_{\text{LSF}} = 1$ indicates that the converted signal is identical to the desired one, whereas $P_{\text{LSF}} = 1$ specifies that the converted signal is not at all similar to the desired one.

In the computation of the performance index, four different converted samples of M1 to F1, F2 to M2, F1 to F2, and M1 to M2 combinations are considered. Comparative performance between cepstrum based VC algorithm and MCEP based VC is shown in Table 1. The results specified that the performance of the complex cepstrum based VC performed better than MCEP based VC algorithm.

Along with performance index, the different objective measures, namely, deviation (D_i), root mean square error (RMSE), and correlation coefficients ($\sigma_{x,y}$), are also calculated for different speaker pairs. Deviation parameter is defined as the percentage variation in the actual (x_k) and predicted (y_k) formant frequencies derived from the speech frames. It corresponds to the percentage of test frames within a specified deviation. Deviation (D_k) is calculated as

$$D_k = \frac{|x_k - y_k|}{x_k} \times 100. \tag{20}$$

TABLE 1: The performance index of complex cepstrum based VC and MCEP based VC.

Type of conversion	Sample 1 Cep. based VC	Sample 1 MCEP based VC	Sample 2 Cep. based VC	Sample 2 MCEP based VC	Sample 3 Cep. based VC	Sample 3 MCEP based VC	Sample 4 Cep. based VC	Sample 4 MCEP based VC
M1-F1	0.7679	0.6230	0.7483	0.6356	0.7127	0.6080	0.8350	0.6768
F2-M2	0.7389	0.5781	0.7150	0.6988	0.7780	0.6908	0.7848	0.6845
F1-F2	0.7921	0.6576	0.6908	0.5954	0.6740	0.6209	0.7946	0.6925
M1-M2	0.7023	0.6490	0.6821	0.6012	0.6432	0.5801	0.7852	0.7012

TABLE 2: Prediction performance of MCEP based for formant frequencies.

Transformation model	Formant frequencies	2%	5%	10%	15%	20%	25%	50%	μ_{RMSE}	$\Upsilon_{X,y}$
M1-F1	F1	51	74	80	81	83	85	90	4.45	0.7235
	F2	45	63	68	78	82	87	89	3.73	0.8182
	F3	57	62	79	86	87	89	92	3.34	0.8703
	F4	69	79	84	89	88	90	100	2.39	0.8629
F2-M2	F1	36	58	67	74	82	86	90	4.28	0.7190
	F2	57	82	86	87	87	89	91	6.30	0.7238
	F3	72	77	89	91	92	94	95	5.23	0.7474
	F4	66	74	89	90	93	95	100	4.91	0.7957

(Columns 2%–50%: % Predicted frame within deviation.)

The root mean square error is calculated as percentage of average of desired formant values obtained from the speech segments:

$$\mu_{\text{RMSE}} = \frac{\sqrt{\sum_k |x_k - y_k|^2}}{\bar{x}} \times 100,$$

$$\sigma = \sqrt{\sum_k d_k^2}, \quad d_k = e_k - \mu, \tag{21}$$

$$e_k = x_k - y_k, \quad \mu = \frac{\sum_k |x_k - y_k|}{N}.$$

The error e_k is the difference between the actual and predicted formant values. N is the number of observed formant values of speech frames. The parameter d_k is the error in the deviation. The correlation coefficient $\Upsilon_{X,y}$ is the parameter which is to be determined from the covariance $\text{COV}(X,Y)$ between the target (x) and the predicted (y) formant values and the standard deviations σ_X, σ_Y of the target and the predicted formant values, respectively. The parameters $\Upsilon_{X,y}$ and $\text{COV}(X,Y)$ are calculated using

$$\Upsilon_{X,y} = \frac{\text{COV}(X,Y)}{\sigma_X \sigma_Y},$$

$$\text{COV}(X,Y) = \frac{\sum_k |(x_k - \bar{x})(y_k - \bar{y})|}{N}. \tag{22}$$

The objective measures, namely, deviation (D_i), root mean square error (RMSE), and correlation coefficients $(\Upsilon_{X,y})$ of M1-F1 and F2-M2 are obtained for MCEP based VC algorithm and shown in Table 2. Similarly, the Table 3 shows the measures obtained for proposed VC system. From the tables it can be observed that the μ_{RMSE} between the desired and the predicted acoustic space parameters for proposed model are less than the baseline model. However, every time RMSE does not give strong information about the spectral distortion. Consequently, scatter plots and spectral distortion are employed additionally as objective evaluation measures. The scatter plots for first, second, third, and fourth formant frequencies for MCEP based VC and complex cepstrum based VC models are shown in Figures 6 and 7, respectively. Figures show that complex cepstrum VC based vocal tract envelope in term of predicted formants closely orient towards the desired speech frames formants as compared to MCEP based predicted formants. The clusters obtained using complex cepstrum based VC are more compact and diagonally oriented than that using MCEP based VC. As perfect prediction means all the data points in scatter plot are diagonally oriented in right side. The compact clusters obtained for proposed method implies its ability to capture the formant structure of desired speaker

The transformed formant patterns for a specific frame of source and target speech signal are obtained using both complex cepstrum and MCEP based VC models and shown in Figures 8(a) and 8(b), respectively. Figure 8(a) depicts that the patterns of particular target signal closely follows the corresponding transformed signal, whereas Figure 8(b) shows that the predicted formant pattern closely follows the target pattern only for lower formants.

Figure 9(a) shows the normalized frequency spectrogram of desired and transformed speech signals obtained from M1 to F1 and F2 to M2 of complex cepstrum based VC model. Similarly, Figure 9(b) shows the spectrogram for M1 to F1 and

Table 3: Prediction performance of complex cepstrum based for formant frequencies.

Transformation model	Formant frequencies	% Predicted frame within deviation 2%	5%	10%	15%	20%	25%	50%	μ_{RMSE}	$\Upsilon_{X,y}$
M1-F1	F1	59	80	89	91	91	93	95	4.45	0.7197
	F2	52	72	85	88	91	92	95	3.55	0.8149
	F3	63	83	90	92	93	95	99	2.65	0.8837
	F4	72	86	91	93	95	97	100	2.049	0.8909
F2-M2	F1	38	60	71	78	80	86	90	8.56	0.757
	F2	60	82	89	92	92	93	98	3.62	0.790
	F3	72	87	92	95	95	95	100	2.93	0.778
	F4	70	86	91	96	97	99	100	2.41	0.756

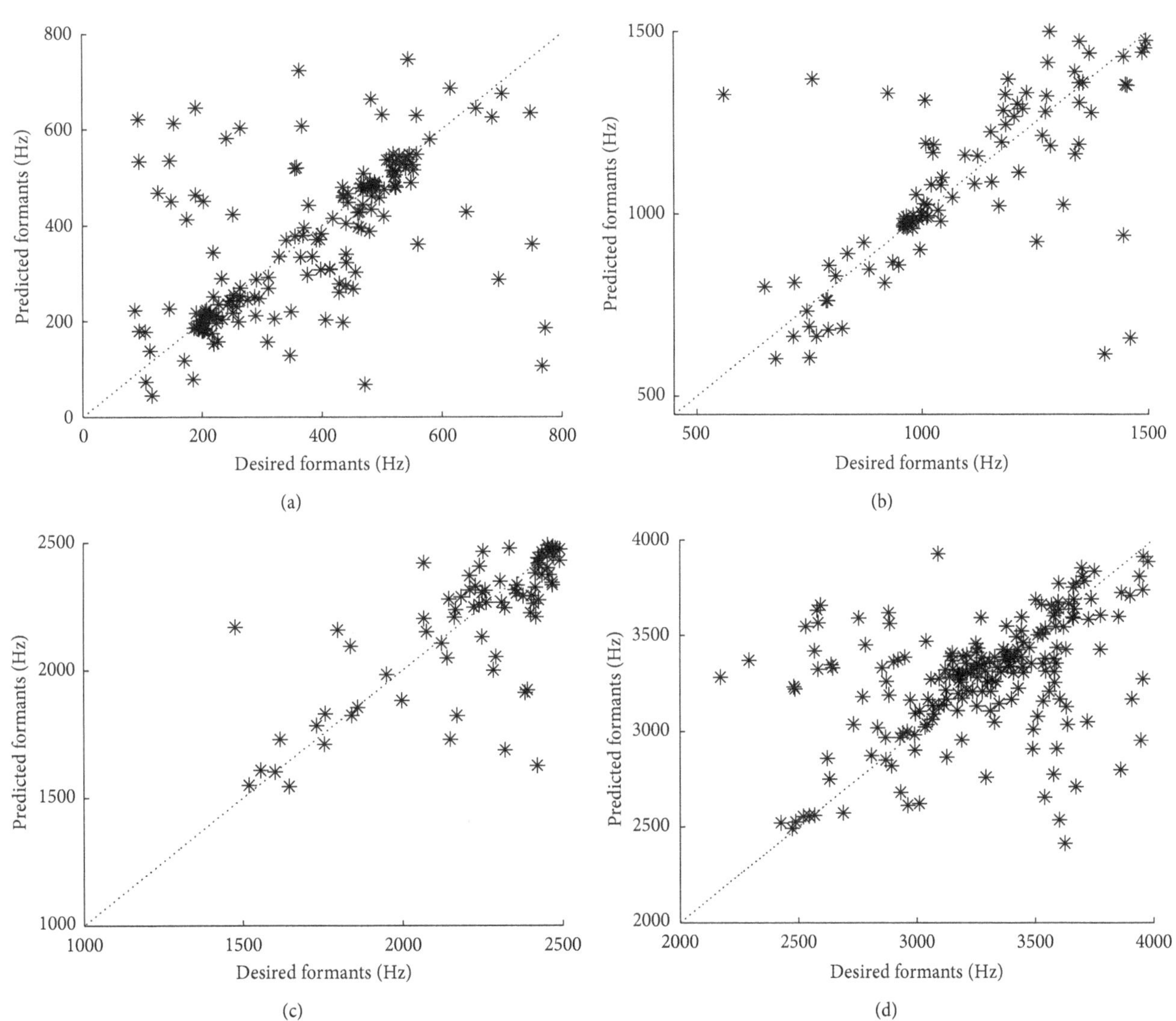

Figure 6: Desired and predicted formant frequencies for F2 to M2 VC using MCEP based approach (a) first formant, (b) second formant, (c) third formant, and (d) fourth formant.

F2 to M2 for the MCEP based VC model. It has been observed that the dynamics of the first three formant frequencies in both the algorithms are closely followed in the target and the transformed speech samples.

5.2. Subjective Evaluation. The effectiveness of the algorithm is also evaluated using listening tests. These subjective tests are used to determine the closeness between the transformed and target speech sample. The mapping functions are developed using 50 parallel utterances of the source and target speakers. Twenty-five different synthesized speech utterances are obtained from the mapping function for inter- and intragender speech conversion and corresponding target utterances are presented to twelve listeners. They are asked to

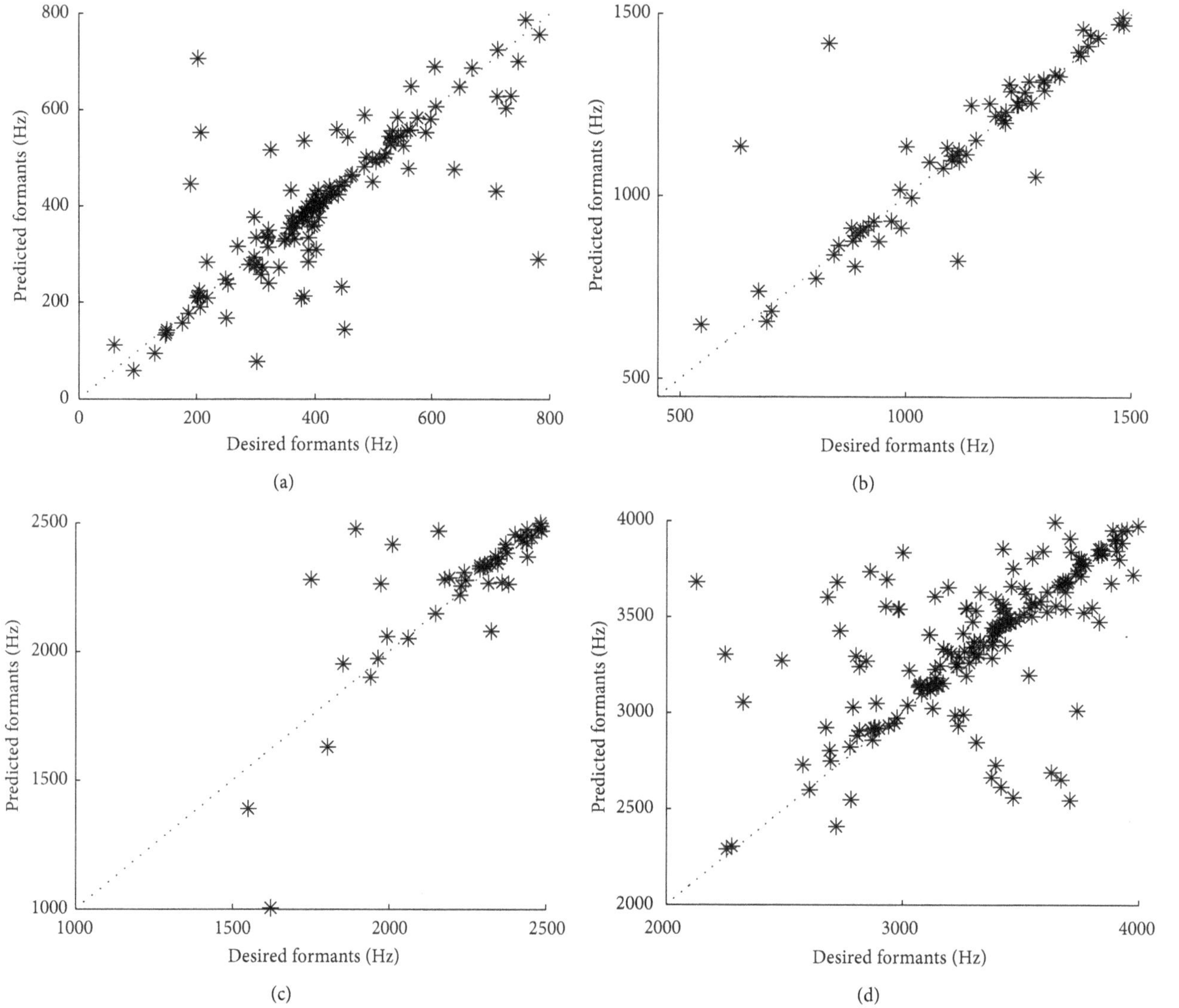

Figure 7: Desired and predicted formant frequencies for F2 to M2 VC using the complex cepstrum based (a) first formant, (b) second formant, (c) third formant, and (d) fourth formant.

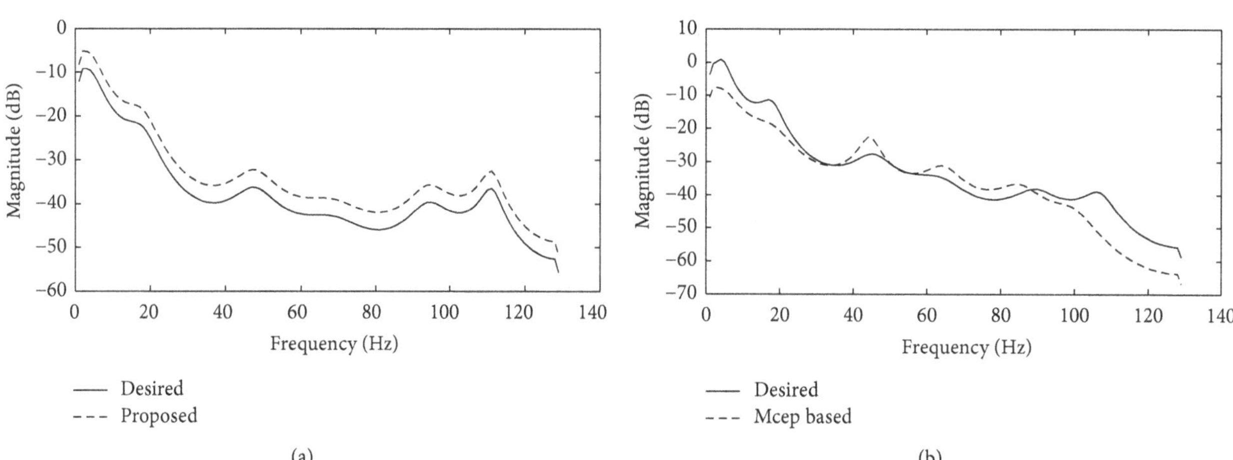

Figure 8: Target and transformed spectral envelopes of the desired speaker using (a) complex cepstrum based VC and (b) MCEP based VC.

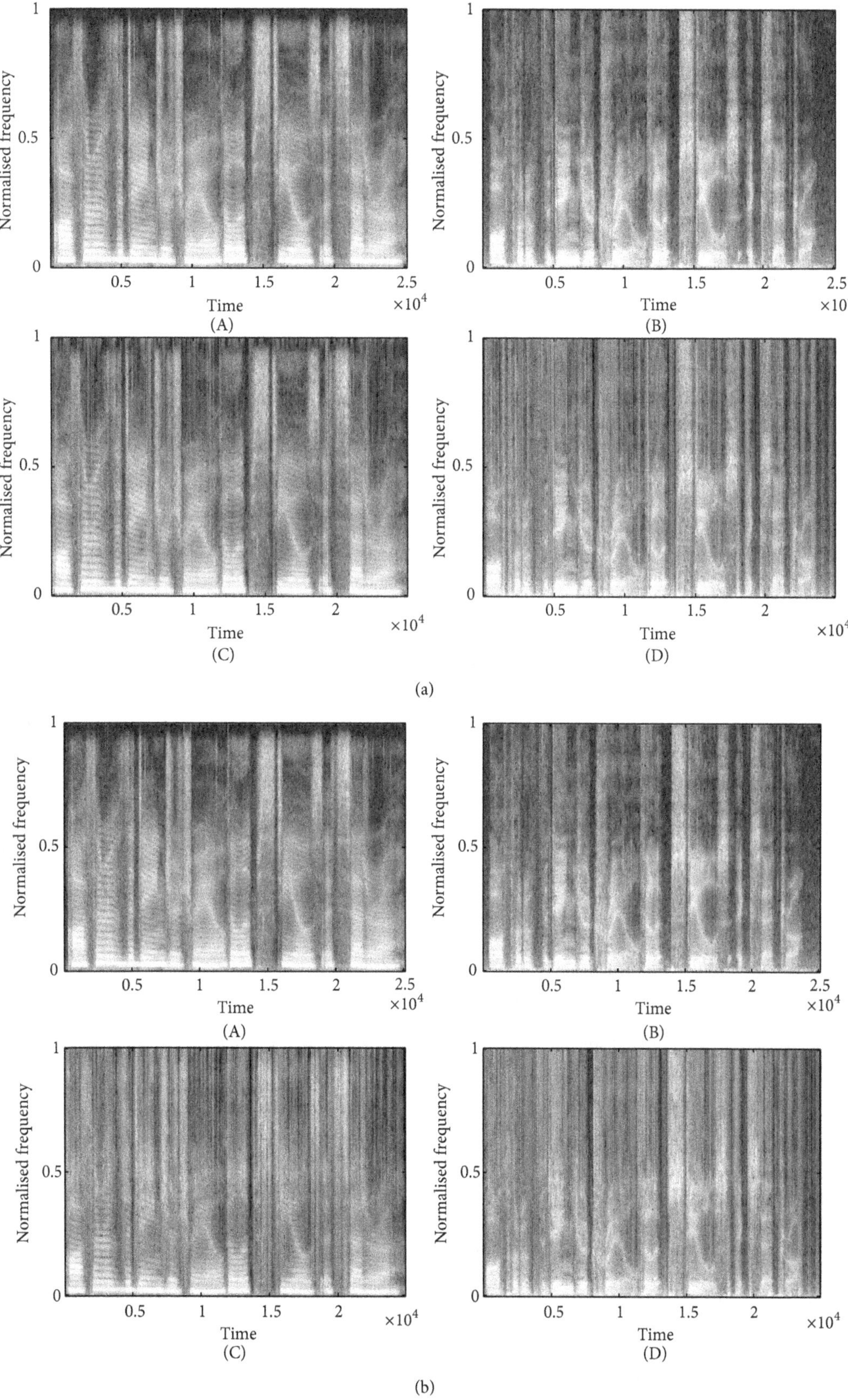

FIGURE 9: Spectrogram of the desired and the transformed signal for M1 to F1 ((A) to (C)) and F2 to M2 ((B) to (D)) using (a) complex cepstrum based VC and (b) MCEP based VC.

Table 4: MOS and ABX evaluations of complex cepstrum and MCEP based VC models.

Conversion data	MOS		ABX	
	Cepstrum based	MCEP based	Cepstrum based	MCEP based
M1-F1	4.64	4.31	4.55	4.25
F2-M1	4.18	3.92	4.34	4.23
M1-M2	4.07	3.88	4.19	4.06
F1-F2	4.24	3.76	4.36	4.13

evaluate their relative performance in term of voice quality (MOS) and speaker identity (ABX) with corresponding source and target speaker speech samples on a scale of 1 to 5, where rating 5 specifies an excellent match between the transformed and target utterances, rating 1 indicates a poor match, and the other ratings indicate different levels of variation between 1 and 5. The ratings given to each set of utterances are used to calculate the MOS for different speaker combinations like M1 to F1, M1 to M2, F1 to F2, and F2 to M2; the results are presented in Table 4. The dissimilarity in the length of the vocal tract and the intonation patterns of different genders is the major reason for variation in the MOS results for source and target utterances of different genders. The ABX (A: Source, B: Target, X: Transformed speech signal) test is also performed using the same set of utterances and speakers. In the ABX test, the listeners are asked to judge whether the unknown speech sample X sounds closer to the reference sample A or B. The ABX is a measure of identity transformation. The higher value of ABX percentage indicates that the transformed speech lies in close proximity of the target utterance. The results of the ABX test are also shown in Table 4.

6. Conclusion

The VC algorithm comprising of complex cepstrum, that preserves the phase related information content of the synthesized speech outcome, is presented. A mixed phase system is designed to yield far better transformed speech signal than the minimum phase systems. The vocal tract and excitation parameters of the speech signal are obtained with the help of low and high time liftering. Radial basis functions are explored to capture the nonlinear mapping function for modifying the real and imaginary parts of the vocal tract and source excitations of the source speaker speech to that of the target speaker speech. In baseline VC algorithm MCEP method is used to interpret the vocal tract whereas, the fundamental frequency (F_0) represent the source excitation. The RBF based mapping function is used to capture the nonlinear relationship between the MCEP of the source speaker to that of the target speaker and statistical mean and standard deviation is used for transformation of fundamental frequency. The proposed complex cepstrum based VC is compared with the MCEP based VC using various objective and subjective measures. The evaluation results reveal that the complex cepstrum based VC performs slightly better than the MCEP based VC model in term of speech quality and speaker identity. The reason may be the fluctuation of MLSA filter parameters with limited margins in Padé approximation. It may be unstable momentarily, when the parameters vary rapidly by contrast the complex cepstrum with finite impulse response is always stable.

Conflict of Interests

The authors declare that there is no conflict of interests regarding the publication of this paper.

References

[1] H. Kuwabara and Y. Sagisak, "Acoustic characteristics of speaker individuality: control and conversion," *Speech Communication*, vol. 16, no. 2, pp. 165–173, 1995.

[2] K.-S. Lee, "Statistical approach for voice personality transformation," *IEEE Transactions on Audio, Speech and Language Processing*, vol. 15, no. 2, pp. 641–651, 2007.

[3] A. Kain and M. W. Macon, "Design and evaluation of a voice conversion algorithm based on spectral envelope mapping and residual prediction," in *Proceedings of the IEEE Interntional Conference on Acoustics, Speech, and Signal Processing*, vol. 2, pp. 813–816, May 2001.

[4] D. Sundermann, "Voice conversion: state-of-the-art and future work," in *Proceedings of the 31st German Annual Conference on Acoustics (DAGA '01)*, Munich, Germany, 2001.

[5] D. G. Childers, B. Yegnanarayana, and W. Ke, "Voice conversion: factors responsible for quality," in *Proceeding of the IEEE International Conference on Acoustics, Speech, and Signal Processing (ICASSP '85)*, vol. 1, pp. 748–751, Tampa, Fla, USA, 1985.

[6] M. Abe, S. Nakanura, K. Shikano, and H. Kuwabara, "Voice conversion through vector quantization," in *Proceeding of the IEEE International Conference on Acoustics, Speech and Signal Processing*, pp. 655–658, 1988.

[7] W. Verhelst and J. Mertens, "Voice conversion using partitions of spectra feature space," in *Proceedings of the IEEE International Conference on Acoustics, Speech, and Signal Processing (ICASSP '96)*, pp. 365–368, May 1996.

[8] N. Iwahashi and Y. Sagisaka, "Speech spectrum conversion based on speaker interpolation and multi-functional representation with weighting by radial basis function networks," *Speech Communication*, vol. 16, no. 2, pp. 139–151, 1995.

[9] Y. Stylianou, O. Cappé, and E. Moulines, "Continuous probabilistic transform for voice conversion," *IEEE Transactions on Speech and Audio Processing*, vol. 6, no. 2, pp. 131–142, 1998.

[10] S. Desai, A. W. Black, B. Yegnanarayana, and K. Prahallad, "Spectral mapping using artificial neural networks for voice conversion," *IEEE Transactions on Audio, Speech and Language Processing*, vol. 18, no. 5, pp. 954–964, 2010.

[11] C. Orphanidou, I. M. Moroz, and S. J. Roberts, "Wavelet-based voice morphing," *WSEAS Journal of Systems*, vol. 10, no. 3, pp. 3297–3302, 2004.

[12] E. Helander, T. Virtanen, N. Jani, and M. Gabbouj, "Voice conversion using partial least squares regression," *IEEE Transcation on Audio, Speech, Language Processing*, vol. 18, no. 5, pp. 912–921, 2010.

[13] L. M. Arslan, "Speaker transformation algorithm using segmental codebooks (STASC)," *Speech Communication*, vol. 28, no. 3, pp. 211–226, 1999.

[14] K. S. Rao, "Voice conversion by mapping the speaker-specific features using pitch synchronous approach," *Computer Speech and Language*, vol. 24, no. 3, pp. 474–494, 2010.

[15] S. Hayakawa and F. Itakura, "Text-dependent speaker recognition using the information in the higher frequency band," in *Proceedings of the International Conference on Acoustics, Speech, and Signal Processing (ICASSP '94)*, pp. 137–140, Adelaide, Australia, 1994.

[16] H. Kawahara, I. Masuda-Katsuse, and A. de Cheveigné, "Restructuring speech representations using a pitch-adaptive time-frequency smoothing and an instantaneous-frequency-based F0 extraction: possible role of a repetitive structure in sounds," *Speech Communication*, vol. 27, no. 3, pp. 187–207, 1999.

[17] R. J. McAulay and T. F. Quatieri, "Phase modelling and its application to sinusoidal transform coding," in *Proceedings of the IEEE International Conference on Acoustics, Speech, and Signal Processing (ICASSP '86)*, vol. 1, pp. 1713–1716, Tokyo, Japan, 1986.

[18] J. Nirmal, P. Kachare, S. Patnaik, and M. Zaveri, "Cepstrum liftering based voice conversion using RBF and GMM," in *Proceedings of the IEEE International Conference on Communications and Signal Processing (ICCSP '13)*, pp. 570–575, April 2013.

[19] A. V. Oppenheim, "Speech analysis and synthesis system based on homomorphic filtering," *The Journal of the Acoustical Society of America*, vol. 45, no. 2, pp. 458–465, 1969.

[20] W. Verhelst and O. Steenhaut, "A new model for the short-time complex cepstrum of voiced speech," *IEEE Transactions on Acoustics, Speech, and Signal Processing*, vol. 34, no. 1, pp. 43–51, 1986.

[21] H. Deng, R. K. Ward, M. P. Beddoes, and M. Hodgson, "A new method for obtaining accurate estimates of vocal-tract filters and glottal waves from vowel sounds," *IEEE Transactions on Audio, Speech and Language Processing*, vol. 14, no. 2, pp. 445–455, 2006.

[22] T. Drugman, B. Bozkurt, and T. Dutoit, "Complex cepstrum-based decomposition of speech for glottal source estimation," in *Proceedings of the 10th Annual Conference of the International Speech Communication Association (INTERSPEECH '09)*, pp. 116–119, Brighton, UK, September 2009.

[23] T. F. Quatieri Jr., "Minimum and mixed phase speech analysis-synthesis by adaptive homomorphic deconvolution," *IEEE Transactions on Acoustics, Speech, and Signal Processing*, vol. 27, no. 4, pp. 328–335, 1979.

[24] T. Drugman, B. Bozkurt, and T. Dutoit, "Causal-anticausal decomposition of speech using complex cepstrum for glottal source estimation," *Speech Communication*, vol. 53, no. 6, pp. 855–866, 2011.

[25] M. Vondra and R. Vích, "Speechmodeling using the complex cepstrum," in *Toward Autonomous, Adaptive, and Context-Aware Ultimodal Interfaces: Theoretical and Practical, Issues*, vol. 6456 of *Lecture Notes in Computer Science*, pp. 324–330, 2011.

[26] R. Maia, M. Akamine, and M. Gales, "Complex cepstrum as phase information in statistical parametric speech synthesis," in *Proceedings of the IEEE International Conference on Acoustics, Speech, and Signal Processing (ICASSP '12)*, pp. 4581–4584, 2012.

[27] K. Shikano, S. Nakamura, and M. Abe, "Speaker adaptation and voice conversion by codebook mapping," in *Proceedings of the IEEE International Symposium on Circuits and Systems*, pp. 594–597, June 1991.

[28] T. Toda, H. Saruwatari, and K. Shikano, "Voice conversion algorithm based on Gaussian mixture model with dynamic frequency warping of straight spectrum," in *Proceedings of the IEEE Interntional Conference on Acoustics, Speech, and Signal Processing*, pp. 841–844, May 2001.

[29] H. Ye and S. Young, "High quality voice morphing," in *Proceedings of the IEEE International Conference on Acoustics, Speech, and Signal Processing*, pp. I9–I12, May 2004.

[30] R. Laskar, K. Banerjee, F. Talukdar, and K. Sreenivasa Rao, "A pitch synchronous approach to design voice conversion system using source-filter correlation," *International Journal of Speech Technology*, vol. 15, pp. 419–431, 2012.

[31] D. Sündermann, A. Bonafonte, H. Ney, and H. Höge, "A study on residual prediction techniques for voice conversion," in *Proceedings of the IEEE International Conference on Acoustics, Speech, and Signal Processing (ICASSP '05)*, pp. I13–I16, March 2005.

[32] K. S. Rao, R. H. Laskar, and S. G. Koolagudi, "Voice transformation by mapping the features at syllable level," in *Pattern Recognition and Machine Intelligence*, vol. 4815 of *Lecture Notes in Computer Science*, pp. 479–486, Springer, 2007.

[33] http://sp-tk.sourceforge.net/.

[34] S. Imai, "Cepstral analysis and synthesis on the mel-frequency scale," in *Proceedings of the IEEE International Conference on Acoustics, Speech, and Signal Processing (ICASSP '83)*, pp. 93–96, Boston, Mass, USA, 1983.

[35] K. Tokuda, T. Kobayashi, T. Masuko, and S. Imai, "Mel-generalized cepstral analysis—a unified approach to speech spectral estimation," in *Proceedings of the International Conference on Spoken Language Processing (ICSLP '94)*, pp. 1043–1046, 1994.

[36] J. Kominek and A. W. Black, "CMU ARCTIC speech databases," in *Proceedings of the 5th ISCA Speech Synthesis Workshop*, pp. 223–224, Pittsburgh, Pa, USA, June 2004.

15

Weighted Least Squares Based Detail Enhanced Exposure Fusion

Harbinder Singh,[1] Vinay Kumar,[2] and Sunil Bhooshan[3]

[1] *Department of Electronics and Communication Engineering, Baddi University of Emerging Sciences and Technology, District Solan, Himachal Pradesh 173205, India*
[2] *Grupo de Procesado Multimedia, Departamento de Teoría de la Señal y Comunicaciones, Universidad Carlos III de Madrid Leganes, Madrid 28911, Spain*
[3] *Department of Electronics and Communication Engineering, Jaypee University of Information Technology, Waknaghat, District Solan, Himachal Pradesh 173215, India*

Correspondence should be addressed to Harbinder Singh; harbinder.ece@gmail.com

Academic Editors: A. Fernandez-Caballero, H. Hu, C. S. Lin, and E. Salerno

Many recent computational photography techniques play a significant role to avoid limitation of standard digital cameras to handle wide dynamic range of the real-world scenes, containing brightly and poorly illuminated areas. In many of these techniques, it is often desirable to fuse details from images captured at different exposure settings, while avoiding visual artifacts. In this paper we propose a novel technique for exposure fusion in which Weighted Least Squares (WLS) optimization framework is utilized for weight map refinement. Computationally simple texture features (i.e., detail layer extracted with the help of edge preserving filter) and color saturation measure are preferred for quickly generating weight maps to control the contribution from an input set of multiexposure images. Instead of employing intermediate High Dynamic Range (HDR) reconstruction and tone mapping steps, well-exposed fused image is generated for displaying on conventional display devices. A further advantage of the present technique is that it is well suited for multifocus image fusion. Simulation results are compared with a number of existing single resolution and multiresolution techniques to show the benefits of the proposed scheme for variety of cases.

1. Introduction

In recent years several new techniques have been developed that are capable of providing precise representation of complete information of shadows and highlights present in the real-world natural scenes. The direct 8-bit gray and 24-bit RGB representation of visual data, with the standard digital cameras in single exposure settings, often causes loss of information in the real-world scenes because the dynamic range of most scenes is beyond what can be captured by the standard digital cameras. Such representation is referred to as low dynamic range (LDR) image. Digital cameras have the aperture setting, exposure time, and ISO value to regulate the amount of light captured by the sensors. It is therefore important to somehow determine exposure setting for controlling charge capacity of the Charge Coupled Device (CCD). In modern digital cameras, Auto Exposure Bracketing (AEB) allows us to take all the images without touching the camera between exposures, provided the camera is on a tripod and a cable release is used. Handling the camera between exposures can increase the chance of missalignment resulting in an image that is not sharp or has ghosting. However, most scenes can be perfectly captured with nine exposures [1], whereas many more are within reach of a camera that allows 5–7 exposures to be bracketed. When the scene's dynamic range exceeds the dynamic range (DR) of camera it is exposure setting that determines which part of the scene will be optimally exposed in the photographed image. The DR of a digital camera is typically defined as the charge capacity divided by the noise [1, 2]. At single exposure setting, either detail in the poorly illuminated area (i.e., shadows) is visible with long exposure or brightly illuminated area (i.e., highlights) with short exposure (see Figure 1). Thus, image captured by the standard digital camera at single exposure setting from a scene containing highlights and shadows is partially over- or underexposed. As a result,

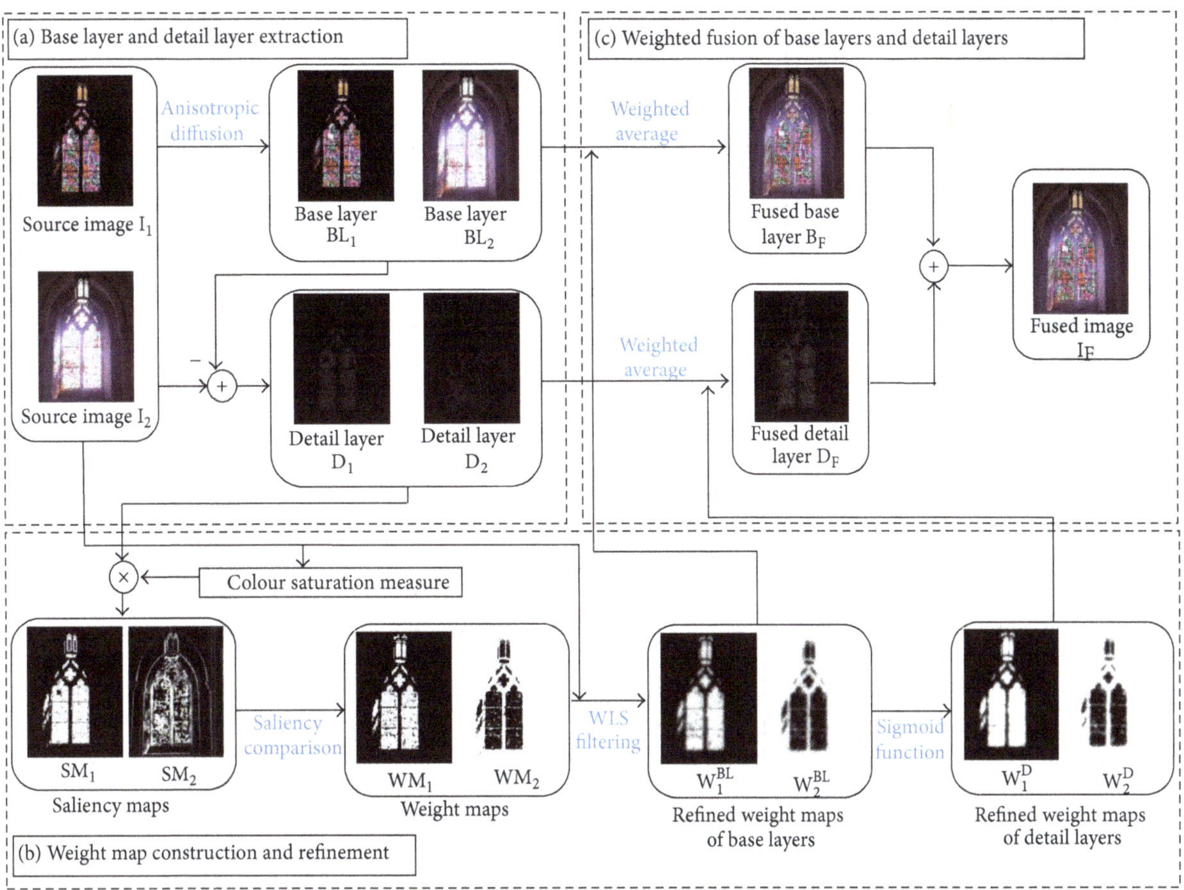

FIGURE 1: Illustration of proposed framework consisting of three principal blocks. (a) Base layer and detail layer extraction. The input images are transformed into two-scale decomposition. (b) Weight map construction and refinement. (c) Weighted fusion of base layers and detail layers. The base layers and detail layers across input image series are fused using simple weighted average approach.

there will always be a need to capture the detail of the entire scene with a sufficient number and value of exposures. The process of collecting complete luminance variations in rapid successions at different exposure settings is known as exposure bracketing.

In principle, there are two major approaches to handle the incapability of existing image capturing devices. The first approach is to develop HDR [3–7] reconstruction from multiexposure images that reconstruct full dynamic range up to 8 orders of magnitude and later tone map these images to adjust their tonal range, to some extent, for depiction on typical display devices. HDR [3–7] imaging is called scene-referred representation which represents the original captured scene values as close as possible. Such representation is sometimes referred to as extrasensory data representation. One of the important applications of HDR capturing techniques for security application is capturing video at entrance of the buildings [1]. Conventional cameras are not able to faithfully capture the interior and exterior of a building simultaneously while HDR camera, which is based on two-phase workflow, would be able to simultaneously record indoor as well as outdoor activities. Other important applications of HDR representation are satellite, scientific, and medical imagery, in which data is analyzed and visualized to record more than what is visible to the naked eye. On the other hand, because of limited contrast ratio, standard displays (LCD, CRT) and printers are unable to reproduce full dynamic range captured by the HDR devices. In such cases, HDR data needs to be remapped [7] with a lower precision for display on conventional devices. Tone mapping algorithms can be either spatially variant or spatially invariant. In particular, spatially variant methods (also called local operators) [7–10] exploit local adaptation properties of human visual system (HVS), while spatially invariant methods [11–13] exploit global adaptation (also called global operators) of HVS.

Higher bit depths are usually not used because the display devices would not be able to reproduce such images at levels that are practical for human viewing [1]. Although for some real-world scenes low bit depth is sufficient to capture entire detail, there are countless situations that are not accommodated by low bit depth. Although HDR display devices will be developed in the near future, conventional printers may lead to inconsistencies which will be responsible for loss of details in the output. Recently Sunnybrook technologies, BrightSide, and Dolby prototypes of HDR display devices have been proposed [1, 14, 15] that can display HDR data directly. As a result, to avoid these inconsistencies, we must use tone mapping operators [7–13] to prepare HDR imagery

(a) (b) (c)

Figure 2: Proposed results for different input multiexposure sequences. (a) Carnival (top: three input exposures, bottom: fusion results), (b) house (top: four input exposures, bottom: fusion results), and (c) bellavita (top: two input exposures, bottom: fusion results). Input image sequences are courtesy of HDRsoft.com, Tom Mertens.

for display on LDR devices. Alternatively, we may directly generate 8-bit low dynamic range (LDR) image that looks like a tone-mapped image [1].

The second approach for the purpose is combining multiexposure images directly into 8-bit single LDR image that does not contain underexposed and overexposed regions [18, 26]. Thus it provides convenient and consistent way for preserving details in both brightly and poorly illuminated areas by skipping the construction of HDR image and the use of tonemapping operators [7–13]. The incorporation of the notion of combining multiple exposures without typical HDR and tone mapping steps is known as "exposure fusion," as shown in Figure 1. The fundamental goal of exposure fusion is often to improve the chance of creating a realistic scene without HDRI representation and tone mapping step. The underlying idea of various exposure fusion approaches [18, 26] is based on the utilization of different local measures for generating weight map to preserve details present in the different exposures. The current manuscript belongs to the second approach. The block diagrammatic representation of the present detail enhanced framework is shown in Figure 1. We have used edge preserving filter based on partial differential equations (PDE) [27] for two-scale decomposition that separates sharp details and fine details across various input images with different exposure levels. The current state-of-the-art method for automatic exposure fusion exploits the capability of edge preserving filter [27, 28] to generate weight function that guides the fusion of different exposures based on two-scale decomposition. We propose WLS filter [20] optimization framework and sigmoid function for weight map refinement of base layers and detail layers, respectively. Farbman et al. [20] has utilized WLS filter to construct a multiscale edge preserving decomposition multiscale tone and detail manipulation. To achieve the optimal contrast in the fused image the current paper develops an appropriate mask based on weak textures and color saturation measure to composite multiexposure images. The method is applicable for the fusion of broad range of textured images. See Figure 2 for an example of our exposure fusion results for typical scene contain artificial light source (i.e., highlights), shadows, reflections, indoor details, and outdoor details.

Texture features [29] refer to the characterization of regions in an image by their spatial arrangement of color or intensities. Image textures are one way that can be used to help in classification of images [30]. Weak edges or texture information are the ideal indicators to detect over (or under) exposed regions in the image [17]. Raman and Chaudhuri [17] employ a Bilateral Filter (BLT) for compositing multi-exposure images, in which weak edges were considered to design weight map. Interestingly, thus an analysis of weak textures seems to be the definition of perceived contrast. We take advantage of such possibility and design the appropriate matting function based on anisotropic diffusion for exposure fusion.

To analyze an image texture, there are primarily two approaches: structural approach and statistical approach. Structural approach uses a set of primitive texture elements in some regular or repeated pattern to characterize spatial relationship. While statistical approach defines an image texture as a quantitative measure of the arrangement of intensities in a region. In general later approach is easier to compute and is more widely used in computer graphics applications, since natural textures are made of patterns of irregular subelements. It has been noticed that simple averaging to fuse details from multiexposure image data yields low contrast in the fused image, especially in brightly and poorly illuminated areas. In the present approach, texture details will decide the contribution of corresponding pixel from different exposures in the fused image. A rich texture details mean a maximum contribution, which tells that image block has higher weight during the fusion process. Such metric is used to quantify the perceived local contrast of an image under different exposure settings and allows discarding underexposed and overexposed pixels. Therefore, to handle underexposed and overexposed regions, we propose a texture feature analysis based on Anisotropic Diffusion (ANI) [27, 28] that has the applicability to design weighting function as shown in Figure 1. Our goal is to exploit the edge preserving property of ANI to produce well-exposed image from input images captured under different exposure settings. The detailed description of ANI based two-layer decomposition and weight map computation is given in the later sections. Our main contributions in this paper are highlighted follows.

(1) Two-scale decomposition based on anisotropic diffusion is proposed for fast exposure fusion, which does not require optimization of number of scales as required in the traditional multiscale techniques.

(2) A novel weight construction approach is proposed to combine texture features and saturation measure for guiding image fusion process. For weight map construction, we seek to utilize the strength of texture details under the change of exposure setting that takes place between an underexposed and an overexposed image. WLS filtering is proposed for weight refinement. Furthermore, fast sigmoid function based weight map generation for detail layers is proposed that reduce computational complexity of the algorithm.

(3) The important contribution of this paper is the advantages including ease of implementation, quality of compositing, and the provision of detail layer enhancement without introducing artifacts.

The remainder of this paper is structured as follows. Section 2 discusses the current available literature. Section 3 discusses description of separation of large scale variations and smaller scale details (i.e., texture details) based on ANI, consideration of smaller scale details and saturation measure for weight map generation, and the WLS and sigmoid based weight map refinement that produce single well-exposed image using simple weighted average approach. Section 4 discusses the utility of proposed approach for multifocus image fusion and the comparison with the popular single resolution exposure fusion, multiresolution exposure fusion and popular tone mapping operators. Section 5 summarizes the paper with future directions and conclusion.

2. Previous Work

2.1. HDR Imaging. There is a tremendous need to record a much wider gamut than standard 24 bit RGB. The practice of assembling HDR image from multiple exposure images recovers true radiance value present in the real-world scenes [2]. The camera response function recovered from differently exposed images is used to create HDR image whose pixel values are equivalent to the true radiance value of a scene. Radiance maps are stored in a file format that can encode recovered HDR data without losing information. "Floating point tiff" format tends to encode dynamic range up to 79 orders of magnitude and has better precision than the radiance format. Reinhard et al. [1] has provided the description and evaluation of formats available to store true radiance values. However, the success of HDR image capture has shown that it is possible to produce an image that exhibits details in poorly and brightly illuminated areas. Moreover, HDR formats have since found widespread applications in the computer graphics and HDR photography.

The prototypes of HDR display devices provide direct HDR display capabilities by means of a projector or Light Emitting Diode (LED) array that lights the Liquid Crystal Diode (LCD) from behind with a spatially varying light pattern [14, 15]. Unfortunately, conventional display devices (i.e., CRT and flat panel display) have dynamic ranges spanning a few orders of magnitude, much lower than those of the real-world scenes, often less than 100 : 1. In order to display HDR images on monitors or print them on paper [31], we must remap the dynamic range of the HDR images to reproduce low dynamic range (LDR) images suitable for human visual system (HVS). In the literature, several tone mapping methods for converting real-world luminances to display luminances have been developed and fulfilling the fast growing demand to be able to display HDR images on low dynamic range (LDR) display devices. Most tone-reproduction algorithms make use of photoreceptor adaptation [32] to achieve visually plausible results. Local operators [7–10] involve the spatial manipulation of local neighboring pixel values based on the observation that HVS system is only sensitive to relative local contrast. Global operators [11–13] do not involve spatial processing. Tonemapping is achieved by applying spatially invariant operator to treat every pixel independently. Both types of techniques have their own advantages and disadvantages in terms of computational cost, easy implementation, halo effects (artifacts), spatial sharpness, and practical application. Reinhard et al. [1] give detailed review of various tone mapping operators.

A simple S-shaped curve (sigmoid function) has been utilized as tone mapping function [33]. The middle portion of such sigmoidal function is nearly linear and thus resembles logarithmic behavior. Moreover, sigmoidal functions have two asymptotes: one for very small values and one for large values. Fattal et al. [9] has introduced gradient based approach to preserve details from HDR image. To simulate the adaptation behavior of human visual system, they have attempted gradient modification at various scales. A reduced, low dynamic range image is then obtained by solving a Poisson equation on the modified gradient field. The algorithm has used local intensity range to reduce the dynamic range in transform domain and preserve local changes of small magnitudes. The method was almost free of artifacts and does not require any manual parameter tweaking.

Recently, dynamic range compression based on two-scale decomposition has been proposed [10]. The base layer was obtained using a nonlinear BLT filter [17] and detail layer was computed by taking difference between the input image and the base layer. Only the contrast of base layer was reduced, thereby preserving fine details.

2.2. Exposure Fusion. In recent years, various fusion algorithms have been developed to combine substantial information from multiple input images into a single composite image. The principal motivation for image fusion is to extend the depth-of-field, extend spatial and temporal coverage, to increase reliability, and extend dynamic range of the fused image and the compact representation of information. Imaging sensor records the time and space varying light intensity information reflected and emitted from object in a three-dimensional observed physical scene. However, image fusion has a fundamental difficulty in preventing artifacts and preserving local contrast when fusing the characteristics recorded from the incident radiations, such as exposure value, focusing, modality, and environmental conditions. The automated procedure of extracting all the meaningful details from the input images to a final fused image is the main motive of image fusion. To facilitate image fusion, it may be necessary to align input images of the same scene captured at different times, or with different sensors, or with different exposure (EV) settings (called bracketing), or from different viewpoint using local and global registration methods [34, 35]. Normally it is assumed that the input images are captured with the help of tripod mounting. Hence, in general, we expect point-by-point correspondence between different input exposures of a scene. From technical stand point, the fused image reveals all details present in the scene without introducing any artifacts or inconsistencies which would distract the human observer or subsequent image processing stages.Ogden et al. [36] has proposed pyramid solution for image fusion. The pyramid becomes a multiresolution sketch pad to fill in the local spatial information at increasingly fine detail (as an artist does when painting). The Laplacian pyramid contains several spatial frequency bands which depicts certain edge information [37]. Image gradient orientation coherence model based fusion has been proposed for blending flash and ambient images [38, 39]. This model seeks to utilize the properties of image gradients that remain invariant under the change of lighting that takes place between a flash and an ambient image. Region segmentation and spatial frequency have been utilized for multifocus image fusion [40]. A fast multifocus algorithm has recently been developed [23] which utilizes weighted nonnegative matrix factorization and focal point analysis for preserving feature information in the fused image.

Raman and Chaudhuri [17] have utilized edge-preserving filter (i.e., bilateral filter) for the fusion of multiexposure images, in which appropriate matte is generated based on local texture details for automatic compositing process. Goshtasby [41] proposed exposure fusion method based on weights determined by blending function. Information metric was considered to design blending function. Smaller weights were assigned to an image block carrying less information while higher weights were assigned to best-exposed image block. Therefore, an image block was considered best-exposed within an area if it carries more information about the area than any other image blocks. To maximize information content in the fused image, a gradient-ascent algorithm was used to determine optimal block size and width of the blending functions. The size of the block automatically varies with image content as the type of image is changed. Szeliski [42] has used multidimensional histogram as postprocessing operator to achieve optimal contrast enhancement in the fused image; simple averaging was performed to smoothly combine the pixels into a fused image. This method was based on the observation that if the average intensity of the image is maintained during the average operation using histogram equalization, then new image can be created with increased contrast.

Mertens et al. [18] have used multiresolution approach [38] for the fusion of multiexposure image series. The technique was designed to create a well-exposed image without extending the dynamic range and tone mapping of the final image. This approach blends multiple exposures in Laplacian-pyramid code based on quality metrics like saturation and contrast. A part of the technique was stitching of flash and no-flash images, which seems to be suitable for detail enhancement in the fused image. The performance of this multiscale technique is dependent on the number of decomposition levels, that is, the pyramid height. However, the presented approach seems to be computationally expensive. Recently, various fast and effective weighted average based exposure fusion approaches have been proposed. Among these guided filtering based two-scale decomposition fusion approach [43], median filter and recursive filtering based fusion approach [44] and global optimization using generalized random walks for fusion [21] are producing fusion results with better quality. These methods were utilizing different image feature for weight calculation and further refined weight was used to control the contribution of pixels from input exposures. Instead, we use anisotropic diffusion which is effective for two-scale decomposition and weight map generation based on image feature such as weak textures. The major advantage of our technique is that it is based on single-resolution weighted average approach. Generally speaking, due to computational simplicity the present approach can be used in various consumer cameras entering the commercial market. Moreover, we noticed that the present approach can be applied for the multifocus image fusion and has much better results than existing multifocus and multiexposure image fusion methods.

3. WLS Based Exposure Fusion

3.1. Overview. A new type of exposure fusion technique is developed to avoid the limitation of conventional digital camera to handle the luminance variation in the entire scene. The primary focus of this paper is the development of a fast and robust exposure fusion approach based on local texture features computed from edge-preserving filter. Unlike most previous multiexposure fusion methods, we build on ANI, a nonlinear filter introduced by Perona and Malik [27] in 1990 that has the ability to preserve large discontinuities (edges). It derives from magnitude of the gradient of the image intensity and controls diffusion strength in the image to prevent blurring across edges. As such, the algorithm implemented (see Figure 4) includes four steps.

(1) A first step, in our algorithm, is two-scale decomposition based on ANI which is used to separate coarser details (base layer) and finer details (detail layer) across each input exposure.

(2) Weak texture details (i.e. detail layer computed from ANI) and saturation measure are utilized to generate weight mask for controlling the contribution of pixels from base layers separated across all the multiple exposures.

(3) WLS and sigmoid function based weight map refinement is performed for coarser details and finer details computed in the first step, respectively.

(4) Weighted average based blending of coarser details and finer details are performed to form a composite seamless image without blurring or loss of detail near large discontinuities.

3.2. Extraction of Coarser Details and Finer Details. Edge preserving filters have received considerable attention in computational photography over the last decade. BLT [45] and ANI [27] are the most popular edge-preserving operators. Standard BLT uses distances of neighboring pixels in space and range. The space varying weighting function is computed at a space of higher dimensionality than the signal being filtered. As a result, such filters have high computational costs [46]. ANI has led to an excellent tool for smoothing fine details of an image while preserving the coarser details (i.e., edges). It is modeled using partial differential equations (PDEs) and based on nonlinear iterative process. The diffusion equation in two dimensions is defined as follows:

$$\frac{\partial I(i,j,t)}{\partial t} = \operatorname{div}\left[g\left(\|\nabla I\|\right)\nabla I\right], \tag{1}$$

where the operator ∇I calculates the image gradient of an input image I, $\|\nabla I\|$ represents the magnitude of the gradient of image intensity, $g(\|\nabla I\|)$ is a spatially varying nonlinear operator that smoothes fine details while avoiding blurring of coarser details, (i, j) specifies spatial position, and t is the iteration parameter.

The diffusion strength in the image is determined by the conduction coefficient which is influenced by the gradient of the image intensity. The principles of conduction coefficient are (i) smoothing the fine textures and (ii) preserving coarser details in the image data. Such type of nonlinear diffusion is achieved by considering image structure. On the other hand, fixed value of conduction coefficient (i.e., $g(\cdot) = 1$) yields isotropic linear diffusion that tends to have constant response for fine textures and strong edges. Therefore to achieve nonlinear diffusion the conduction coefficient is chosen to satisfy $g(i) \rightarrow 0$ when $i \rightarrow \infty$ so that the diffusion process is "stopped" across the region boundaries (i.e., edges) at locations of high gradients.

A diffusion functions $g(\cdot)$ used in our approach can be defined as follows:

$$g(\nabla I) = e^{(-(\|\nabla I\|/K)^2)}, \tag{2}$$

where K is a scale parameter that is determined by empirical constant and the selection of scale parameter may be different for a particular application [27]. In our algorithm the value of $K = 1/7$ was fixed for all cases, which is determined empirically to yield optimally diffused image for fine details extraction.

Anisotropic diffusion [27] for discrete signal is computed as follows:

$$I_s^{t+1} = I_s^t \frac{\gamma}{|\eta_s|} \sum_{p\in\eta_s} g\left(\nabla I_{s,p}\right)\nabla I_{s,p}, \tag{3}$$

where I_s^t is a discrete version of input signal, s determine the sample position in the discrete signal, and t determines iterations. The constant γ is a scalar that determines the rate of diffusion, η_s represents the spatial neighborhoods of current sample position s, and $|\eta_s|$ is the number of neighbors.

For discrete image, North (g_N), South (g_S), East (g_E), and West (g_W) spatial locations are considered for the computation of conduction coefficients. In our case, local window of size (3×3) from input image (I) is chosen, which intuitively appears most suitable for the computation of conduction coefficient at low computation cost, but other window sizes are possible as well. So after computing all the possible values of the conduction coefficients for pixel position (s) in the discrete image, the diffused image is obtained as follows:

$$I_s^{t+1} = I_s^t + \frac{\gamma}{|\eta_s|}\left[g_N \cdot \nabla_N I + g_S \cdot \nabla_S I + g_E \cdot \nabla_E I + g_W \cdot \nabla_W I\right]_s^t, \tag{4}$$

where ∇_N, ∇_S, ∇_E, and ∇_W indicate the difference of North, South, East, and West neighbor for pixel position s, respectively.

Let I_n be the nth source image which needs to be operated by an ANI filter. In order to separate coarser and finer details, we first decompose source images into two-scale representations by using anisotropic diffusion. The base layer (i.e., the diffused image defined in (4)) of each source image is obtained as follows:

$$\mathrm{BL}_n = \mathrm{ANI}_{K,t,\gamma}\left(I_n\right). \tag{5}$$

Once the base layer is obtained for each nth input image, the detail layer (D_n) can be directly calculated by subtracting the base layer from the source image as follows:

$$D_n = I_n - \mathrm{BL}_n. \tag{6}$$

3.3. Weight Estimation. The motivation behind weight map computation is to yield nonlinear adaptive function for controlling the contribution of pixels from base layers and detail layers computed across all input exposures.

Interestingly, the detail layer (D_n) computed for nth source image in (6) yields analysis of weak textures that seems to be indicator of contrast variation in the image. We adopted such metric to quantify the perceived local contrast of an image under different exposure settings and allow discarding underexposed and overexposed pixels.

Furthermore, in order to accomplish optimal contrast and color details in the fused image, we additionally incorporate the color saturation measure (SAT_n) to our weighting function. In practice for nth RGB source image, SAT_n is computed for nth source image as the standard deviation within the R', G', and B' channel, at each pixel

$$\mathrm{SAT}_n = \sqrt{\frac{1}{3}\left[\left(R'-\mu\right)^2 + \left(G'-\mu\right)^2 + \left(B'-\mu\right)^2\right]}, \tag{7}$$

where, $\mu = (1/3)(R' + G' + B')$.

As shown in Figure 1, in order to remove the influences of underexposed and overexposed pixels for producing well-exposed image, the two image features, that is, D_n and SAT_n, are combined together by multiplication to estimate combined features (FR_n):

$$\mathrm{FR}_n = D_n \times \mathrm{SAT}_n. \quad (8)$$

Then FR_n is convolved with the symmetric Gaussian low pass kernel (U) having 5×5 kernel size (r_U) with standard deviation (σ_U) 5 to construct the saliency maps SM_n:

$$\mathrm{SM}_n = \left|\mathrm{FR}_n\right| * U_{r_U,\sigma_U}. \quad (9)$$

Next, the saliency maps are compared to determine the weight maps as follows:

$$\mathrm{WM}_n^k = \begin{cases} 1, & \text{if } \mathrm{SM}_n^k = \max\left(\mathrm{SM}_1^k, \mathrm{SM}_2^k, \ldots, \mathrm{SM}_M^k\right) \\ 0, & \text{otherwise,} \end{cases} \quad (10)$$

where M is number of source images, SM_n^k is the saliency value of the pixel k in the nth image.

3.4. WLS Based Weight Refinement and Weighted Fusion of Coarser Details and Finer Details. In this section, we propose WLS optimization framework [20] and sigmoid function [28] based weight map refinement approach to obtain noiseless and smooth weight maps. First, WLS filtering is performed on each weight map WM_n with the corresponding source image serving as the source image for the affinity matrix [20]. The motivation behind weight maps refinement is as follows. The fusion rules (weight map) computed in (10) are hard (the value of weight maps are changing abruptly), noisy, and not aligned with the object boundaries. Weight maps need to be as smooth as possible, since rapid changes in the weight maps will introduce seam and artifacts in fused image.

3.4.1. WLS Optimization Framework. WLS [20] based edge-preserving operator may be viewed as a compromise between two possible contradictory goals. Given an input image v, we seek a new image w, which, on the one hand, is as close as possible to v, and at the same time, is as smooth as possible everywhere, except across significant gradients in v. To achieve these objectives we seek to minimize the following quadratic functional:

$$\sum_{p'} \left(\left(w_{p'} - v_{p'}\right)^2 + \lambda \left(q_{x,p'}(v) \left(\frac{\partial w}{\partial x}\right)_{p'}^2 + q_{y,p'}(v) \left(\frac{\partial w}{\partial y}\right)_{p'}^2 \right) \right), \quad (11)$$

where the subscript p' denotes the spatial location of a pixel. The goal of the data term $(w_{p'} - v_{p'})^2$ is to minimize the distance between w and v, while the second (regularization) term strives to achieve smoothness by minimizing the partial derivatives of w. The smoothness requirement is enforced in a spatially varying manner via the smoothness weights q_x and q_y, which depend on v. Finally, λ is responsible for the balance between the two terms, increasing the value of λ results in progressively smoother images w.

Using matrix notation we may rewrite (11) in following quadratic form:

$$(w - v)^T (w - v) + \lambda \left(w^T O_x^T V_x O_x w + w^T O_y^T V_y O_y w \right). \quad (12)$$

Here V_x and V_y are diagonal matrices containing the smoothness weights $q_x(v)$ and $q_y(v)$, respectively, and the matrices O_x and O_y are discrete differentiation operators.

The vector w that minimizes (12) is uniquely defined as the solution of the linear system

$$\left(I_d + \lambda L_v\right) w = v, \quad (13)$$

where I_d is the identity matrix and $L_v = O_x^T V_x O_x + O_y^T V_y O_y$. Modulo the difference in notation, this is exactly the linear system used in [47], where it was primarily used to drive piecewise smooth adjustment maps from a sparse set of constraints.

In the present approach, O_x and O_y are forward difference operators, and hence O_x^T and O_y^T are backward difference operators, which means that L_v is a five-point spatially inhomogeneous Laplacian matrix. As for the smoothness weights, we define in the same manner as in [47]:

$$\begin{aligned} q_{x,p'}(v) &= \left(\left| \frac{\partial \ell}{\partial x}(p') \right|^{\alpha} + \varepsilon \right)^{-1}, \\ q_{y,p'}(v) &= \left(\left| \frac{\partial \ell}{\partial y}(p') \right|^{\alpha} + \varepsilon \right)^{-1}, \end{aligned} \quad (14)$$

where ℓ is the log-luminance channel of the input image v and the exponent α (typically between 1.2 and 2.0) determines the sensitivity to the gradients of v, while ε is a small constant (typically 0.0001) that prevents division by zero in areas where v is constant.

Equation (13) tells us that w is obtained from v by applying a nonlinear operator Z_λ, which depends on v:

$$w = Z_\lambda(v) = \left(I_d + \lambda L_v\right)^{-1} v. \quad (15)$$

In the present approach, $\mathrm{WLS}_{\lambda,\alpha,\varepsilon}(v)$ represents the WLS filtering operation. Where λ,α, and ε are the parameters which decide the degree of smoothness, sensitivity to the gradients and small constant of the WLS filter, respectively. In our case WM_n computed in (10) serves as the input image to WLS filter (i.e., $v = \mathrm{WM}_n$). More specifically, the coarser version of weight map WM_n will serve as refined weight map for nth base layer W_n^{BL}:

$$W_n^{\mathrm{BL}} = \mathrm{WLS}_{\lambda,\alpha,\varepsilon}(v). \quad (16)$$

Once the resulting weight maps for base layer are obtained, sharp and edge-aligned weights are computed based on 1-D sigmoid function for fusing the detail layers. As shown in Figure 1, the spatially smoothed weight maps of base layer are utilized to compute sharp weight mask of detail layer which preserve texture details in the fused image. Therefore, unlike [43], the proposed solution attempts computationally simple approach to estimate the best possible weight maps for detail layer fusion.

Let W_n^D denote refined weight map for nth detail layer and $\text{SIG}_{a,t',\theta}(W_n^{\text{BL}})$ is the 1-D sigmoid function [28] applied on W_n^{BL}, where $a \in \mathfrak{R}$, $t' \in \mathfrak{R}$, and θ are the weight parameter, independent variable, and the parameter which decide the threshold to further control the degree of sharpness, respectively. Then W_n^D is computed as

$$W_n^D = \text{SIG}_{a,t',\theta}\left(W_n^{\text{BL}}\right). \quad (17)$$

In theory, the 1-D sigmoid is computed as:

$$\text{SIG}_{a,t',\theta}(t) = \frac{1}{1 + e^{(-at'+\theta)}}, \quad (18)$$

where $t' \in \mathfrak{R}$ is the independent variable, $a \in \mathfrak{R}$ is a weight parameter of the sigmoid function, and θ be a fixed threshold to further control the sharpness of sigmoid function.

Once the resulting weight maps W_n^{BL} and W_n^D are obtained, the pixelwise weighted composition of base layers (i.e., fused base layer BL_F), detail layers (i.e., fused detail layer D_F), and the resulting fused image I_F can be directly calculated as follows:

$$\text{BL}_F = \sum_{n=1}^{N} W_n^{\text{BL}} \text{BL}_n, \quad (19)$$

$$D_F = \sum_{n=1}^{N} W_n^D D_n, \quad (20)$$

$$I_F = \text{BL}_F + D_F. \quad (21)$$

4. Experimental Results and Analysis

In order to evaluate the performance and effectiveness of the proposed image fusion approach, we have summarized the comparison of our fusion approach with different exposure fusion, tone mapping, and multifocus image fusion methods. Two objective evaluation metrics (i.e., quality score $Q^{AB/F}$ [48] and visual information fidelity for fusion VIFF [49]) were employed to access the fusion quality and to analyze the effect of free parameters used in the approach. Currently, all experimental results are generated by the MATLAB implementation. Furthermore, to measure distortion in the fused image and strengthen the evaluation capability of $Q^{AB/F}$ and VIFF, we incorporate the Dynamic Range Independent Visible Difference Predictor (DRIVDP) [22]. DRIVDP metric is sensitive to three types of structural changes for distortions measurement (i.e., loss of visible contrast, amplification of invisible contrast, and reversal of visible contrast) between images under a specific viewing condition.

4.1. Comparison with Other Exposure Fusion, Multifocus Image Fusion and Tone Mapping Methods. Figures 2, 3, and 4 depict examples of fused images from the source multi-exposure images. It is noticed that the proposed approach enhances texture details while preventing halos near strong edges. In order to check effectiveness and robustness of present approach, the algorithm is tested on variety of multiexposure image series. The proposed approach is computationally simple and the results are comparable to several exposure fusion and tone mapping techniques. As shown in Figures 2(a)–2(c) the details from all of the source images are perfectly combined and reveal fine textures while preserving local contrast and natural color. In Figures 3(a)–3(d) we compare our results to the recently proposed approaches. Figure 3(d) depicts the results of optimization framework [16] and Figure 3(e) shows the mate-based fusion results using the edge preserving filter such as BLT [17]. It can be observed that other fusion methods perform well in preserving image details while they fail to reconstruct texture and color details in the brightly illuminated areas. The result of Mertens et al. [18] (see Figure 3(f)) appears blur and loses texture details while in our results (see Figure 3(c)) the fine texture are emphasized that are difficult to be visible in Figure 3(f). This is because of utilization of Gaussian kernel for pyramid generation as it removes Pixel-to-pixel correlations by subtracting a low-pass filtered copy of the image from the image itself to generate Laplacian pyramid and result is a texture and edge details reduction in the fused image. The results produced in Figures 3(d)–3(f) lose visibility in brightly illuminated areas and details are lost in the tree leaves, and the texture on the wall is washed out. Although the results of Raman and Chaudhuri [17] (see Figure 3(e)) exhibit better color details in tree leaves, they appear slightly blurry. In our results (Figure 3(c)) details are preserved in the brightly illuminated areas, yet at the same time fine details are well preserved (tree leaves, wall texture, and lizard).

To further compare our results visually with Mertens et al. [18], iCAM06 [19], WLS [20], and GRW [21], respectively, Figures 4(a), 4(b), 4(c), 4(d), and 4(e) depict experimental results for National Cathedral sequence (1024 × 768 × 3). Proposed fusion results shown in Figure 4(a) illustrate the ability of enhancing fine texture details. As well as having the ability to produce good color information with natural contrast. This can bring an increased illusion of depth to an image textures. Therefore, enhanced texture details in the fused image let you get everything sharp and yield an accurate exposure that is entirely free of halo artifacts. Although tone mapped results of iCAM06 [19] and WLS [20] have produced comparable results, but they do not preserve contrast from input LDR image series. Figures 4(b) and 4(e) show the results of pyramid approach [18] and GRW optimization framework [21], respectively, which preserve global contrast but losses color information. GRW [21] based exposure fusion is shown in Figure 4(e) which depicts less texture and color details in brightly illuminated regions (i.e., lamp and window glass). Note that Figure 4(a) retains colors, sharp edges, and details while also maintaining an overall reduction in high frequency artifacts near strong edges. The results produced in Figures 4(b)–4(d) were generated by the programs provided by their respective authors. The HDR images for iCAM06 [19] and WLS [20] were generated using HDR reconstruction [2]. The results of GRW [21] shown in Figure 4(e) are taken from its

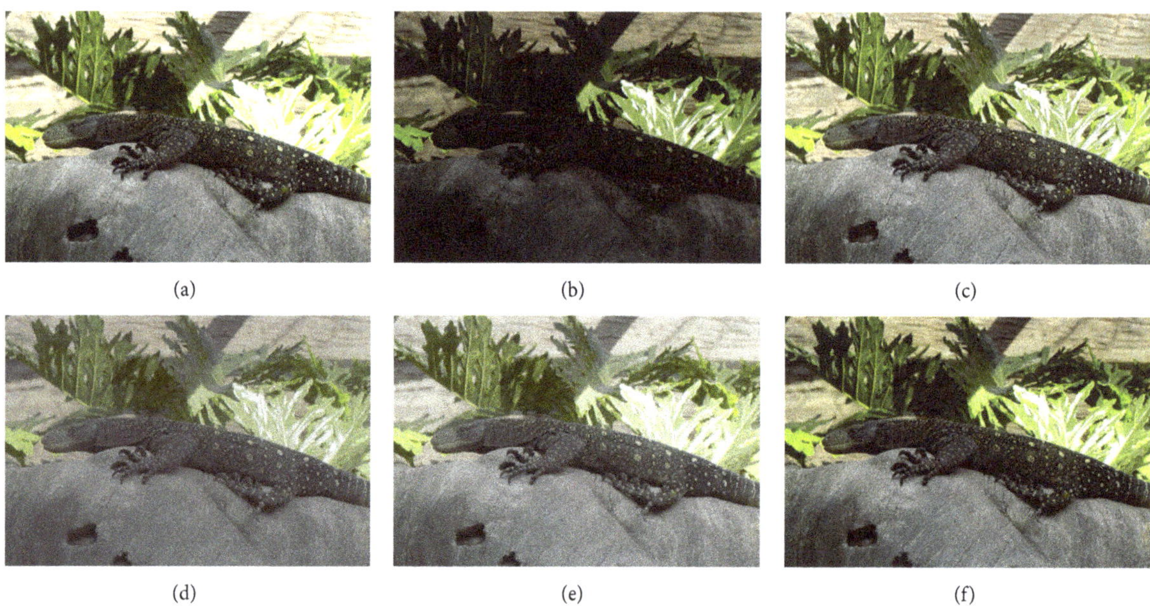

FIGURE 3: Lizard: comparison with popular exposure fusion approaches. (a), (b) Source images, (c) results of our new exposure fusion method. Tree leaves and wall texture appear overexposed and blurry in (d) Kotwal and Chaudhuri [16], (e) Raman and Chaudhuri [17], and (f) Mertens et al. [18]. In the proposed results, it becomes almost possible to preserve texture and strong edge features simultaneously. Moreover, by applying the proposed approach, the fine textures are accurately enhanced. Input image sequence is Eric Reinhard, University of Bristol.

FIGURE 4: National Cathedral: comparison with popular exposure fusion and tone mapping methods. (a) Results of our new exposure fusion method, (b) Mertens et al. [18], (c) iCAM06 [19], (d) WLS [20], and (e) GRW [21]. Note that our method yields enhanced texture and edge features with better color appearance. Input image sequence is courtesy of Max Lyons.

paper. In order to give a relatively fair comparison in our experiments, we have used default sets of parameters for tone mapping [19, 20] and exposure fusion [18] methods.

Figure 5 shows the distortion maps computed from DRIVDP metric proposed by Aydin et al. [22]. This quality assessment metric detects loss of visible contrast (green) and amplification of invisible contrast (blue). The main advantage of this metric is that it yields meaningful results even if the input images have different dynamic ranges. Though we consider here DRIVDP based quality assessment to compare proposed method with one exposure fusion [18] method and two tone mapping methods [19, 20]. We assume that the LDR images are shown in a typical LCD display with maximum luminance 100 and gamma 2.2. We also assume that for all the LDR images, the viewing distance is 0.5 metres and the number of pixels per visual degree is 30 and peak contrast is 0.0025. Significance of the choice of these parameters can be found in [22]. Figures 5(a)–5(v) show a side-by-side comparison of the loss of visible contrast (green), and amplification of invisible contrast (blue) of proposed results with others methods. To compute visible contrast loss illustrated in Figures 5(d), 5(i), 5(n), and 5(s), respectively, for fused images in Figures 5(c), 5(h), 5(m), and 5(r), the underexposed image (i.e., Figure 5(a)) is used as reference image. Similarly, to compute visible contrast loss illustrated in Figures 5(e), 5(j), 5(o), and 5(t), respectively, for fused

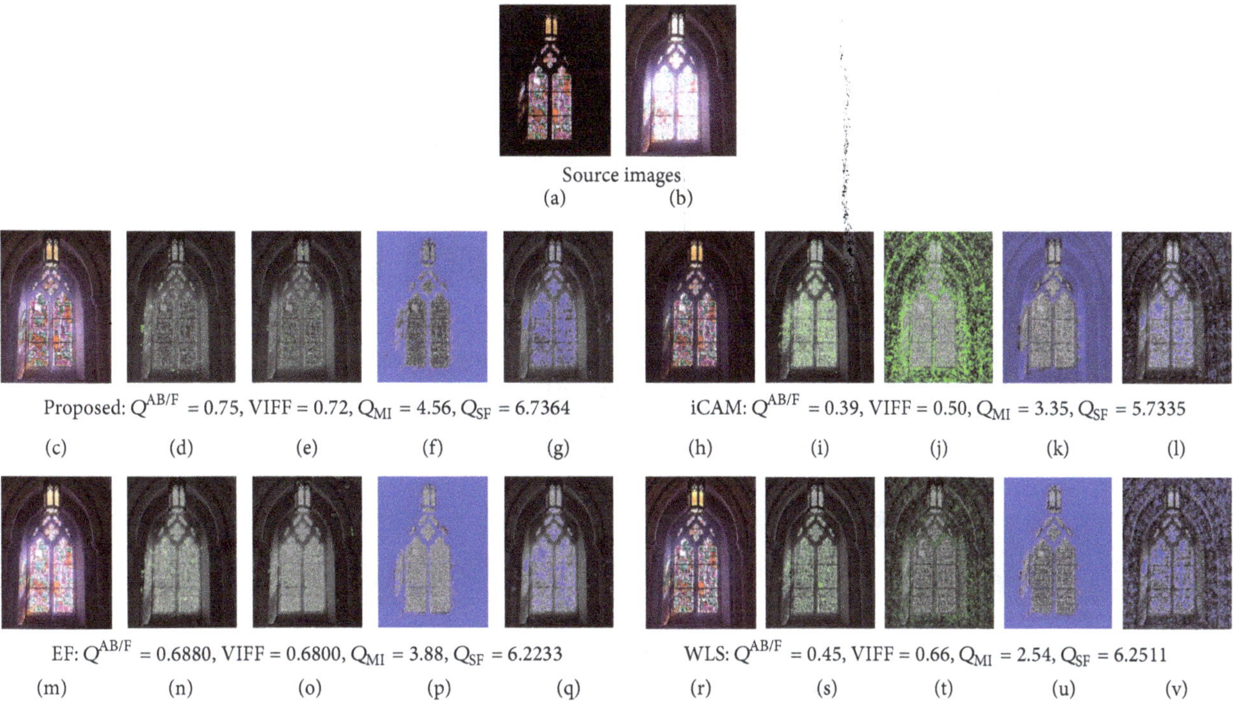

FIGURE 5: Comparison of proposed results with iCAM06, EF, and WLS on the National Cathedral sequence using DRIVDP [22]. (a), (b) The two source images give good exposures for the paintings on window glass and lamp and the wall, respectively. In a distortion map, green, blue, and gray pixels indicate visible contrast loss, amplification, and no distortion, respectively. Proposed results are more effective in preserving local details and colors than the others. For the paintings on window glass and lamp and the wall, proposed results (see (c)–(g)) depict the least distortion, followed by iCAM06 (see (h)–(l)), Mertens et al. EF (see (m)–(q)), and WLS (see (r)–(v)).

images in Figures 5(c), 5(h), 5(m), and 5(r), the overexposed image (i.e., Figure 5(b)) is used as reference image. We ran invisible amplification metrics on fused images, which are generated using a similar procedure as used for loss of visible contrast metric. The two source images with good exposures, respectively, for the brightly illuminated region (i.e., window) and the poorly illuminated region (i.e., wall) are given in Figures 5(a) and 5(b). The distortion maps for proposed method, iCAM06 [19], Mertens [18], and WLS [20] are given in Figures 5(c)–5(g), Figures 5(h)–5(l), Figures 5(m)–5(q), and Figures 5(r)–5(v), respectively, along with the fused images. In a distortion map, green, blue, and gray pixels indicate contrast loss, amplification, and no distortion, respectively. It can be noticed that the proposed results are more effective in preserving local contrast and color information than the other methods. Please note that visible contrast loss and distortions are the least using the proposed approach. Moreover, to compare the performance of the proposed approach, iCam06, Mertens et al., and WLS, we have employed four fusion quality metrics, that is, $Q^{AB/F}$, VIFF, Mutual Information (Q_{MI}) [50], and Spatial Frequency (Q_{SF}) [51].

$Q^{AB/F}$ [48] evaluates the amount of edge information transferred from input images to the fused image. A Sobel operator is applied to yield the edge strength and orientation information for each pixel. For two input images A and B, and a resulting fused image F (i.e., I_F computed in (21)), the Sobel edge operator is applied to yield the edge strength $e(n',m')$ and orientation $\beta(n',m')$ information for each pixel as

$$e_A(n',m') = \sqrt{sb_A^x(m',n')^2 + sb_A^y(m',n')^2},$$
$$\beta_A(n',m') = \tan^{-1}\left[\frac{sb_A^x(m',n')}{sb_A^y(m',n')}\right], \tag{22}$$

where $sb_A^x(m',n')$ and $sb_A^y(m',n')$ are horizontal and vertical Sobel template cantered on pixel $P_A(n',m')$ and convolved with the corresponding pixels of image A. The relative strength and orientation values of an input image A with respect to F are formed as

$$\left(G_{n',m'}^{AF}, A_{n',m'}^{AF}\right) = \left(\left(\frac{e_{n',m'}^F}{e_{n',m'}^A}\right)^{\Psi}, 1 - \frac{\left|\beta_A(n',m') - |\beta_F(n',m')|\right|}{\frac{\pi}{2}}\right),$$
$$\text{where}\quad \Psi = \begin{cases} 1 & \text{if } e_A(n',m') > e_F(n',m') \\ -1 & \text{otherwise.} \end{cases} \tag{23}$$

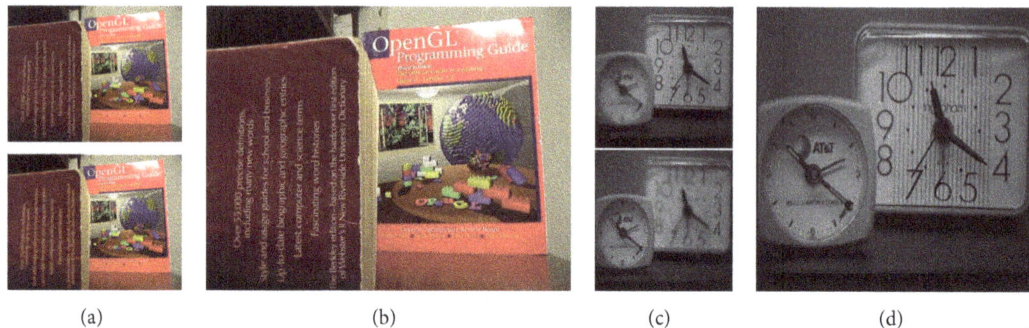

FIGURE 6: Multifocus image fusion results: (a), (c) demonstrates the effect of selective focus to capture more details from a particular part of a scene and (b), (d) images generated by the proposed approach. Note how in the fused image all objects appear all-in-focus and enhance the color and texture details present in the foreground and background extracted from the original input images (input sequence is courtesy of Adu and Wang [23]).

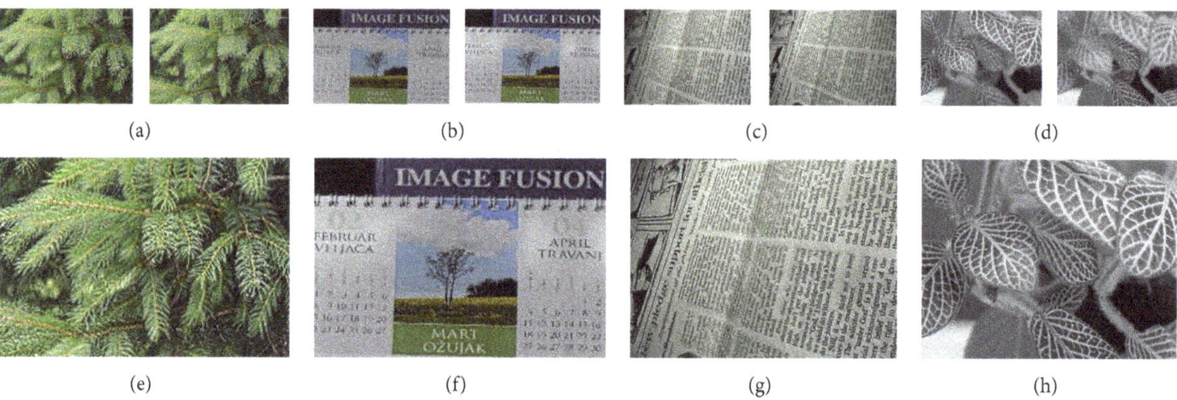

FIGURE 7: Multifocus image fusion results: (a), (b), (c), and (d) demonstrate the effect of selective focus to capture more details from a particular part of a scene and (e), (f), (g), and (h) images generated by the proposed approach. These results demonstrate that present approach has helped to handle typical situations of foreground and background present in the scene (input sequence is courtesy Slavica Savic).

The process of edge information preservation values is defined in [48].

Finally, the $Q^{AB/F}$ is defined as

$$Q^{AB/F} = \frac{\sum_{\forall n',m'} Q^{AF}_{n',m'} w^{A}_{n',m'} + Q^{AF}_{n',m'} w^{B}_{n',m'}}{\sum_{\forall n',m'} w^{A}_{n',m'} + w^{B}_{n',m'}}, \quad (24)$$

which evaluates the sum of edge information preservation values for both inputs Q^{AF} and Q^{BF} weighted by local importance perceptual factors w^A and w^B. We defined $w^A(n',m') = [e_A(n',m')]^L$ and $w_B(n',m') = [e_B(n',m')]^L$. L is a constant. For the "ideal fusion," $Q^{AB/F} = 1$.

VIFF [49] first decomposes the source and fused images into blocks. Then, VIFF utilizes the models in visual information fidelity (VIF) (i.e., Gaussian Scale Mixture (GSM) model, Distortion model, and human visual system (HVS) model) to capture visual information from the two source-fused pairs. With the help of an effective visual information index, VIFF measures the effective visual information of the fusion in all blocks in each subband. Finally, the assessment result is calculated by integrating all the information in each subband:

$$\text{VIFF}(I_1,\dots,I_n,I_F) = \sum_k P'_k \cdot \text{VIFF}_k(I_1,\dots,I_n,I_F), \quad (25)$$

where P'_k is a weighting coefficient. According to VIF theory, a high VIF yields a high quality test image. Therefore, as VIFF increases, the quality of the fused image improves.

The quality metric Q_{MI} measures how well the original information from source images is preserved in the fused image:

$$Q_{\text{MI}} = 2\left(\frac{\text{MI}(A,F)}{H(A)+H(F)} + \frac{\text{MI}(B,F)}{H(B)+H(F)}\right), \quad (26)$$

where $H(A)$, $H(B)$, and $H(F)$ are the marginal entropy of A, B, and F and $\text{MI}(A,F)$ is the mutual information between the source image A and the fused image F:

$$\text{MI}(A,F) = H(A) + H(F) - H(A,F), \quad (27)$$

where $H(A,F)$ is the joint entropy between A and F, $H(A)$ and $H(F)$ are the marginal entropy of A and F, respectively, and $\text{MI}(B,F)$ is similar to $\text{MI}(A,F)$.

FIGURE 8: Comparison of Clock image results with recently proposed conventional multifocus image fusion methods. (a) Proposed results, (b) Adu et al. [23], (c) DWT [24], and (d) Tian et al. [25]. It has been found that the proposed approach helps the viewer to observe enhanced texture and edge features simultaneously without depicting visible artifacts. Input image sequence is courtesy of Adu and Wang.

The fourth criterion is Q_{SF}. The spatial frequency, which originated from the human visual system, indicates the overall active level in an image and has led to an effective objective quality index for image fusion [51]. The total spatial frequency of the fused image is computed from row (RF) and column (CF) frequencies of the image block and Q_{SF} is defined as

$$Q_{SF} = \sqrt{RF^2 + CF^2},$$

$$RF = \sqrt{\frac{1}{M'N'} \sum_{m'=1}^{M'} \sum_{n'=1}^{N'} \left(I_F(m',n') - I_F(m',n'-1)\right)^2}, \tag{28}$$

where $I_F(m',n')$ is the gray value of pixel at position (m',n') of image I_F:

$$CF = \sqrt{\frac{1}{M'N'} \sum_{n'=1}^{N'} \sum_{m'=1}^{M'} \left(I_F(m',n') - I_F(m'-1,n')\right)^2} \tag{29}$$

The quantitative performance analysis using the aforesaid evaluation indices are shown in the caption of Figure 5. The present approach has outperformed the other methods. We can see that the proposed method can preserve more useful information compared with iCam06, Mertens et al., and WLS fusion methods. In particular, evaluation results in Figure 5 have demonstrated that $Q^{AB/F}$, VIFF, Q_{MI}, and Q_{SF} have correspondence with the DRIVDP-based evaluation.

Furthermore, to check the applicability of proposed approach for other image fusion applications, we have presented the experimental results for multifocus image fusion. In Figures 6, 7, and 8, it is demonstrated that proposed method is also suitable for multifocus image fusion to yield rich contrast and texture details. One of the key characteristics of present approach for multifocus image fusion application is illustrated in Figure 6(b): the color details are preserved in the fused image with better visualization of texture details. It can also be noticed in Figure 6(d) that the edges and textures are relatively better than that of

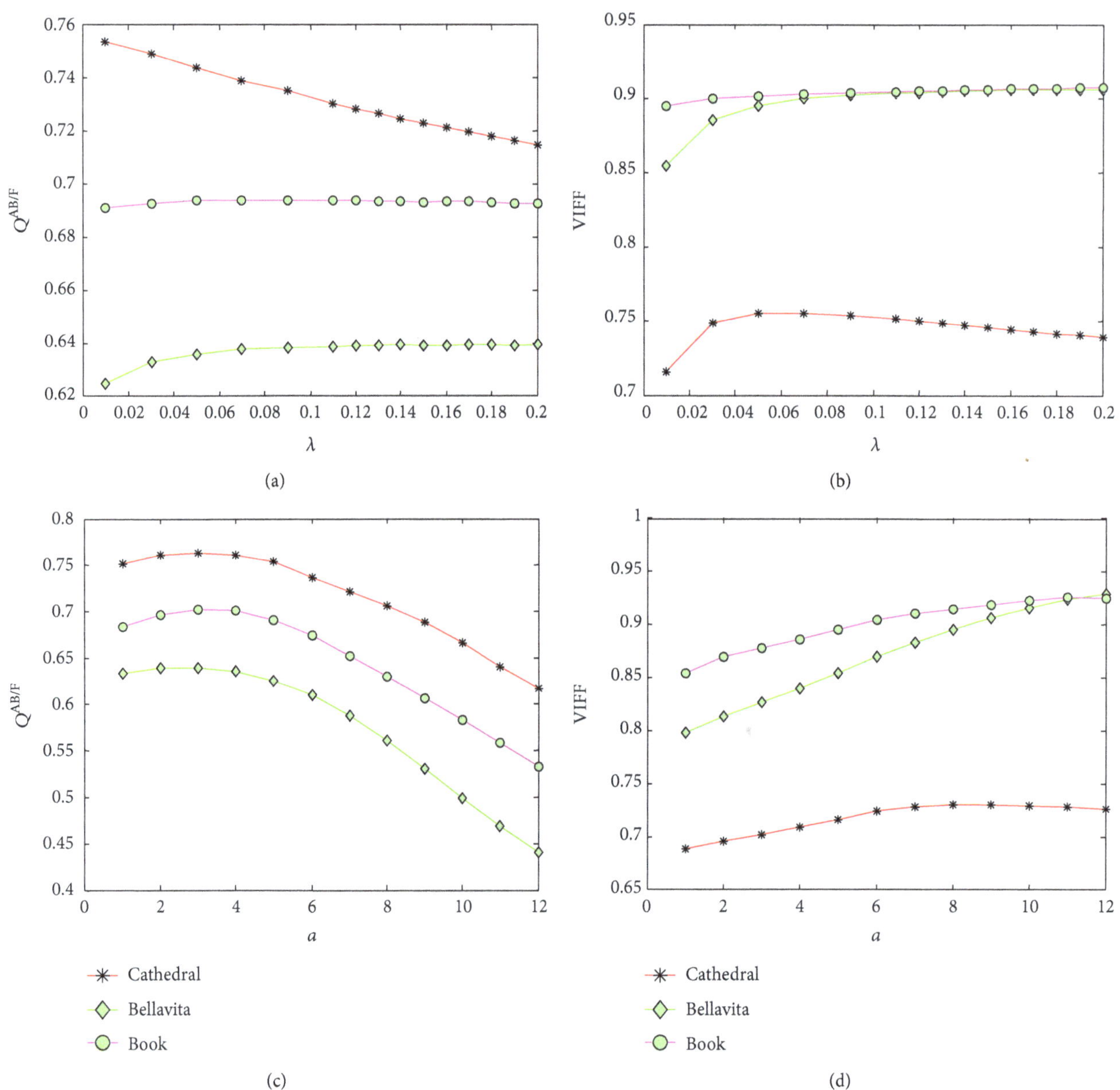

Figure 9: Sensitivity analysis of lambda (λ) and free parameter (a)which, respectively, control the smoothness in weight maps computed for base layers and detail layers. (a) Effectiveness of λ on $Q^{AB/F}$, (b) effectiveness of λ on VIFF, (c) effectiveness of a on $Q^{AB/F}$, and (d) effectiveness of a on VIFF.

source images. Fusion results of proposed method on the four standard test scenes (see Figures 7(a)–7(d)) are shown in Figures 7(e)–7(h). Note that, the strong edges and fine texture details are accurately preserved in the fused image without introducing halo artifacts. The halo artifacts may stand out if the detail layer undergoes a substantial boost. Comparisons of Adu [23], DWT [24], Tian et al. [25], and our approach for multifocus image fusion are illustrated in Figures 8(a)–8(d). The result produced in Figure 8(b) is taken from its paper [23]. Results of DWT [24] shown in Figure 8(c) were generated by the MATLAB Wavelet toolbox. For the DWT-based methods, the low-pass subband coefficients and the high-pass subband coefficients are simply merged by the averaging scheme and the choose-max selection scheme, respectively. The DWT-based fusion algorithm is performed using five-level decomposition and db3 wavelets are used in scale decomposition. The results of Jing et al. shown in Figure 8(d) are generated from the MATLAB code provided by the author. Note that our method (see Figure 8(a)) yields enhanced texture and edge features. We can significantly preserve and enhance fine details separately because our approach excludes fine textures from the base layers.

4.2. Analysis of Free Parameters and Fusion Performance Metrics. Proposed method has eight free parameters, that is, t, K, γ, a, θ, λ, α, ε. We fix $t = 5$, $K = 30$, $\gamma = 1/7$,

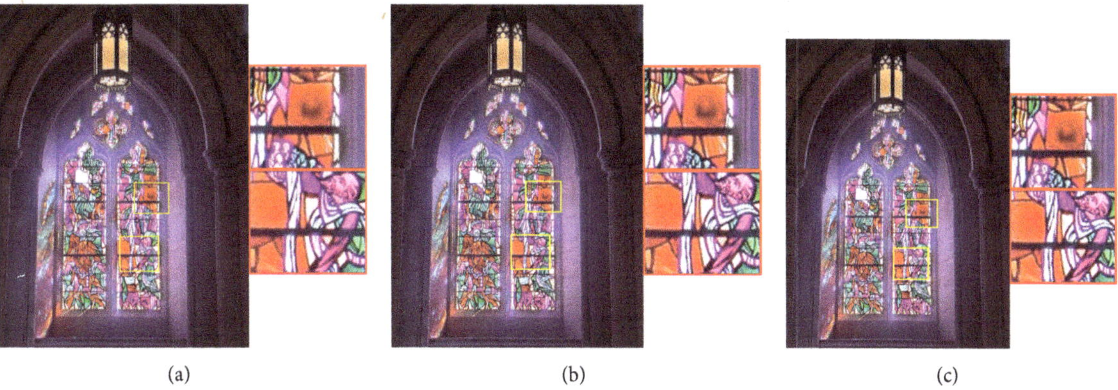

(a) (b) (c)

FIGURE 10: Visual inspection of fine detail enhancement in typical lighting situations: The free parameter a in (18) controls detail enhancement and sharpening. We have found that variation of "a" between 3 and 6 is sufficient for fine details extraction and give better results for various typical situations. By choosing higher value of a,the texture details are accurately enhanced. It also shows that selection of much higher value for detail enhancement does not introduce artifacts near object boundaries due to detail enhancement. (a) $a = 1$, (b) $a = 6$, and (c) $a = 12$.

$a = 3$, $\theta = 0.001$, $\lambda = 0.1$, $\alpha = 1.2$, and $\varepsilon = 0.0001$ in all experiments and they are set as default parameters. It is preferred to have a small number of iterations (t) to reduce computational time. The fusion performance is not affected when $t \leq 5$ because present method does not depend much on the exact parameter choice of t. The parameters selection criterion for K, γ, and θ is given in [27, 28], respectively. We have set $K = 30$, $\gamma = 1/7$, and $\theta = 0.001$ as default parameters for all experiments. In the present approach, the fusion performance is dependent on two free parameters, that is, λ and a. To analyze the effect of lambda (λ) and free parameter (a) on $Q^{AB/F}$ [48] and VIFF [49], we have illustrated four plots (see Figures 9(a)–9(d)) for input image sequence of Cathedral ($1024\times768\times 3$), Bellavita ($800\times535\times 3$), and Book ($569 \times 758 \times 3$). The detailed description of $Q^{AB/F}$ and VIFF is given in the previous subsection. To assess the effect of lambda (λ) and free parameter (a) on fusion performance, the $Q^{AB/F}$ and VIFF are experimented.

To analyze the influence of λ and a on $Q^{AB/F}$ and VIFF, other parameters are set to $t = 5$, $K = 30$, $\gamma = 1/7$, $\theta = 0.001$, $\alpha = 2$, and $\varepsilon = 0.0001$. As shown in Figures 9(a) and 9(b), the fusion performance will be worse when the values of λ and a are too large or too small. It should be noticed in Figure 9 that the $Q^{AB/F}$ and VIFF decrease when the λ and a are too large or too small. The visual inspection of effect of a's on "Cathedral" sequence is depicted in Figures 10(a)–10(c). It can easily be noticed in Figures 10(a)–10(c) that as a increases, the strong edges and textures get enhanced and therefore leads to a detail preserving fusion results. In order to obtain optimal detail enhancement and low computational time, we have concluded that the best results were obtained with $t = 5$, $K = 30$, $\gamma = 1/7$, $a = 3$, $\theta = 0.001$, $\lambda = 0.1$, $\alpha = 1.2$, and $\varepsilon = 0.0001$, which yield reasonably good results and satisfactory subjective performance for all cases.

To further demonstrate the analysis of errors introduced by the free parameter a, four fundamental error performance metrics are adopted, that is, Root Mean Squared Error (RMSE), Normalized Absolute Error (NAE), Laplacian Mean Squared Error (LMSE), and Peak Signal to Noise Ratio (PSNR). The RMSE measure the differences between resulting image and reference image. The error in a pixel $P_{i'}^{*}$ is calculated using Euclidean distance between a pixel in a resulting image with $a > 1$ and the corresponding pixel in the reference image $P_{i'}^{\text{ref}}$ with $a = 1$, that is, $E(P_{i'}^{*}) = \|P_{i'}^{*} - P_{i'}^{\text{ref}}\|$. The total error in a resulting fused image is computed using square root of the Mean Square Error (MSE), that is, $\sqrt{(1/M)\sum_{i'} E(P_{i'}^{*})^2}$.

NAE is a measure of how far is the resulting fused image (when $a > 1$) from the reference fused image (when $a = 1$) with the value of zero being the perfect fit. Large value of NAE indicates poor quality of the image [51]. NAE is computed as follows:

$$\text{NAE} = \frac{\sum_{m'=1}^{M'} \sum_{n'=1}^{N'} \left| I_F(m',n') - \widehat{I_F}(m',n') \right|}{\sum_{m'=1}^{M'} \sum_{n'=1}^{N'} \left| I_F(m',n') \right|}. \quad (30)$$

LMSE is based on the importance of edges detail measurement, which is also the most critical feature for image quality assessment. The large value of LMSE means that image is poor in quality. LMSE is defined as follows:

$$\text{LMSE} = \frac{\sum_{m'=1}^{M'} \sum_{n'=1}^{N'} \left[L\left(I_F(m',n')\right) - L\left(\widehat{I_F}(m',n')\right) \right]^2}{\sum_{m'=1}^{M'} \sum_{n'=1}^{N'} \left[L\left(I_F(m',n')\right) \right]^2}, \quad (31)$$

where $L(I_F(m',n'))$ is laplacian operator:

$$L\left(I\left(m',n'\right)\right) = I_F\left(m'+1,n'\right) + I_F\left(m'-1,n'\right) + I_F\left(m',n'+1\right) + I_F\left(m',n'-1\right) - 4I_F\left(m',n'\right). \quad (32)$$

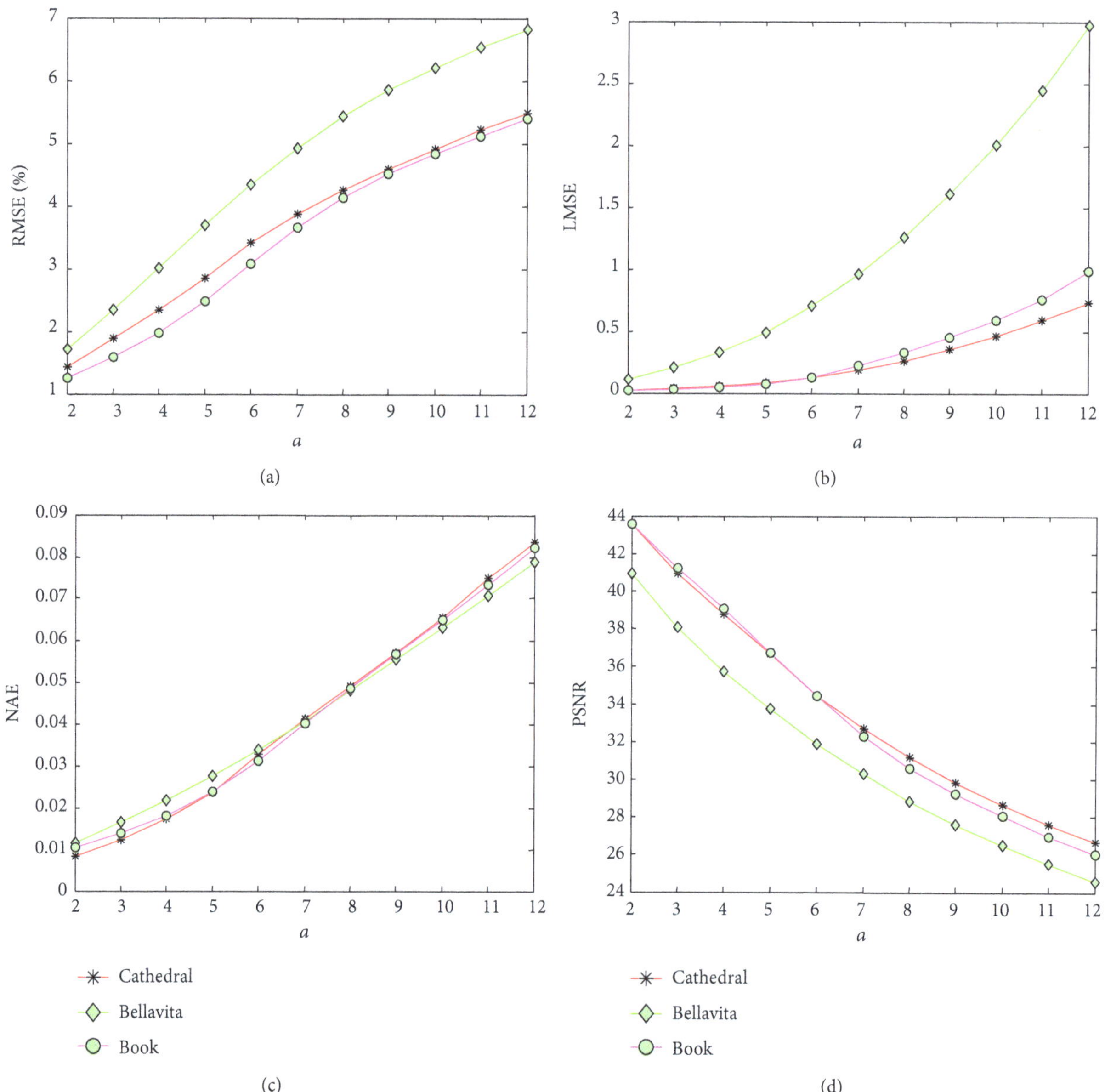

FIGURE 11: Sensitivity analysis of errors introduced by the free parameter a. The errors increases dramatically as a becomes too large but increases slowly when $a \leq 6$. It is observed in (b) and (c) that error increases as a increases but still the deviation range is less. (a) Effectiveness of a on RMSE, (b) effectiveness of a on LMSE, (c) effectiveness of a on NAE, and (d) effectiveness of a on PSNR.

PSNR is the ratio between the maximum possible pixel value of the fused image (MAX_{I_F}) and the MSE, which is computed as follows:

$$\mathrm{PSNR} = 10 \cdot \log_{10}\left(\frac{\mathrm{MAX}_{I_F}^2}{\mathrm{MSE}}\right). \tag{33}$$

As shown in Figures 11(a)–11(c) the RMSE, LMSE, and NAE increase as parameter a increases. It should be noticed in Figure 11(a) that when $a = 12$, the total error introduced is still less than 7%. It is seen from the computed values of objective measures like NAE and LMSE (see Figures 11(b)-11(c)) for $a = 2, 3, 4, 5, 6, 7, 8, 9, 10, 11$, and 12, for input image sequence of Cathedral, Bellavita, and Book, the errors increase dramatically as free parameter a in (18) becomes too large but increase slowly when $a \leq 6$. Using graph presented in Figure 11(d), we want to illustrate what can happen if PSNR is used as distortion measure. It has been found that PSNR decreases gradually as free parameter a increases and can be seen that the proposed approach is performing consistently for different value of free parameter (i.e., a) proposed for detail enhancement.

Another interesting interactive tool for manipulating the detail and contrast in the multifocus image fusion has been experimented. Figures 12(a)–12(f) show the results generated for Clock and Leaves image series with the parameters $\lambda =$

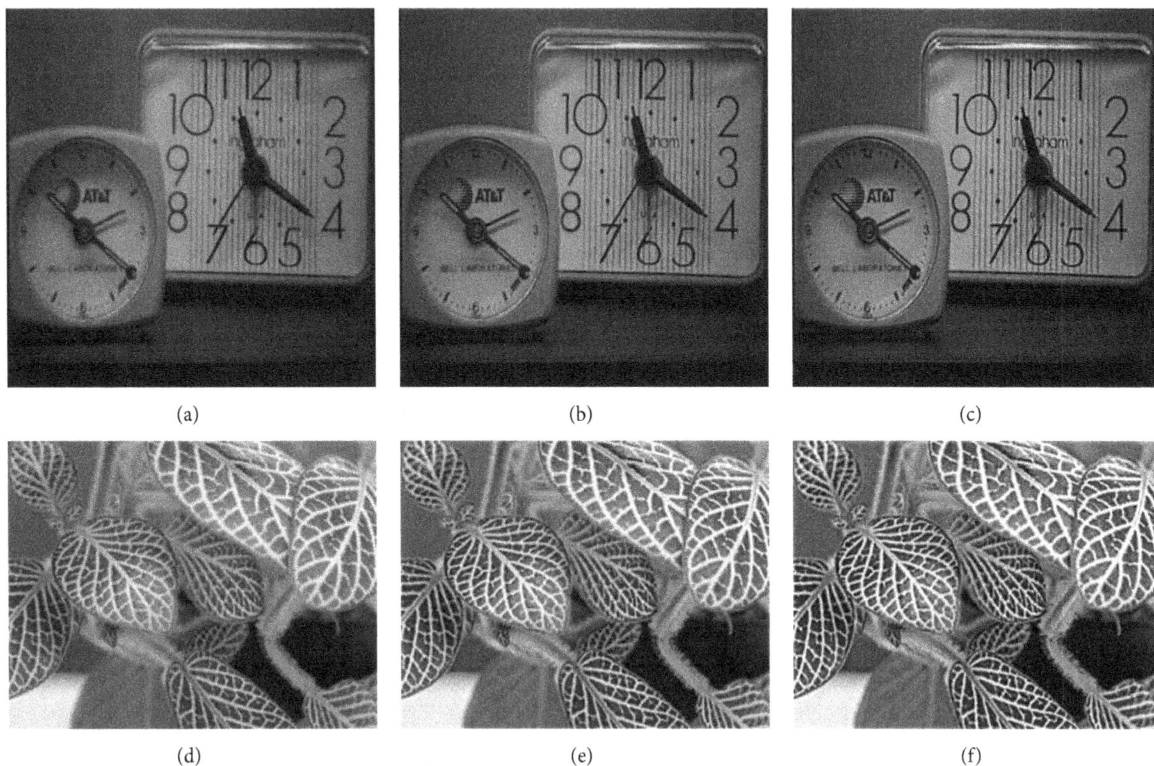

FIGURE 12: Visual inspection of contrast and detail enhancement in multifocus image fusion. The free parameter λ in (15) and a in (18) controls detail enhancement and sharpening, respectively. We are able to achieve aggressive detail enhancement and exaggeration, while avoiding artifacts. Clock image: (a) $\lambda = 0.5$, $a = 7$, (b) $\lambda = 0.5$, $a = 12$, (c) $\lambda = 0.5$, and $a = 17$. Leaves image: (d) $\lambda = 0.5$, $a = 7$, (e) $\lambda = 0.5$, $a = 12$, (f) $\lambda = 0.5$, and $a = 17$.

0.5, $a = 7$ (see Figure 12(a)), $\lambda = 0.5$, $a = 12$ (see Figure 12(b)), $\lambda = 0.5$, $a = 17$ (see Figure 12(c)), $\lambda = 0.5$, $a = 7$ (see Figure 12(d)), $\lambda = 0.5$, $a = 12$ (see Figure 12(e)), and $\lambda = 0.5$, $a = 17$ (see Figure 12(f)), respectively. Here, we demonstrate that we can generate highly detail enhanced fused image from multifocus image series, before objectionable artifacts appear. We found that present approach is very effective for boosting the amount of local contrast and fine details. The effective manipulation range is very wide and will vary in accordance with the texture details present in the input image series: it typically takes a rather extreme manipulation to cause artifacts near strong edges to appear.

5. Conclusion and Future Scope

Our proposed technique constructs a detail enhanced fused image from a set of multiexposure images by using WLS optimization framework. When compared with the existing techniques which use multiresolution and single resolution analysis for exposure fusion the present method perform better in terms of enhancement of texture details in the fused image. Our research was motivated by the edge-preserving property of anisotropic diffusion that has nonlinear response for fine textures and coarser details. The two-layer decomposition based on anisotropic diffusion is used to extract fine textures for detail enhancement. Furthermore, it is interesting to note here that our approach can also be applied for multifocus image fusion problem. More importantly, the information in the resultant fused image can be controlled with the help of proposed free parameters. At last, the future work involves improvement of this method for adaptively choosing the parameters of the WLS filter and checking the utilization for different kinds of image fusion applications.

Conflict of Interests

The authors declare that there is no conflict of interests regarding the publication of this paper.

Acknowledgments

The authors would like to thank Jacques Joffre, Adu and Wang, Max Lyons, Slavica Savic, and Eric Reinhard for the permission to use their images. They would like to thank Rui Shen for providing experimental results for analysis purpose. They would like to express gratitude to the anonymous reviewers for providing valuable suggestions and proposed corrections to improve the quality of the paper.

References

[1] E. Reinhard, G. Ward, S. Pattanaik, and P. Debvec, *High Dynamic Range Imaging Acquisition, Manipulation, and Display*, Morgan Kaufmann, 2005.

[2] P. E. Debevec and J. Malik, "Recovering high dynamic range radiance maps from photographs," in *Proceedings of the Conference on Computer Graphics (SIGGRAPH '97)*, pp. 369–378, August 1997.

[3] S. Mann and R. W. Picard, "Being "undigital" with digital cameras: extending dynamic range by combining differently exposed pictures," in *Proceedings of IST's 48th Annual Conference*, pp. 442–448, May 1995.

[4] K. Jacobs, C. Loscos, and G. Ward, "Automatic high-dynamic range image generation for dynamic scenes," *IEEE Computer Graphics and Applications*, vol. 28, no. 2, pp. 84–93, 2008.

[5] G. Ward, "Fast, robust image registration for compositing high dynamic range photographs from hand-held exposures," *Journal of Graphics Tools*, vol. 8, no. 2, pp. 17–30, 2003.

[6] A. Tomaszewska and R. Mantiuk, "Image registration for multi-exposure high dynamic range image acquisition," in *Proceedings of the International Conference on Computer Graphics, Visualization and Computer Vision*, Plzen, Czech Republic, 2007.

[7] E. Reinhard, M. Stark, P. Shirley, and J. Ferwerda, "Photographic tone reproduction for digital images," *ACM Transactions on Graphics*, vol. 21, no. 3, pp. 267–276, 2002.

[8] Y. Li, L. Sharan, and E. H. Adelson, "Compressing and Companding high dynamic range images with subband architectures," *ACM Transactions on Graphics*, vol. 24, no. 3, pp. 836–844, 2005.

[9] R. Fattal, D. Lischinski, and M. Werman, "Gradient domain high dynamic range compression," *ACM Transactions on Graphics*, vol. 21, no. 3, pp. 249–256, 2002.

[10] F. Durand and J. Dorsey, "Fast bilateral Filtering for the display of high dynamic range images," *ACM Transaction on Graphics*, vol. 21, no. 3, pp. 257–266, 2002.

[11] G. W. Larson, H. Rushmeier, and C. Piatko, "A visibility matching tone reproduction operator for high dynamic range scenes," *IEEE Transactions on Visualization and Computer Graphics*, vol. 3, no. 4, pp. 291–306, 1997.

[12] F. Drago, K. Myszkowski, T. Annen, and N. Chiba, "Adaptive logarithmic mapping for displaying high contrast scenes," *Computer Graphics Forum*, vol. 22, no. 3, pp. 419–426, 2003.

[13] E. Reinhard and K. Devlin, "Dynamic range reduction inspired by photoreceptor physiology," *IEEE Transactions on Visualization and Computer Graphics*, vol. 11, no. 1, pp. 13–24, 2005.

[14] H. Seetzen, W. Heidrich, W. Sturzlinger et al., "High dynamic range display system," *ACM Transaction on Graphics*, vol. 23, no. 3, pp. 760–768, 2004.

[15] H. Seetzen, L. A. Whitehead, and G. Ward, "A high dynamic range display using low and high resolution modulator," *In the Society For InFormation Display International Symposium*, vol. 34, no. 1, pp. 1450–1453, 2003.

[16] K. Kotwal and S. Chaudhuri, "An optimization-based approach to fusion of multi-exposure, low dynamic range images," in *Proceedings of the 14th International Conference on Information Fusion*, Chicago, Ill, USA, July 2011.

[17] S. Raman and S. Chaudhuri, "Bilateral filter based compositing for variable exposure photography," in *Proceedings of Eurographics*, Munich, Germany, 2009.

[18] T. Mertens, J. Kautz, and F. Van Reeth, "Exposure fusion: a simple and practical alternative to high dynamic range photography," *Computer Graphics Forum*, vol. 28, no. 1, pp. 161–171, 2009.

[19] J. Kuang, G. M. Johnson, and M. D. Fairchild, "iCAM06: a refined image appearance model for HDR image rendering," *Journal of Visual Communication and Image Representation*, vol. 18, no. 5, pp. 406–414, 2007.

[20] Z. Farbman, R. Fattal, D. Lischinski, and R. Szeliski, "Edge-preserving decompositions for multi-scale tone and detail manipulation," *ACM Transactions on Graphics*, vol. 27, no. 3, article 67, 2008.

[21] R. Shen, I. Cheng, J. Shi, and A. Basu, "Generalized random walks for fusion of multi-exposure images," *IEEE Transactions on Image Processing*, vol. 20, no. 12, pp. 3634–3646, 2011.

[22] T. O. Aydin, R. Mantiuk, K. Myszkowski, and H.-P. Seidel, "Dynamic range independent image quality assessment," *ACM Transactions on Graphics*, vol. 27, no. 3, article 69, 2008.

[23] J. H. Adu and M. Wang, "Multi-focus image fusion based on WNMF and focal point analysis," *Journal of Convergence Information Technology*, vol. 6, no. 7, pp. 109–117, 2011.

[24] P. M. Zeeuw, "Wavelet and image fusion, CWI," Amsterdam, The Netherlands, March 1998, http://www.cwi.nl/~pauldz/.

[25] J. Tian, L. Chen, L. Ma, and W. Yu, "Multi-focus image fusion using a bilateral gradient-based sharpness criterion," *Optics Communications*, vol. 284, no. 1, pp. 80–87, 2011.

[26] J. Shen, Y. Zhao, and Y. He, "Detail-preserving exposure fusion using subband architecture," *The Visual Computer*, vol. 28, no. 5, pp. 463–473, 2012.

[27] P. Perona and J. Malik, "Scale-space and edge detection using anisotropic diffusion," *IEEE Transactions on Pattern Analysis and Machine Intelligence*, vol. 12, no. 7, pp. 629–639, 1990.

[28] H. Singh, V. Kumar, and S. Bhooshan, "Anisotropic diffusion for details enhancement in multiexposure image fusion," *ISRN Signal Processing*, vol. 2013, Article ID 928971, 18 pages, 2013.

[29] G. Linda Shapiro and C. George, *Stockman, Computer Vision*, Prentice-Hall, Upper Saddle River, NJ, USA, 2001.

[30] K. Laws, *Textured image segmentation [Ph.D. dissertation]*, University of Southern California, 1980.

[31] G. Qiu and J. Duan, "An optimal tone reproduction curve operator for the display of high dynamic range images," in *IEEE International Symposium on Circuits and Systems (ISCAS '05)*, pp. 6276–6279, May 2005.

[32] M. H. Kim, T. Weyrich, and J. Kautz, "Modeling human color perception under extended luminance levels," *ACM Transactions on Graphics*, vol. 28, no. 3, article 27, 2009.

[33] J. Tumblin, J. K. Hodgins, and B. K. Guenter, "Two methods for display of high contrast images," *ACM Transactions on Graphics*, vol. 18, no. 1, pp. 56–94, 1999.

[34] A. Ardeshir Goshtasby, *2-D and 3-D Image Registration for Medical, Remote Sensing, and Industrial Applications*, Wiley-Interscience, 2005.

[35] G. Ward, "Fast, robust image registration for. Compositing high dynamic range photographs from hand-held exposures," *Journal of Graphics Tools*, vol. 8, pp. 17–30, 2003.

[36] J. M. Ogden, E. H. Adelson, J. R. Bergen, and P. J. Burt, "Pyramid based computer graphics," *RCA Engineer*, vol. 30, no. 5, pp. 4–15, 1985.

[37] P. J. Burt and E. H. Adelson, "The laplacian pyramid as a compact image code," *IEEE Transactions on Communications*, vol. 31, no. 4, pp. 532–540, 1983.

[38] A. Agrawal, R. Raskar, S. K. Nayar, and Y. Li, "Removing photography artifacts using gradient projection and flash exposure sampling," *ACM Transaction on Graphics*, vol. 24, no. 3, pp. 828–835, 2005.

[39] G. Petschnigg, R. Szeliski, M. Agrawala, M. F. Cohen, H. Hoppe, and K. Toyama, "Digital photoy with flash and no-flash image pairs," *ACM Transactions on Graphics*, vol. 23, no. 3, pp. 664–672, 2004.

[40] S. T. Li and B. Yang, "Multifocus image fusion using region segmentation and spatial frequency," *Image and Vision Computing*, vol. 26, no. 7, pp. 971–979, 2008.

[41] A. A. Goshtasby, "Fusion of multi-exposure images," *Image and Vision Computing*, vol. 23, no. 6, pp. 611–618, 2005.

[42] R. Szeliski, "System and process for improving the uniformity of the exposure and tone of a digital image," U. S. Patent No. 6687400, 2004.

[43] S. Li, X. Kang, and J. Hu, "Image fusion with guided filtering," *IEEE Transaction on Image Processing*, vol. 22, no. 7, pp. 2864–2875, 2013.

[44] S. Li and X. Kang, "Fast multi-exposure image fusion with median filter and recursive filter," *IEEE Transactions on Consumer Electronics*, vol. 58, no. 2, 2012.

[45] C. Tomasi and R. Manduchi, "Bilateral filtering for gray and color images," in *Proceedings of the IEEE 6th International Conference on Computer Vision*, pp. 839–846, January 1998.

[46] D. Barash, "A fundamental relationship between bilateral filtering, adaptive smoothing, and the nonlinear diffusion equation," *IEEE Transactions on Pattern Analysis and Machine Intelligence*, vol. 24, no. 6, pp. 844–847, 2002.

[47] D. Lischinski, Z. Farbman, M. Uyttendaele, and R. Szeliski, "Interactive local adjustment of tonal values," *ACM Transaction on Graphics*, vol. 25, pp. 646–653, 2006.

[48] C. S. Xydeas and V. Petrović, "Objective image fusion performance measure," *Electronics Letters*, vol. 36, no. 4, pp. 308–309, 2000.

[49] Y. Han, Y. Cai, Y. Cao, and X. Xu, "A new image fusion performance metric based on visual information fidelity," *Information Fusion*, vol. 14, no. 2, pp. 127–135, 20131.

[50] G. Qu, D. Zhang, and P. Yan, "Information measure for performance of image fusion," *Electronics Letters*, vol. 38, no. 7, pp. 313–315, 2002.

[51] R. Sakuldee and S. Udomhunsakul, "Objective performance of compressed image quality assessments," in *Proceedings of World Academy of Science, Engineering and Technology*, vol. 26, pp. 434–443, 2007.

16

Statistically Matched Wavelet Based Texture Synthesis in a Compressive Sensing Framework

Mithilesh Kumar Jha, Brejesh Lall, and Sumantra Dutta Roy

Department of Electrical Engineering, Indian Institute of Technology Delhi, Hauz Khas, New Delhi 110016, India

Correspondence should be addressed to Mithilesh Kumar Jha; jham73@gmail.com

Academic Editors: C.-W. Kok, S. Kwong, A. M. Peinado, and A. Rubio Ayuso

This paper proposes a statistically matched wavelet based textured image coding scheme for efficient representation of texture data in a compressive sensing (CS) frame work. Statistically matched wavelet based data representation causes most of the captured energy to be concentrated in the approximation subspace, while very little information remains in the detail subspace. We encode not the full-resolution statistically matched wavelet subband coefficients but only the approximation subband coefficients (LL) using standard image compression scheme like JPEG2000. The detail subband coefficients, that is, HL, LH, and HH, are jointly encoded in a compressive sensing framework. Compressive sensing technique has proved that it is possible to achieve a sampling rate lower than the Nyquist rate with acceptable reconstruction quality. The experimental results demonstrate that the proposed scheme can provide better PSNR and MOS with a similar compression ratio than the conventional DWT-based image compression schemes in a CS framework and other wavelet based texture synthesis schemes like HMT-3S.

1. Introduction

Texture data contain spatial, temporal, statistical, and perceptual redundancies. Representing texture data using standard compression schemes like MPEG-2 [1] and H.264 [2] is not efficient, as they are based on Shannon-Nyquist sampling [3] and do not account for perceptual redundancies. They are often resource consuming (as they acquire too many samples) due to its fine details in textured image and high frequency content. Variety of applications in computer vision, graphics, and image processing (such as robotics, defence, medicine, and geosciences) demands better compression with good perceptual reconstruction quality, instead of bit accurate (high PSNR) reconstruction. This is because the human brain is able to decipher important variations in data at scales smaller than those of the viewed objects. Ndjiki-Nya et al. [4–8], Bosch et al. [9, 10], Byrne et al. [11, 12], and Zhang et al. [13, 14] have proposed techniques to reconstruct visually similar texture from sample data. Statistically matched wavelet [15] is aimed at designing a filter bank that matches a given pattern in the image and can better represent the corresponding image as compared to other wavelet families.

Compressive sensing (CS) technique [16] has proved that it is possible to achieve a sampling rate lower than the Nyquist rate [3] with acceptable reconstruction quality. Leveraging the concept of transform coding, compressive sensing enables a potentially large reduction in sampling and computation costs for sensing signals that have sparse or compressible representation (by a sparse representation, we mean that for a signal of length N, we can represent it with $K \ll N$ nonzero coefficients). Compressive sensing framework opens a new research dimension in which most of the sparse signals can be reconstructed from a small number of measurements (M), using algorithms like convex optimization, greedy methods, and iterative thresholding [17, 18]. Significant theoretical contributions have been published on the compressive sensing in recent years [16, 17, 19] for image processing applications [20–24]. Compressive sensing framework mainly consists of three stages, that is, sparsification by transformation, measurement (projection), and optimization (reconstruction). Designing

Table 1: Comparative study of texture analysis and synthesis schemes.

Proposed work	Texture analysis and synthesis technique	Texture type	Quality assessment	Limitation	Complexity
Wang and Adelson [29]	Affine warping	Rigid textures	None	Not suitable for structural texture	Medium
Dumitras and Haskell [30]	Steerable pyramids	Rigid textures	None	Not suitable for structural texture	Medium
Ndjiki-Nya and Wiegand [8]	Perspective warping	Rigid and nonrigid textures	Yes	Prone to propagation error	High
Zhang and Bull [14]	ARMA	Rigid and nonrigid textures	Yes	Not suitable for structural texture	Medium
Portilla and Simoncelli [37]	Wavelet based (joint statistics)	Rigid and nonrigid textures	None	Not suitable for structural texture	Medium
Fan and Xia [42]	Wavelet based (HMT-3S)	Rigid and nonrigid textures	None	Not suitable for structural texture	Medium
Our scheme (SMWT-CS)	Compressive sensing and statistically matched wavelet	Rigid and nonrigid textures	Yes (PSNR and MOS)	CS optimization still evolving	Medium

a good measurement matrix with large compression effects and designing a good signal recovery algorithm are the two major challenges for applying CS technique in image compression.

1.1. The Prior Works and Motivation. Existing methods of textured image compression scheme can be broadly classified as parametric or nonparametric. Nonparametric approaches can be applied to a wide variety of textures (with irregular texture patterns) and provide better perceptual results. However, these schemes are often computationally more complex. Parametric approaches can achieve very high compression at low computational cost. However, these techniques are not effective for structured textures such as those with primarily nonstationary data content.

Nonparametric approaches are pixel-based or patch-based. Efros and Leung [25] proposed pixel-based nonparametric sampling to synthesize texture. Wei and Levoy [26] further improve the above using a multiresolution image pyramid based on a hierarchical statistical method. A limitation of the above pixel-based methods is an incorrect synthesis owing to incorrect matches in searching for similar statistics. Patch-based methods overcome this limitation by considering features matching patch boundaries with multiple pixel statistics. People generally use Markov Random Field (MRF) models for texture analysis [7, 26]. The popular choice for texture synthesis is a patch-based graph cut [27]. Region-based texture representation and synthesis algorithms [28–30] have been explored recently to address the limitations of block-based representation in handling the homogenous texture and blocking artifacts. Byrne et al. [11, 12] have demonstrated region-based synthesis structure using a morphological and spectral image representation technique [31]. Region-based representation has also been explored using multiresolution wavelet based decomposition as reported in [32–34]. However that work is limited to document class of images.

Recent work in the parametric approach is typically based on Auto Regressive (AR) [38] or Auto Regressive Moving Average- (ARMA-) based modelling [39–41]. AR- and ARMA-based models in texture synthesis enable blocks to be selectively removed at the encoding stage and reconstructed at the decoder with acceptable perceptual quality. AR- and ARMA-based approaches are suitable for the textures with stationary data, like steady water, grass, and sky; however, they are not suitable for structured texture with nonstationary data as blocks with nonstationary which data are not amenable to AR modelling. Further, they are block-based approaches, and blocking artifacts can appear in the synthesised image. Portilla and Simoncelli [37] propose a statistical model for texture images based on joint statistical constraint on the wavelet coefficients. In the above work, authorspro-posed an algorithm for synthesizing textures using sequential projections on to the constraints surfaces; however, the model is the choice of the statistical constraints and not suitable for texture with structural pattern. Wavelet domain hidden Markov tree (HMT-3S) [42] was recently used for texture analysis and synthesis, where it was assumed that three subbands of the two-dimensional DWT, that is, HL, LH, and HH, are independent. The HMT-3S adds dependencies across subbands by treating jointly the three hidden elements across the three orientations; however, HMT-3S is established in the nonredundant DWT and is inferior to redundant DWT for statistical modelling. Also, for structural textures with regular patterns, both HMT and HMT-3S fail to reproduce the regular patterns.

In this paper, we propose a statistically matched wavelet based texture data representation and a compressive sensing based texture synthesis scheme (*henceforth we refer the proposed scheme in this paper as SMWT-CS*). Statistically matched wavelet based representation [43] causes most of the captured energy to be concentrated in the approximation subspace, while very little information is retained in the detail subspace. We encode not the full-resolution statistically matched wavelet subband coefficients (as normally done in a standard wavelet based image compression) but

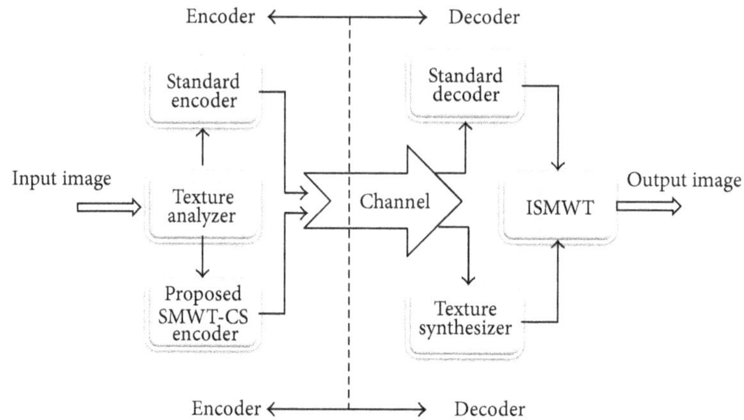

FIGURE 1: Overview of analysis and synthesis based image compression.

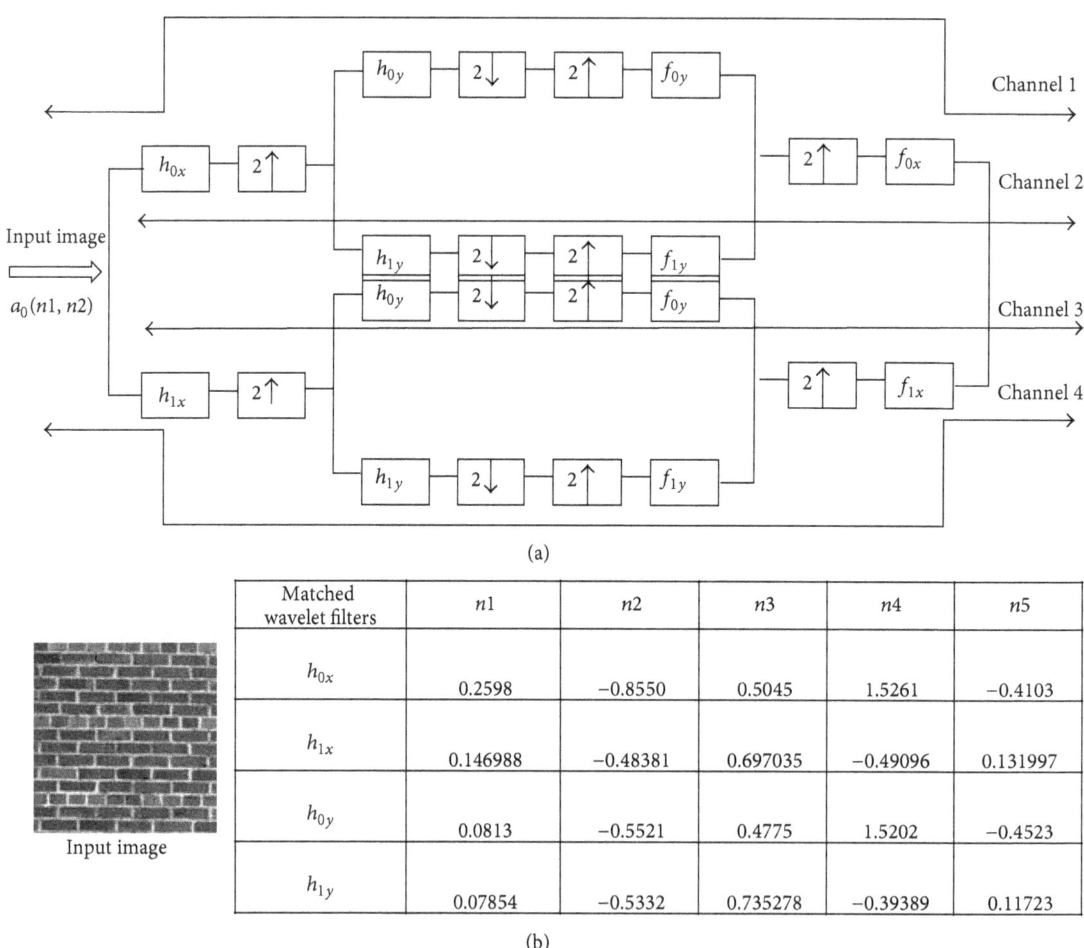

Matched wavelet filters	$n1$	$n2$	$n3$	$n4$	$n5$
h_{0x}	0.2598	−0.8550	0.5045	1.5261	−0.4103
h_{1x}	0.146988	−0.48381	0.697035	−0.49096	0.131997
h_{0y}	0.0813	−0.5521	0.4775	1.5202	−0.4523
h_{1y}	0.07854	−0.5332	0.735278	−0.39389	0.11723

(b)

FIGURE 2: Statistically matched wavelet estimation. (a) Matched Wavelet filter bank (output of Channel 1 represents approximation subspace while output of Channels 2, 3, and 4 is detailed subspace). (b) Estimated analysis filter coefficients for brick wall test image.

only the approximation subband coefficients (LL) using standard image compression scheme like JPEG2000 [44] (which accounts for 1/4th of the total coefficients and can be represented using fewer bits). The detail subband coefficients, that is, HL, LH, and HH (which account for 3/4th of the total coefficients), are jointly encoded in a compressive sensing framework and can therefore be represented with fewer measurements. Quality matrix is an essential component for assessing the performance of an image compression system. We have computed Mean Opinion Score (MOS) for

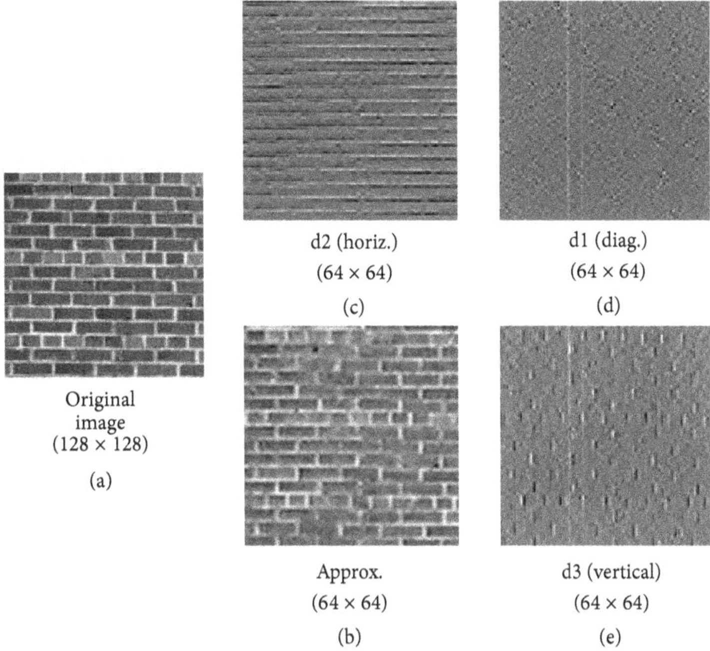

FIGURE 3: Illustration of statistically matched wavelet decomposition (a) input image, (b) approximation subspace image, and (c) to (e) detailed subspace image respective to horizontal, diagonal, and vertical subbands.

subjective assessment as it captures the visual perception of human subjects. We have also computed PSNR for objective assessment of the quality for comparative study.

Table 1 provides a comparison of existing texture analysis and synthesis scheme. We have included our proposed scheme to put the work in perspective. As can be seen the proposed scheme is more efficient than the existing schemes. The main difference between a conventional wavelet based compression scheme in a CS framework (DWT-CS) [21, 35, 45] and our approach is the use of Statistically Matched Wavelet based sparsification [15, 43], as opposed to generic DWT used in previous works. The existing schemes use regular DWT that is not able to fully exploit the signal properties (orthogonal/nonredundant DWT cannot well preserve the regularity and/or periodicity [46]) and hence do not fully exploit the sparsity. CS optimization that is done over detail subspace allows for scalability in choosing measurement vector size providing a tradeoff between compression and perceptual reconstruction quality. The other novelty is that we apply standard coding on the low resolution part (LL subband) of the information, because of the nature of wavelet decomposition (i.e., true especially for multilevel wavelet decomposition); the low frequency information is not sparse and hence not amenable for CS based encoding. Our proposed scheme can be easily integrated into an existing encoding framework [38] with texture analysis and synthesis being performed by the proposed scheme. So far, multiresolution based representation has been used in the domain of documentation class of image; however, to the best of our knowledge no work has reported using statistically matched wavelet and compressive sensing based analysis and synthesis technique for a generic texture class images for compression with subjective reconstruction quality.

1.2. Overview of the Scheme. Figure 1 gives an overall view of the proposed scheme. In this figure, the texture analyzer block decomposes the input textured image into approximation (LL) and detail (HL, LH, and HH) subbands using statistically matched wavelet based image representation [15]. The basic idea is to design a statistically matched wavelet filter bank using source data and decompose the input textured image into approximation and detail subspace [43]. Standard image encoder like JPEG2000 [44] is used to encode approximation subband coefficients (LL). The detail subband coefficients, that is, HL, LH, and HH are jointly encoded in a compressive sensing framework. The proposed SMWT-CS encoder block performs compressive measurement using Noiselet transform [47] over detail subspace coefficients (joint representation for HL, LH, and HH subband coefficients) along with quantization and entropy coding. For a K-sparse signal x, we compute the measurement vector $M \times 1$, which is much smaller in length than the length of signal ($N \times 1$), and therefore the compression is guaranteed. Texture synthesis block does the compressive sensing optimization to synthesize the samples from detail subspace measurements, using convex optimization [16, 19], that is, l_1-norm minimization with equality constraint. Combining the decoded samples from the standard decoder and the texture synthesizer and doing the inverse of statistically matched wavelet transform, a synthesized image is reconstructed.

The rest of the paper is organized as follows. Section 2 provides an overview of statistically matched wavelet based texture representation scheme. Section 3 presents the design and implementation of the compressive measurements and encoding framework, while Section 4 provides a detailed view of synthesis framework. The experimental results are discussed in Section 5 followed by conclusion in Section 6.

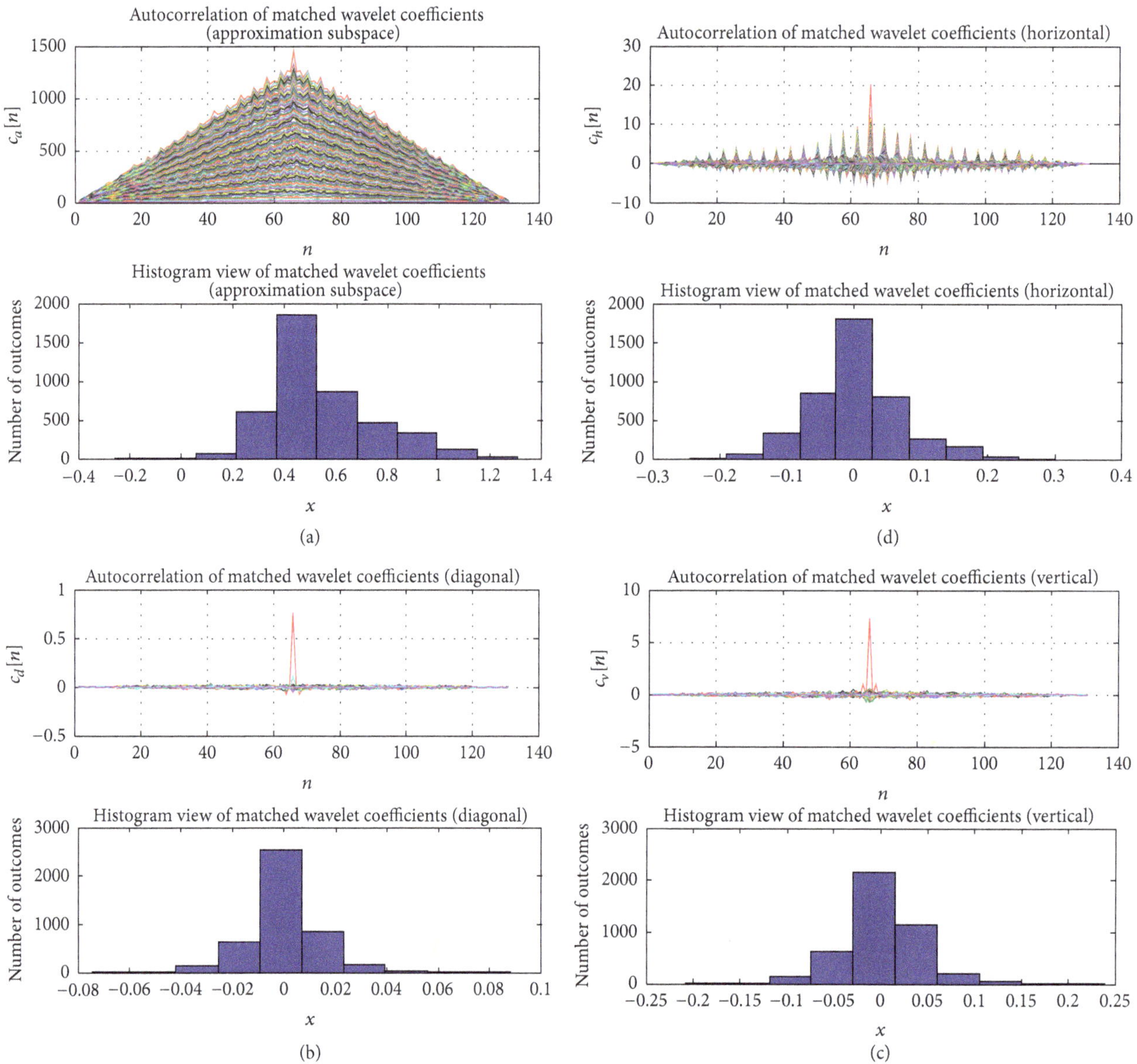

FIGURE 4: Illustration of statistically matched wavelet coefficients distribution in approximation and detail subspace, (a) Histogram and auto corelation plot of Matched wavelet coefficients in approximation sub-space, (b) to (d) Histogram and auto corelation plot of matched wavelet coefficients in detail subspace respective to diagonal, vertical, and horizontal subbands.

2. Statistically Matched Wavelet Based Texture Representation

Gupta et al. [15] propose statistically matched wavelet for the estimation of wavelets that is matched to a given signal in the statistical sense. This concept is further extended for image data and Figure 2 shows a 2D two-band separable kernel wavelet system (with an example of estimated matched wavelet filters for brick wall image). In this figure, x and y represent the horizontal and vertical directions, respectively. The scaling filter is represented by f_{0x} and f_{0y} while the wavelet filter is represented by f_{1x} and f_{1y} corresponding to the horizontal and vertical direction. The dual of them is represented by h_{0x}, h_{0y}, h_{1x}, and h_{1y}. The input to this system is a 2D signal (an image in our framework). The output of Channel 1 in Figure 2 is called approximation subspace or scaling subspace while the outputs of the other three channels (Channels 2, 3, and 4 in Figure 2) are called detail subspace. This system is designed as biorthogonal wavelet system so that it satisfies the condition (1) for perfect reconstruction of the two-band filter bank:

$$h_{1i}(n) = (-1)^n f_{0i}(d-n),$$

$$f_{1i}(n) = (-1)^n h_{0i}(d-n),$$

$$\sum_n h_{0i}(n-2m1)\, f_{0i}(n-2m2)$$

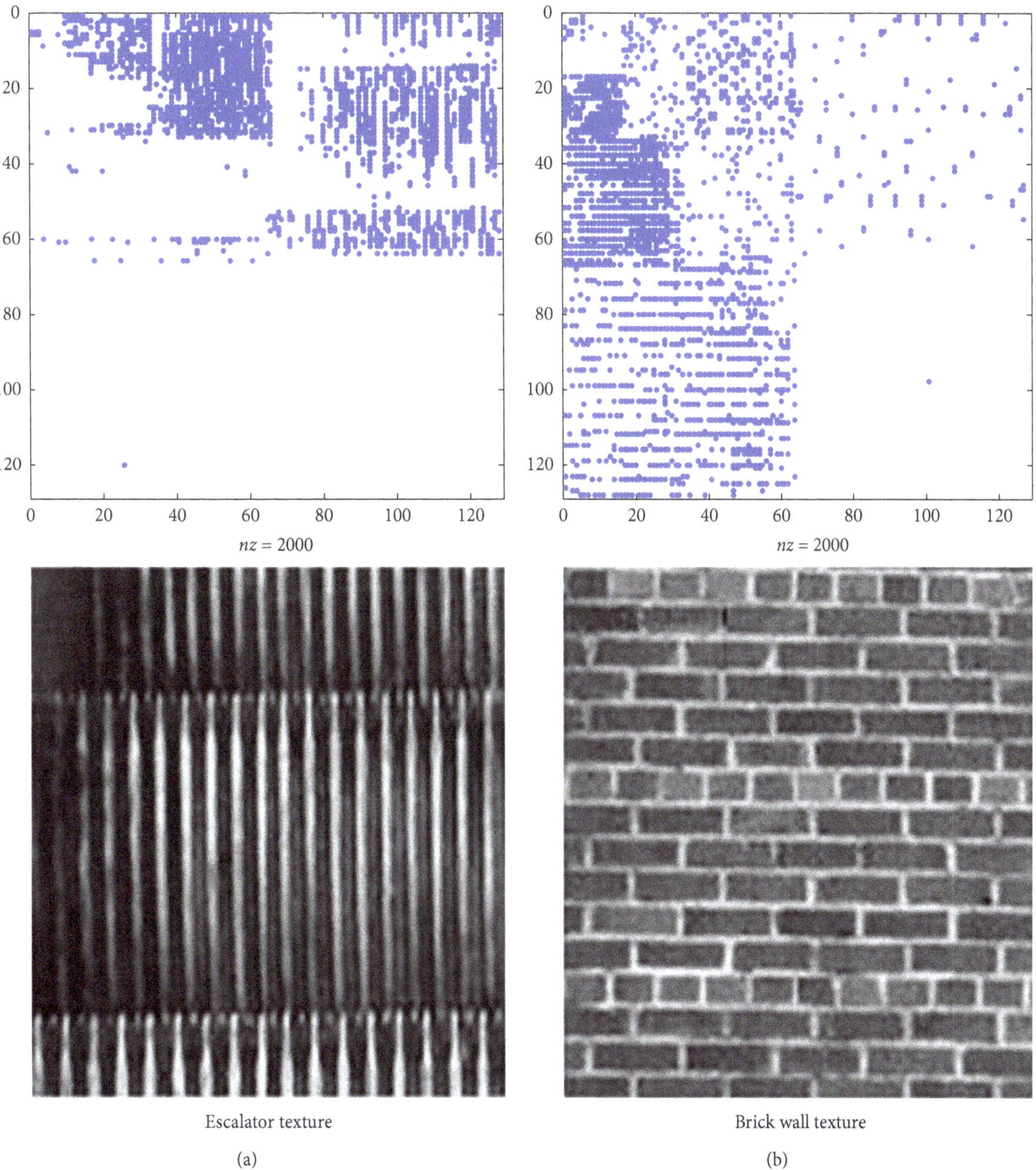

(a) (b)

FIGURE 5: Sparse representation of statistically matched wavelets subbands coefficients for brickwall and escalator texture.

$$= \delta(m1 - m2), \quad \forall m1, m2 \in Z,$$

$$\sum_n h_{0i}(n) h_{1i}(n) = 0, \tag{1}$$

where i can be x or y, respective to horizontal or vertical direction, and d is any odd delay. f_{0i} is the scaling filter while f_{1i} is the wavelet filter. h_{0i} and h_{1i} are the dual of scaling and wavelet filters, respectively.

The optimization criterion used for matched wavelet is the minimization of the energy in the detail subspace [43]. If $a(x,y)$ is a 2D image signal and $\hat{a}(x,y)$ represents the 2D image reconstructed using only the output of Channels 2, 3, and 4 (detail subspace) in Figure 2, then the error function $e(x,y)$ can be defined as

$$e(x, y) = a(x, y) - \hat{a}(x, y). \tag{2}$$

To ensure that the maximum input signal energy moves to approximation subspace, the energy E captured in the difference signal $e(x,y)$ should be maximized with respect to

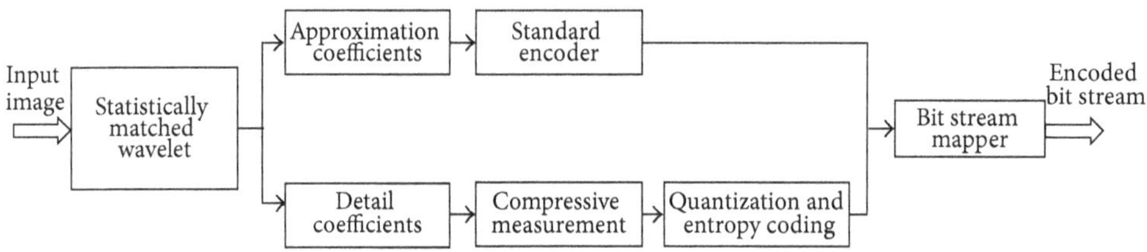

FIGURE 6: Encoder framework.

(1) Design a 2-D separable kernel filter bank using Statistically Matched Wavelet [Detailed in Section 2]
  (a) Decompose input image into approximation and detail subbands coefficients
    (i) Use Wavelet function from MATLAB "wavecut" to represent approximation and detail coefficients
(2) Design a CS measurement matrix Φ using Noiselet Transform [47] and detail subbands coefficients
(3) Do quantization of the CS measurements using (7)
(4) Do Entropy coding of the quantized CS measurements
(5) Combine standard and CS encoded bit streams

ALGORITHM 1: CS measurement and encoding.

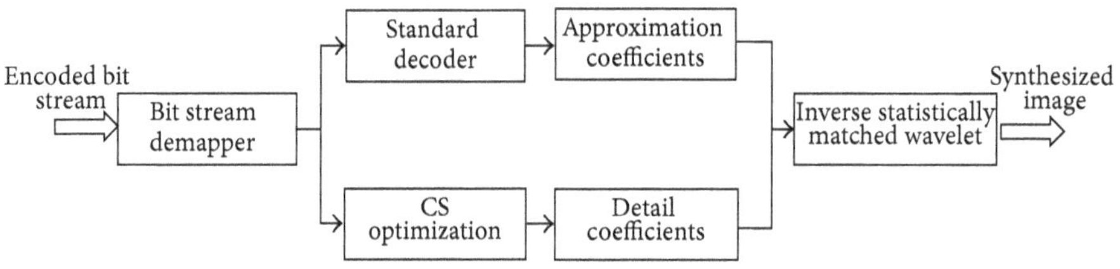

FIGURE 7: Decoder framework.

Inputs: Initial Reference Point, Observation Vector, Optimization Parameter

Initialize: $s_a = N \times 1$ vector(guess point), $y = k \times 1$ measurement vector,
T (tolerance for primal-dual algorithm) = 5, I (Maximum primal dual iterations) = 20,
T_g (Tolerance for conjugate gradients) = $1e - 8$, I_g (Maximum conjugate gradients) = 300

If (Valid Starting Reference, s_a)
  Minimize $f(x)$ subject to $\Theta x = s_a$ (Where, f-convex, rank$\Theta = M < N$)
    (Use Primal Dual Interior Point Methods with Equality Constraint)

  While (Surrogate duality gap < T OR Iterations > I) do
    (1) Optimality Condition: $\nabla f(x^*) + \Theta^T V^* = 0, \Theta(x^*) = s_a$
    (2) Compute the Newton step and decrement Δx_{nt}, $\lambda(x)$
    (3) Solve $K \times K$ Positive definite system of equations from Newton step
      (3a) Use Conjugate Gradients Method, MATLAB command, cgsolve
      (3b) Stop Conjugate Gradient algorithm, If (norm($\Theta_x - s_a$)/ norm(s_a) < T_g)
    (4) Do line Search. Choose step size **t** by backtracking line search
    (5) Update. $x = x + t\Delta x_{nt}$
    (6) Update Central and Dual residuals
  end while
end If
Output: Sparse representation Wavelet Coefficients $\hat{\mathbf{s}}$

ALGORITHM 2: Texture synthesis in a CS framework.

TABLE 2: Estimate of statistically matched wavelet filters.

Test sequence	Matched wavelet filters	n_1	n_2	n_3	n_4	n_5
Brick wall	h_{1x}	0.1441	−0.4869	0.6997	−0.4855	0.1300
	h_{1y}	0.0862	−0.5230	0.7329	−0.4099	0.1180
Escalator	h_{1x}	0.1564	−0.4931	0.6894	−0.4827	0.1553
	h_{1y}	0.1248	−0.3860	0.7252	−0.5496	0.0860
Floor box	h_{1x}	0.1256	−0.4708	0.7171	−0.4852	0.1134
	h_{1y}	0.1018	−0.3761	0.7148	−0.5668	0.1264
Black hole	h_{1x}	0.1276	−0.4873	0.7047	−0.4808	0.1360
	h_{1y}	0.1058	−0.4367	0.7065	−0.5257	0.1503
D20	h_{1x}	0.1197	−0.4864	0.7371	−0.4459	0.0832
	h_{1y}	0.1080	−0.4851	0.7307	−0.4558	0.1062
D36	h_{1x}	0.1012	−0.4673	0.7495	−0.4516	0.0753
	h_{1y}	0.0661	−0.4249	0.7732	−0.4633	0.0517
D75	h_{1x}	0.1301	−0.4776	0.7051	−0.4896	0.1342
	h_{1y}	0.1366	−0.5195	0.7060	−0.4446	0.1242
D68	h_{1x}	0.1341	−0.4849	0.7026	−0.4832	0.1405
	h_{1y}	0.1171	−0.4321	0.7241	−0.5139	0.1057
D87	h_{1x}	0.1354	−0.4858	0.7052	−0.4811	0.1303
	h_{1y}	0.1529	−0.4996	0.6894	−0.4813	0.1418
Fish fabric	h_{1x}	0.0942	−0.4654	0.7448	−0.4607	0.0870
	h_{1y}	0.0687	−0.4580	0.7686	−0.4373	0.0583

both x and y direction filters. This leads to the following set of equations [43]:

$$\sum_k h_{1x}\left[\sum_m a_{0x}(2m+k)\,a_{0x}(2m+r)\right]=0 \quad \text{for } r=0,1,2,\ldots,j-1,j+1,\ldots,N-1, \tag{3}$$

$$\sum_k h_{1y}\left[\sum_m a_{0y}(2m+k)\,a_{0y}(2m+r)\right]=0 \quad \text{for } r=0,1,2,\ldots,j-1,j+1,\ldots,N-1. \tag{4}$$

Here the jth filter weight is kept constant leading to a close form expression. These is a set of $N-1$ linear equations in filter weight that can be solved simultaneously. All rows of image are placed adjacent to each other to form a 1D signal having variations in horizontal direction only and represented as a_{0x} in (3). Similarly all the columns of image are placed together to form 1D signal having variations in vertical direction only and represented as a_{0y} in (4). The solution of (3) and (4) gives the analysis high pass filters (wavelet filters) h_{1x} and h_{1y}. From these, other filters (analysis low pass, synthesis high pass, and synthesis low pass) are computed using finite impulse response perfect reconstruction biorthogonal filter bank design as presented with (1).

The estimation of scaling and wavelet function is done separately for the variations in the horizontal and vertical directions of the input image. Passing the input image through the 2D statistically matched wavelet filter bank, we get the output as subsampled image corresponding to approximation and detail subspace. Figure 3 demonstrates the decomposition of one of the input tests sequences in approximation and detail subspace and Figure 4 gives the distribution of the matched wavelet coefficients in approximation and detail subspace (horizontal, vertical, and diagonal subspace). As one can observe from Figures 3 and 4, most of the information is represented by approximation subspace alone and very little information is present in the detail subspace. In the rest of the paper we refer to the approximation or scaling subspace (output of Channel 1 in Figure 2) as *the candidate data for host encoding* and the detail subspace (sum of output of Channels 2, 3, and 4 in Figure 2) as *the candidate data for CS measurements and synthesis.*

3. Compressive Measurement and Encoding

Compressive sensing exploits the fact that most natural or artificial signals are sparse in some domain and hence compressible. A real valued signal $x \in \mathbf{R}^N$, can be represented as a function of basis vectors as follows [16, 17]:

$$\mathbf{x}=\sum_i^N s_i\psi_i \quad \text{or} \quad \mathbf{x}=\Psi\mathbf{s}, \tag{5}$$

where $\mathbf{x}$ and $\mathbf{s}$ are $N\times1$ column vectors and Ψ is an $N\times N$ sparsifying basis matrix. The signal x is called K-sparse if it can be represented as a linear combination of only K-basis vectors, that is, only K elements of the vector $\mathbf{s}$ are nonzero. The signal x can be treated as compressible if it can be well approximated by a signal with only K ($K \ll N$) nonzero coefficients. Figure 5 shows sparse representation of subbands coefficients corresponding to brick wall and escalator test sequences.

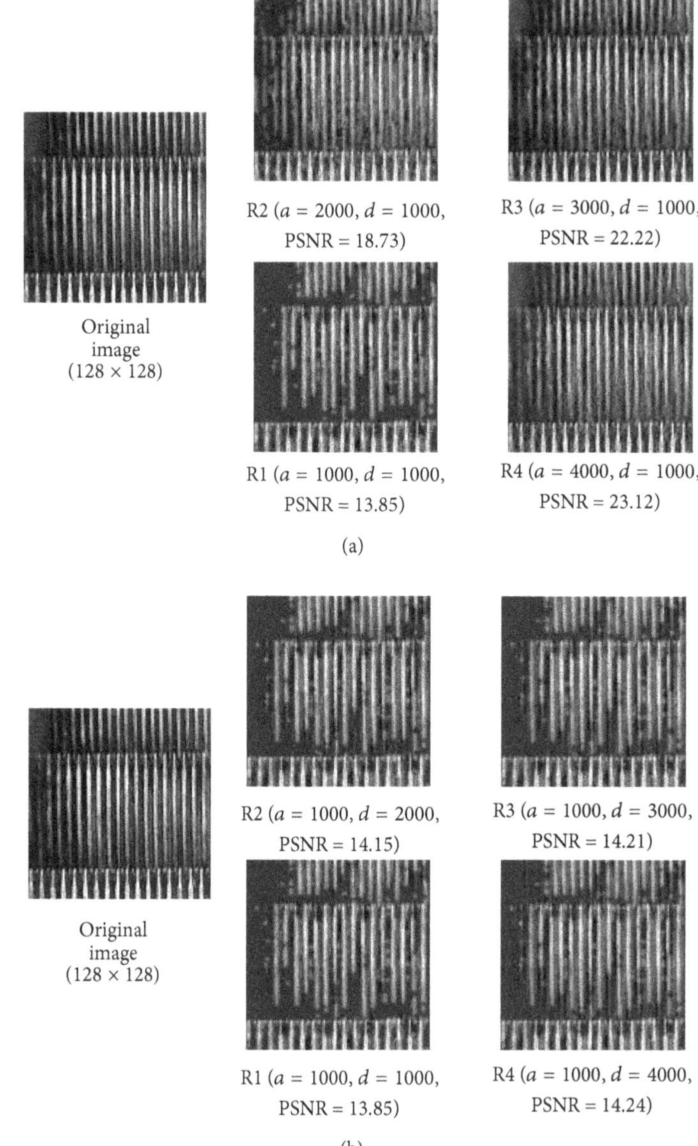

FIGURE 8: Illustrating the significance of approximation and detail subbands coefficients in image reconstruction. R1, R2, R3, and R4 are the reconstructed images using different approximation (indicated as a) and detail subbands coefficients (indicated as d in the figure). The upper part shows reconstruction for varying a with constant d, while lower part shows reconstruction for constant a with varying d.

Significant wavelet coefficients in the figure are represented by blue pixels while all other nonsignificant coefficients are shown in white. One can observe that most of the detail subbands coefficients are close to zero, and therefore CS can be best utilised to exploit this sparsity for encoding detailed subbands coefficients. Compressive measurement is computed through linear projection as shown in (6):

$$\mathbf{y} = \Phi\mathbf{x} = \Phi\Psi\mathbf{s} = \Theta\mathbf{s}. \tag{6}$$

Here $\mathbf{y}$ is an $M \times 1$ measurement vector where $M < N$. Θ is an $M \times N$ measurement matrix. The matrix Φ represents a dimensionality reduction matrix; that is, it maps $\mathbf{R}^N$ to $\mathbf{R}^M$, where M is typically much smaller than N. The main challenge with CS theory is that how should we design the sensing matrix Φ so that it preserves the information in the signal x. Several theories have been proposed in the literature for reconstructing x from y, if Θ satisfies a Restricted Isometry Property (RIP) [16]. We have used statistically matched wavelet for the sparsifying matrix (Ψ) and Noiselet [47] for sensing matrix (Φ) for compressive measurements according to (6). Measurements from Noiselet have been chosen because they are highly incoherent with the considered sparse domain and RIP tends to hold for reasonable values of M. In addition, noiselet comes with very fast algorithms and just like the Fourier

TABLE 3: Illustrating the PSNR and MOS data of the proposed model ((a) floor box, (b) black hole, (c) escalator, and (d) brick wall) using JPEG2000, conventional wavelet based CS scheme (DWT-CS) [21, 35], and our proposed CS scheme (SMWT-CS).

(a) Sequence: floor box (128 × 128).

CS measurements (Number of coefficients for JPEG2000)	(PSNR in db) (JPEG2000)	(PSNR in db) (DWT-CS)	(PSNR in db) (Our scheme, SMWT-CS)	(MOS) (Our scheme, SMWT-CS)
2000	39.19	30.47	42.86	5.0
4000	41.30	32.93	44.02	5.0
6000	41.85	36.39	45.40	5.0
8000	41.90	38.19	46.96	5.0
10000	41.90	42.27	48.98	5.0
12000	41.90	45.01	51.16	5.0

(b) Sequence: black hole (128 × 128).

CS measurements (Number of coefficients for JPEG2000)	(PSNR in db) (JPEG2000)	(PSNR in db) (DWT-CS)	(PSNR in db) (Our scheme, SMWT-CS)	(MOS) (Our scheme, SMWT-CS)
2000	38.88	31.18	46.8	5.0
4000	41.51	34.43	47.86	5.0
6000	41.97	36.72	49.06	5.0
8000	41.98	39.63	50.58	5.0
10000	41.98	42.86	52.24	5.0
12000	41.98	46.30	54.16	5.0

(c) Sequence: escalator (128 × 128).

CS measurement (Number of coefficients for JPEG2000)	(PSNR in db) (JPEG2000)	(PSNR in db) (DWT-CS)	(PSNR in db) (Our scheme, SMWT-CS)	(MOS) (Our scheme, SMWT-CS)
2000	30.86	13.93	17.22	3.0
4000	31.08	16.13	19.63	3.5
6000	34.76	18.26	22.13	4.0
8000	38.39	20.88	24.75	4.2
10000	40.07	23.75	27.84	4.4
12000	40.56	27.48	31.76	4.8

(d) Sequence: brick wall (128 × 128).

CS measurements (Number of coefficients for JPEG2000)	(PSNR in db) (JPEG2000)	(PSNR in db) (DWT-CS)	(PSNR in db) (Our scheme, SMWT-CS)	(MOS) (Our scheme, SMWT-CS)
2000	30.47	18.52	21.52	3.5
4000	31.41	20.49	22.79	3.5
6000	33.25	22.34	24.50	4.1
8000	35.19	24.50	26.17	4.1
10000	37.48	27.14	28.35	4.4
12000	38.95	30.50	31.04	4.8

transform, the noiselet matrix does not need to be stored to be applied to a vector. This is of crucial importance for efficient numerical computations without which applying CS can be very complex. The CS measurement matrix created is orthogonal and self-adjoint, thus being easy to manipulate. Scalar quantization over the measurement vector is proposed to achieve better compression ratio.

It is important to note that (6) is ill-conditioned since there are more unknowns than the number of equations as $M < N$; however, it has been shown that if the signal x is K-sparse and the locations of the K nonzero elements are known then the problem can be solved provided $M \geq K$ through a simplified linear equation by deleting all those columns and elements corresponding to zero or nonsignificant elements. Figure 6 gives an overall block diagram of the proposed encoder. Bit mapper block is responsible for generating a final bit stream encompassing encoded data from standard codecs for approximation subspace and CS measurements

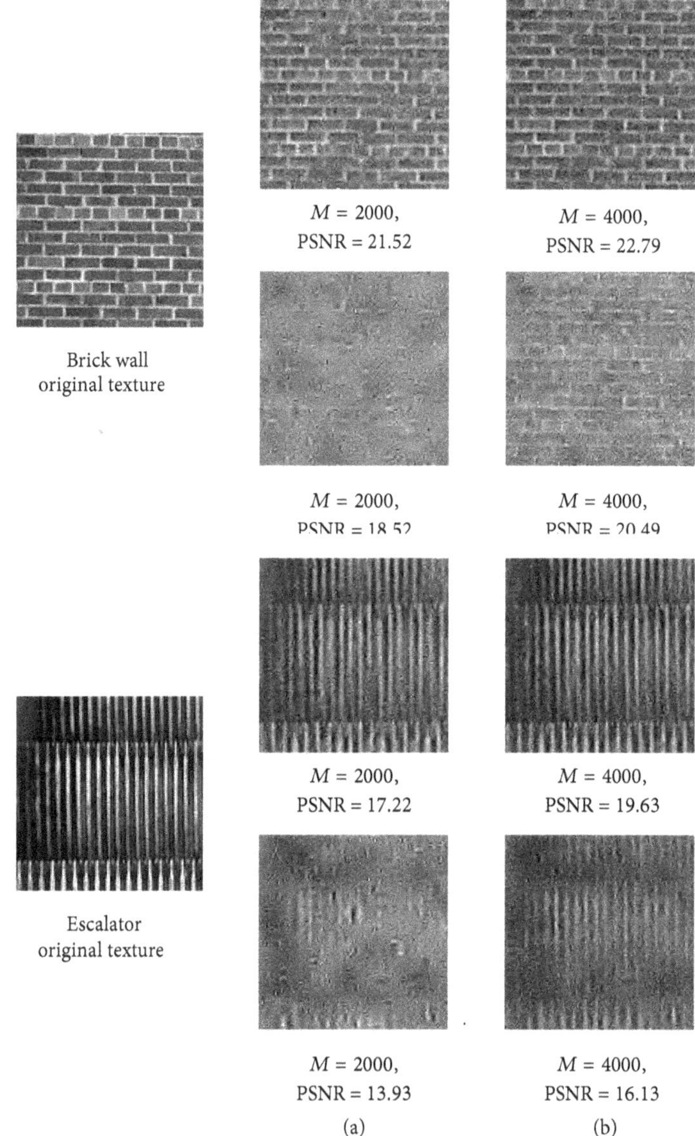

FIGURE 9: Synthesis results of structural texture with pseudoperiodic patterns (brick wall and escalator textures). For each original texture, the lower image pair is the synthesized texture using conventional DWT-based CS scheme for image synthesis [21, 35], and the upper image pair is the synthesized texture using our proposed scheme. (a) Synthesis results for CS measurements (M) = 2000. (b) Synthesis results for CS measurements (M) = 4000.

for detail subspace. Algorithm 1 provides the implementation summary of the proposed encoder framework:

$$y = \frac{y}{Q}; \quad y = \text{round}(y); \quad y = y * Q, \tag{7}$$

where Q is the quantization factor.

4. Texture Synthesis Framework

In this section, we present the overall texture synthesis framework. Figure 7 gives an overall block diagram of the proposed decoder.

At the decoder, the detail subbands coefficients are synthesized from the CS measurements using the compressive optimization, while the approximation coefficients are decoded using standard JPEG2000. In all our experiments, we have used convex optimization (l_1-norm minimization with equality constraint) for good recovery in a CS framework. This matches the l_0 norm as RIP is ensured due to noiselet measurements. Because of nondifferentiability of the l_1-norm, this optimization principle leads to sparser decompositions [17] and ensures fast and stable resolution better than other methods from the class of greedy algorithms. The l_1 norm also provides a computationally viable approach to sparse signal recovery. In all our experiments

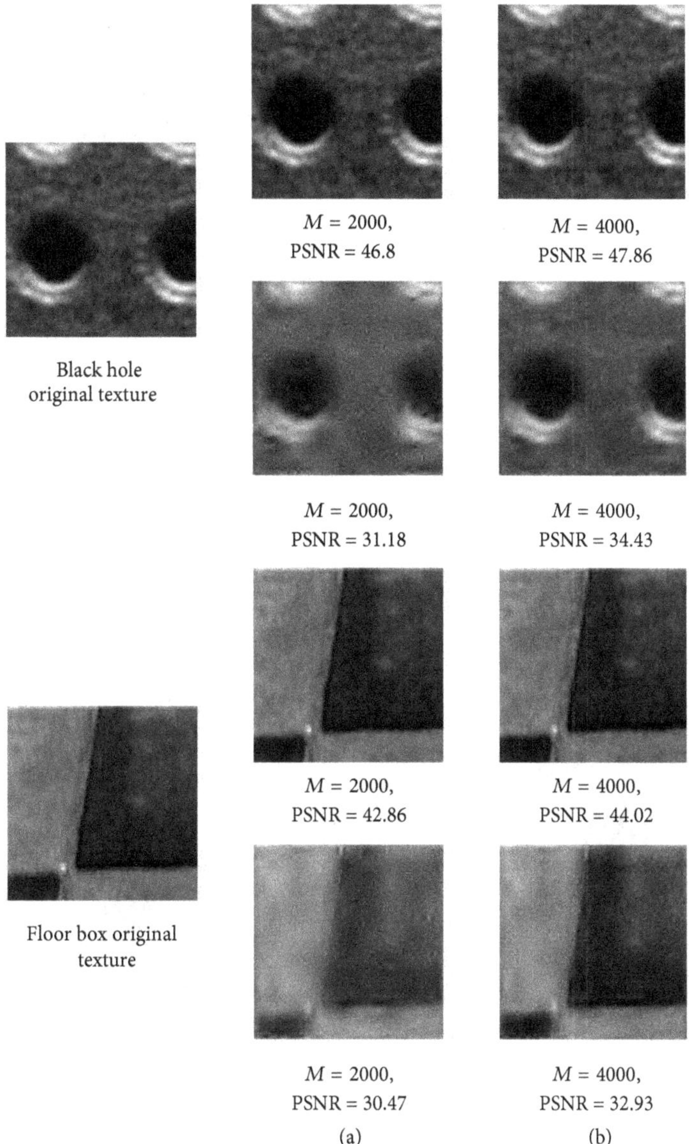

FIGURE 10: Synthesis results of statistical texture (black hole and floor box textures). For each original texture, the lower image pair is the synthesized texture using conventional DWT-based CS scheme for image synthesis [21, 35], and the upper image pair is the synthesized texture using our proposed scheme. (a) Synthesis results for CS measurements (M) = 2000. (b) Synthesis results for CS measurements (M) = 4000.

we have used standard basis pursuit using a primal-dual algorithm [48] which finds the vector with smallest l_1-norm (8). Algorithm 2 provides the implementation summary of the convex optimization and proposed decoder framework. Combining the decoded samples from the standard decoder and the texture synthesizer (CS optimisation) and doing the inverse of statistically matched wavelet transform, a synthesized image is reconstructed:

$$\begin{aligned} &\min && \|x\|_1 \\ &\text{subject to} && \Theta x = s_a, \quad \|x\|_1 = \sum_i |x_i| . \end{aligned} \tag{8}$$

5. Results and Discussion

In this section, we present the experimental results. For our experiments, we have used texture database from Brodatz album [36] and Portilla and Simoncelli [37] website to select different class of structural textured images (periodic, pseudoperiodic, and aperiodic) and complex structured photographic textures. All the test sequences are 128 × 128, 8-bit, and gray scale texture. PSNR and MOS have been used as the quality metrics. Mean Opinion Score (MOS) computation was done by collecting responses of various students and staff working in the lab and averaging them. All the experiments are carried out in MATLAB, running on windows XP PC with P4 CPU and 1 GB RAM.

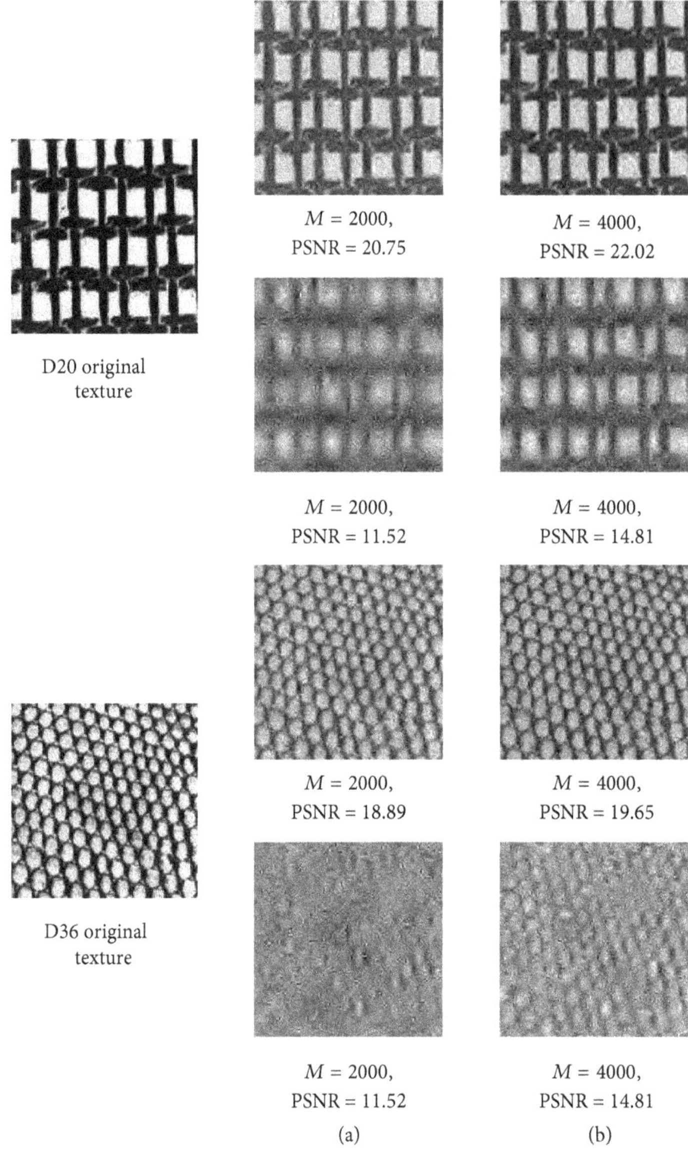

FIGURE 11: Synthesis results of structural texture with periodic patterns (D20 and D36 textures from Brodatz album [36]). For each original texture, the lower image pair is the synthesized texture using conventional DWT-based CS scheme for image synthesis [21, 35], and the upper image pair is the synthesized texture using our proposed scheme. (a) Synthesis results for CS measurements (M) = 2000. (b) Synthesis results for CS measurements (M) = 4000.

5.1. Statistically Matched Wavelet Filter Estimation. In this section we present the simulation results of statistically matched wavelet based 2-D separable kernel wavelet filters used to input data decomposition and subbands representation in the proposed framework. Table 2 gives the result of statistically matched analysis wavelet filters estimated using input data with the filter length set to 5. Using the analysis filters we have computed all other wavelet filters as described in Section 2, to construct the filter bank. Figure 8 shows the reconstruction results of one of our test sequences (escalator), using different set of subbands coefficients for approximation and detail subbands. As one can see that by increasing the number of approximation subband coefficients (a = 1000 to 4000) while keeping the number of detail subbands coefficients constant (d = 1000) improves the reconstruction quality and PSNR significantly. As opposed to this if we keep the approximation subband coefficients at constant (a = 1000) and increase the number of detail subbands coefficients (d = 1000 to 4000), the reconstruction quality and PSNR remain almost unchanged. This experiment demonstrates the correctness of our theoretical claim that statistically matched wavelet based texture representation ensures that maximum energy is captured in the approximation subspace with very little information left in detail subspace. One can also observe from Figure 8 that we need smaller number of coefficients from detail subbands (we select 1000 coefficients out of total

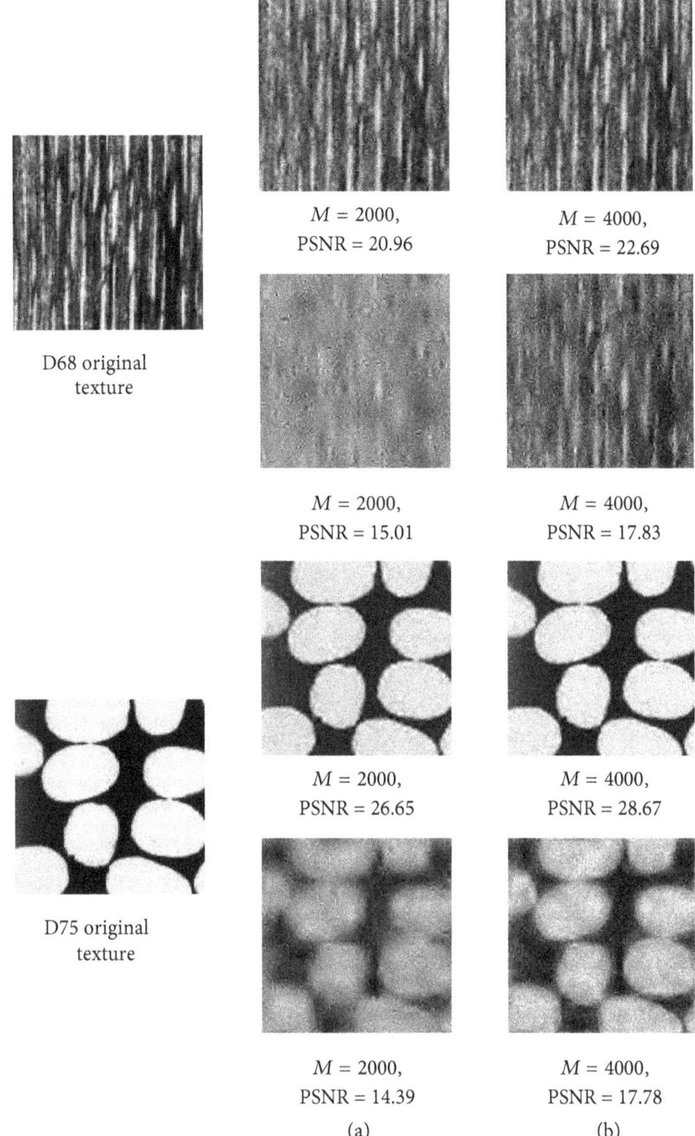

FIGURE 12: Synthesis results of structural texture with irregular and regular patterns (D68 and D75 textures from Brodatz album [36]). For each original texture, the lower image pair is the synthesized texture using conventional DWT-based CS scheme for image synthesis [21, 35], and the upper image pair is the synthesized texture using our proposed scheme. (a) Synthesis results for CS measurements (M) = 2000. (b) Synthesis results for CS measurements (M) = 4000.

13872 coefficients (3∗68∗68)) as compared to those from approximation subband. This experiment was performed to illustrate the significance of approximation and detail subbands coefficients for good quality reconstruction.

5.2. Texture Synthesis Results. In this section we present the texture synthesis results of our proposed scheme (SMWT-CS) and conventional DWT-based texture synthesis scheme in a CS framework (DWT-CS) [21, 35]. We have compared our simulation results with the conventional DWT based image synthesis scheme in a CS framework as presented in [21, 35]. In addition, we have done a comparative study of our synthesis results with joint statistics based statistical model for texture synthesis schemes as presented in [37, 42] and standard JPEG2000. The texture synthesis performance is measured using varying CS measurement samples such as $M = 2000$ and $M = 4000$.

(i) Figures 9 and 10 present the synthesis results for structural texture such as brickwall, escalator, and statistical texture such as floorbox and balckhole, using conventional DWT-CS and our proposed scheme. Table 3 gives the PSNR values and MOS scores of the synthesized texture. As one can observe from Table 3, the proposed scheme outperforms the conventional DWT-CS scheme and can provide significantly better PSNR (5 to 10 dB gain) for the same compressive measurements ($M = 2000$ or $M = 4000$) or better compression

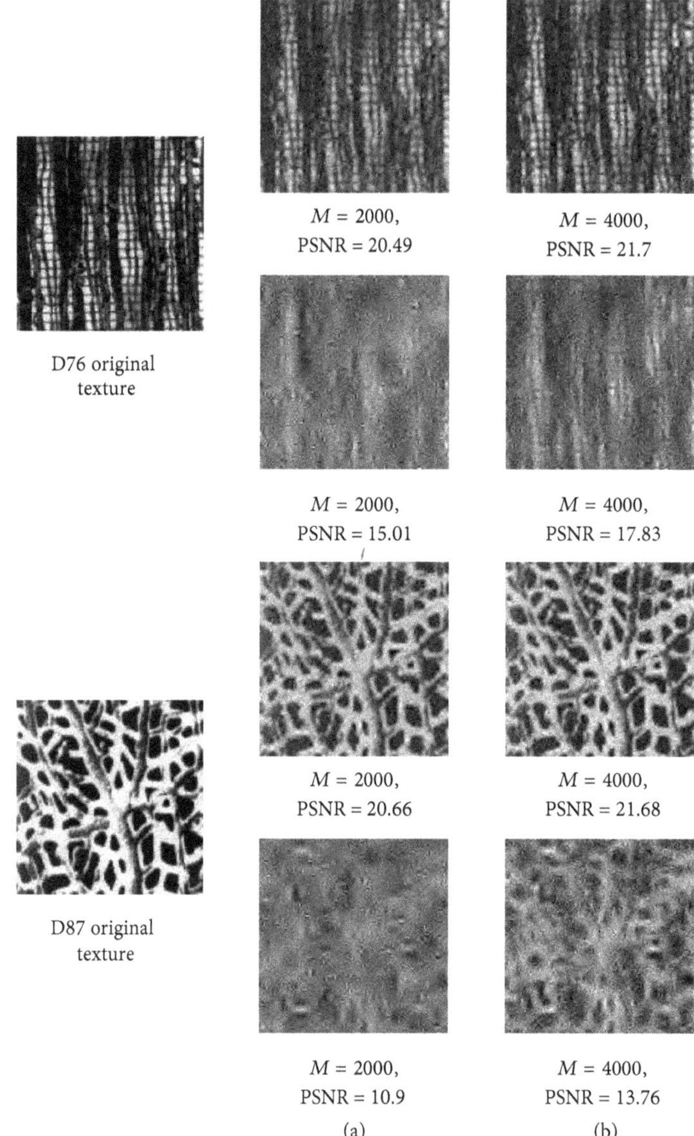

FIGURE 13: Synthesis results of structural texture with irregular patterns (D76 and D87 textures from Brodatz album [36]). For each original texture, the lower image pair is the synthesized texture using conventional DWT-based CS scheme for image synthesis [21, 35], and the upper image pair is the synthesized texture using our proposed scheme. (a) Synthesis results for CS measurements (M) = 2000. (b) Synthesis results for CS measurements (M) = 4000.

at the same PSNR. It is important to note that SMWT-CS is able to reconstruct the textural pattern smoother and sharper as compared to conventional DWT-CS based scheme for the same CS measurements (Figure 9). In addition, we can observe that the texture synthesis quality can be improved by increasing the CS measurement. This can be observed both subjectively (MOS assessment) as well as objectively (PSNR data), hinting at the scalability in the proposed framework.

(ii) Figures 11, 12, and 13 show the synthesis results for structural textures with regular patterns such as D20, D36, and D75 and structural textures with irregular patterns such as D68, D76, and D87 from Brodatz album [36]. As one can observe, the proposed scheme can synthesize the regular and irregular structural patterns with better PSNR and perceptual quality as compared to conventional DWT-CS schemes for the same CS measurements. When compared with the synthesis results for the same textures with other wavelet based texture synthesis schemes such as HMT-3S [42], the proposed scheme outperforms them in terms of perceptual synthesis quality. In fact, such joint statistics based statistical texture model cannot handle the regular and periodic structures as reported by the authors in [42].

(iii) Figure 14 shows the synthesis results for complex structured photographic textures such as stone wall and fish fabric from Portilla website [37]. As one can observe, the proposed scheme results in better PSNR and perceptual

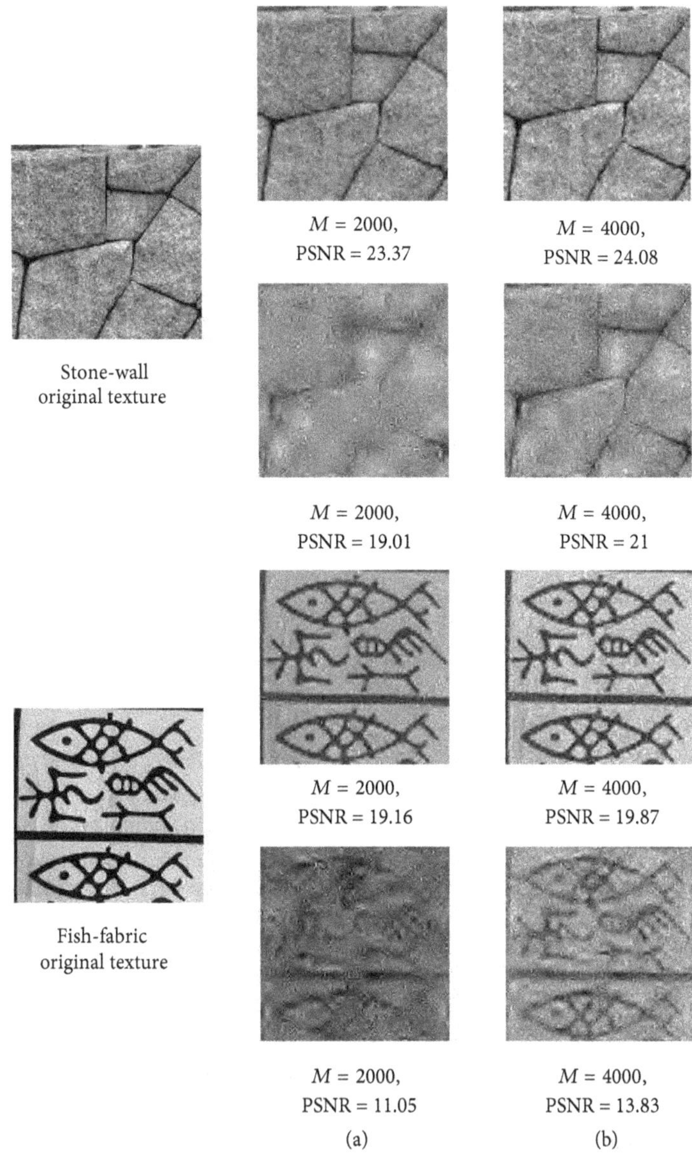

FIGURE 14: Synthesis results of complex structural texture with irregular patterns (stone-*wall* and fish-fabric textures from portilla website [37]). For each original texture, the lower image pair is the synthesized texture using conventional DWT based CS scheme for image synthesis [21, 35], and the upper image pair is the synthesized texture using our proposed scheme. (a) Synthesis results for CS measurements (M) = 2000. (b) Synthesis results for CS measurements (M) = 4000.

quality as compared to conventional DWT-CS schemes for the same CS measurements. Also, when compared with the synthesis results based on joint-statistical models for texture synthesis as suggested by Portilla and Simoncelli [37] for the same textures, the proposed scheme provides much better synthesis quality and the capability to synthesise all the patterns types.

When compared with JPEG2000, we obtain approx 4 dB PSNR gain for statistical textures such as blackhole and floorbox (Table 3). For structural textures such as brickwall and escalator (Figure 9), JPEG2000 performs better at the same bit rate. The reason for this is the presence of significant energy in the high frequency region for structured textures. The coding efficiency loss in a scene with very high frequency transition is due to the fact that compressive sensing recovery of sparse signals typically requires a number of measurements to be larger than the number of nonzero samples.

6. Conclusion

In this paper, we propose statistically matched wavelet based texture data representation and synthesis in a compressive sensing framework (SMWT-CS). Statistically matched wavelet based representation causes most of the captured energy to be concentrated in the approximation subspace, while very little information is retained in the detail subspace. We encode not the full-resolution statistically matched wavelet subband coefficients but only the approximation subband coefficients (LL) using standard image compression scheme like JPEG2000. The detail subband coefficients, that is, HL, LH, and HH, are jointly encoded in a compressive sensing framework using compressive projection and measurements. The experimental results demonstrate that the proposed scheme can provide significantly better PSNR for the same compressive measurements or better compression at the same PSNR as compared to conventional DWT-based image compression scheme in a CS framework. It can also be observed that performing linear compression over the approximation subspace provides better reconstruction quality for the same number of samples as compared to CS measurements over approximation subspace. This indicates that performing linear compression over approximation subspace and CS measurements over detail subspace provides optimal compression and reconstruction quality as against using only standard linear compression or using only CS measurements over both approximation and detail subspace.

Conflict of Interests

The authors declare that there is no conflict of interests regarding the publication of this paper.

References

[1] ITU-T Rec. H. 262 and ISO/IEC 13818-2 MPEG-2, Generic Coding of Moving Pictures and Associated Audio Information—Part 2 Video.

[2] ITU-T Rec. H. 264 and ISO/IEC 4496-10 (MPEG-4/AVC), Advanced video coding for generic audio visual services, Standard version 7, ITU-T and ISO/IEC JTC 1.

[3] H. Nyquist, "Certain topics in telegraph transmission theory," *Transactions of the American Institute of Electrical Engineers*, vol. 47, no. 2, pp. 617–644, 1928.

[4] P. Ndjiki-Nya, T. Hinz, C. Stiuber, and T. Wiegand, "A content-based video coding approach for rigid and non-rigid textures," in *Proceedings of the IEEE International Conference on Image Processing (ICIP '06)*, pp. 3169–3172, Atlanta, Ga, USA, October 2006.

[5] P. Ndjiki-Nya, C. Stüber, and T. Wiegand, "Texture synthesis method for generic video sequences," in *Proceedings of the IEEE International Conference on Image Processing (ICIP '07)*, pp. III397–III400, San Antonio, Tex, USA, September 2007.

[6] P. Ndjiki-Nya, T. Hinz, and T. Wiegand, "Generic and robust video coding with texture analysis and synthesis," in *Proceedings of the IEEE International Conference onMultimedia and Expo (ICME '07)*, pp. 1447–1450, Beijing, China, July 2007.

[7] P. Ndjiki-Nya, M. Köppel, D. Doshkov, and T. Wiegand, "Automatic structure-aware inpainting for complex image content," in *Advances in Visual Computing*, vol. 5358 of *Lecture Notes in Computer Science*, pp. 1144–1156, 2008.

[8] P. Ndjiki-Nya and T. Wiegand, "Video coding using closed-loop texture analysis and synthesis," *IEEE Journal of Selected Topics in Signal Processing*, 2011.

[9] M. Bosch, F. Zhu, and E. J. Delp, "Spatial texture models for video compression," in *Proceedings of the IEEE International Conference on Image Processing (ICIP '07)*, pp. 193–196, San Antonio, Tex, USA, September 2007.

[10] M. Bosch, F. Zhu, and E. J. Delp, "Video coding using motion classification," in *Proceedings of the 15th IEEE International Conference on Image Processing (ICIP '08)*, pp. 1588–1591, San Diego, Calif, USA, October 2008.

[11] J. Byrne, S. Ierodiaconou, D. Bull, D. Redmill, and P. Hill, "Unsupervised image compression-by-synthesis within a JPEG framework," in *Proceedings of the 15th IEEE International Conference on Image Processing (ICIP '08)*, pp. 2892–2895, San Diego, Calif, USA, October 2008.

[12] S. Ierodiaconou, J. Byrne, D. R. Bull, D. Redmill, and P. Hill, "Unsupervised image compression using graphcut texture synthesis," in *Proceedings of the 16th IEEE International Conference on Image Processing (ICIP '09)*, pp. 2289–2292, Cairo, Egypt, November 2009.

[13] F. Zhang, D. R. Bull, and N. Canagarajah, "Region-based texture modelling for next generation video codecs," in *Proceedings of the 17th IEEE International Conference on Image Processing (ICIP '10)*, pp. 2593–2596, Hong Kong, China, September 2010.

[14] F. Zhang and D. R. Bull, "Enhanced video compression with region-based texture models," in *Proceedings of the Picture Coding Symposium (PCS '10)*, pp. 54–57, Nagoya, Japan, December 2010.

[15] A. Gupta, S. D. Joshi, and S. Prasad, "A new approach for estimation of statistically matched wavelet," *IEEE Transactions on Signal Processing*, vol. 53, no. 5, pp. 1778–1793, 2005.

[16] D. L. Donoho, "Compressed sensing," *IEEE Transactions on Information Theory*, vol. 52, no. 4, pp. 1289–1306, 2006.

[17] R. G. Baraniuk, "Compressive sensing," *IEEE Signal Processing Magazine*, vol. 24, no. 4, pp. 118–121, 2007.

[18] S. S. Chen, D. L. Donoho, and M. A. Saunders, "Atomic decomposition by basis pursuit," *SIAM Journal on Scientific Computing*, vol. 20, no. 1, pp. 33–61, 1998.

[19] E. J. Candès, J. Romberg, and T. Tao, "Robust uncertainty principles: exact signal reconstruction from highly incomplete frequency information," *IEEE Transactions on Information Theory*, vol. 52, no. 2, pp. 489–509, 2006.

[20] Y. Zhang, S. Mei, Q. Chen, and Z. Chen, "A novel image/video coding method based on compressed sensing theory," in *Proceedings of the IEEE International Conference on Acoustics, Speech and Signal Processing (ICASSP '08)*, pp. 1361–1364, Las Vegas, Nev, USA, April 2008.

[21] D. Venkatraman and A. Makur, "A compressive sensing approach to object-based surveillance video coding," in *Proceedings of the IEEE International Conference on Acoustics, Speech, and Signal Processing (ICASSP '09)*, pp. 3513–3516, Taipei City, Taiwan, April 2009.

[22] J. Prades-Nebot, Y. Ma, and T. Huang, "Distributed video coding using compressive sampling," in *Proceedings of the Picture Coding Symposium (PCS '09)*, pp. 1–4, Chicago, Ill, USA, May 2009.

[23] M. B. Wakin, J. N. Laska, M. F. Duarte et al., "Compressive imaging for video representation and coding," in *Proceedings of the Picture Coding Symposium (PCS '06)*, pp. 1289–1306, Beijing, China, April 2006.

[24] Y. Yang, O. C. Au, L. Fang, X. Wen, and W. Tang, "Perceptual compressive sensing for image signals," in *Proceedings of the IEEE International Conference on Multimedia and Expo (ICME '09)*, pp. 89–92, New York, NY, USA, July 2009.

[25] A. A. Efros and T. K. Leung, "Texture synthesis by non-parametric sampling," in *Proceedings of the 7th IEEE International Conference on Computer Vision (ICCV'99)*, vol. 2, pp. 1033–1038, September 1999.

[26] L. Y. Wei and M. Levoy, "Fast texture synthesis using tree-structured vector quantization," in *Proceedings of SIGGRAPH '00*, pp. 479–488, New Orleans, La, USA, July 2000.

[27] V. Kwatra, A. Schödl, I. Essa, G. Turk, and A. Bobick, "Graphcut textures: image and video synthesis using graph cuts," in *Proceedings of the SIGGRAPH '03*, pp. 277–286, San Diego, Calif, USA, July 2003.

[28] P. Hill, *Wavelet based texture analysis and segmentation for imgae retrieval and fusion [Ph.D. thesis]*, University of Bristol, 2002.

[29] J. Y. A. Wang and E. H. Adelson, "Representing moving hands with layers," *IEEE Transactions on Image Processing*, vol. 3, no. 5, pp. 625–638, 1994.

[30] A. Dumitraş and B. G. Haskell, "An encoder-decoder texture replacement method with application to content-based movie coding," *IEEE Transactions on Circuits and Systems for Video Technology*, vol. 14, no. 6, pp. 825–840, 2004.

[31] R. J. O'Callaghan and D. R. Bull, "Combined morphological-spectral unsupervised image segmentation," *IEEE Transactions on Image Processing*, vol. 14, no. 1, pp. 49–62, 2005.

[32] H. Choi and R. G. Baraniuk, "Multiscale image segmentation using wavelet-domain hidden Markov models," *IEEE Transactions on Image Processing*, vol. 10, no. 9, pp. 1309–1321, 2001.

[33] J. Li and R. M. Gray, "Context-based multiscale classification of document images using wavelet coefficient distributions," *IEEE Transactions on Image Processing*, vol. 9, no. 9, pp. 1604–1616, 2000.

[34] M. Acharyya and M. K. Kundu, "An adaptive approach to unsupervised texture segmentation using M-band wavelet transform," *Signal Processing*, vol. 81, no. 7, pp. 1337–1356, 2001.

[35] A. Schulz, L. Velho, and E. A. B. da Silva, "On the empirical rate-distortion performance of compressive sensing," in *Proceedings of the 16th IEEE International Conference on Image Processing (ICIP '09)*, pp. 3049–3052, Cairo, Egypt, November 2009.

[36] P. Brodatz, *Textures—A Photographic Album for Artists and Designers*, Dover, New York, NY, USA, 1999.

[37] J. Portilla and E. P. Simoncelli, "A parametric texture model based on joint statistics of complex wavelet coefficients," *International Journal of Computer Vision*, vol. 40, no. 1, pp. 49–71, 2000.

[38] A. Khandelia, S. Gorecha, B. Lall, S. Chaudhury, and M. Mathur, "Parametric video compression scheme using ar based texture synthesis," in *Proceedings of the 6th Indian Conference on Computer Vision, Graphics and Image Processing (ICVGIP '08)*, pp. 219–225, Bhubaneswar, India, December 2008.

[39] A. Stojanovic, M. Wien, and J. R. Ohm, "Dynamic texture synthesis for H.264/AVC inter coding," in *Proceedings of the IEEE International Conference on Image Processing (ICIP '08)*, pp. 1608–1611, San Diego, Calif, USA, October 2008.

[40] A. Stojanovic, M. Wien, and T. K. Tan, "Synthesis-in-the-loop for video texture coding," in *Proceedings of the 16th IEEE International Conference on Image Processing (ICIP '09)*, pp. 2293–2296, Cairo, Egypt, November 2009.

[41] H. Chen, R. Hu, D. Mao, R. Zhong, and Z. Wang, "Video coding using dynamic texture synthesis," in *Proceedings of the IEEE International Conference on Multimedia and Expo (ICME '10)*, pp. 203–208, Singapore, July 2010.

[42] G. Fan and X. Xia, "Wavelet-based texture analysis and synthesis using hidden Markov models," *IEEE Transactions on Circuits and Systems I*, vol. 50, no. 1, pp. 106–120, 2003.

[43] S. Kumar, R. Gupta, N. Khanna, S. Chaudhury, and S. D. Joshi, "Text extraction and document image segmentation using matched wavelets and MRF model," *IEEE Transactions on Image Processing*, vol. 16, no. 8, pp. 2117–2128, 2007.

[44] I.T. R. T. 800, "Jpeg-2000:core coding system," Tech. Rep., International Telecommunication Union, 2000.

[45] C. Deng, W. Lin, B. Lee, and C. T. Lau, "Robust image compression based on compressive sensing," in *Proceedings of the IEEE International Conference on Multimedia and Expo (ICME '10)*, pp. 462–467, Singapore, July 2010.

[46] A. Mojsilovic, M. V. Popovic, and D. M. Rackov, "On the selection of an optimal wavelet basis for texture characterization," *IEEE Transactions on Image Processing*, vol. 9, no. 12, pp. 2043–2050, 2000.

[47] R. Coifman, F. Geshwind, and Y. Meyer, "Noiselets," *Applied and Computational Harmonic Analysis*, vol. 10, no. 1, pp. 27–44, 2001.

[48] S. Boyd and L. Vandenberghe, *Convex Optimization*, Cambridge University Press, New York, NY, USA, 2004.

Radar Coincidence Imaging under Grid Mismatch

Dongze Li, Xiang Li, Yongqiang Cheng, Yuliang Qin, and Hongqiang Wang

School of Electronic Science and Engineering, National University of Defense Technology, Changsha 410073, China

Correspondence should be addressed to Xiang Li; lixiang01@vip.sina.com

Academic Editors: I. Buciu and B. K. Gunturk

Radar coincidence imaging is an instantaneous imaging technique which does not depend on the relative motion between targets and radars. High-resolution, fine-quality images can be obtained using a single pulse either for stationary targets or for complexly maneuvering ones. There are two image-reconstruction algorithms used for radar coincidence imaging, that is, the correlation method and the parameterized method. In comparison with the former, the parameterized method can achieve much higher resolution but is seriously sensitive to grid mismatch. In the presence of grid mismatch, neither of the two algorithms can obtain recognizable high-resolution images. The above problem largely limits the applicability of radar coincidence imaging in actual imaging scenes where grid mismatch generally exists. This paper proposes a joint correlation-parameterization algorithm, which uses the correlation method to estimate the grid-mismatch error and then iteratively modifies the results of the parameterized method. The proposed algorithm can achieve high resolution with fine imagery quality under the grid mismatch. Examples are provided to illustrate the improvement of the proposed method.

1. Introduction

Radar coincidence imaging is developed as the extension of classical coincidence imaging in microwave radar systems [1]. Classical coincidence imaging, which is realized in thermal optical systems, is a method to nonlocally image an object with high resolution by producing sharply fluctuating signals in the imaging plane [1, 2]. Similarly, the essential principle of radar coincidence imaging is to produce time-space independent signals in the detecting area. Then radar signals at different positions have mutually independent waveforms. Illuminated by such signals, target-scattering centers within a radar beam will also reflect echoes of independent waveforms associated with their respective positions. Therefore, the echo component of each scattering center can be extracted from the receiving signal and then correlated to their respective positions. Consequently, scattering centers within a beam can be resolved and the spatial distribution of the target scattering centers can be obtained. Obviously, it is quite different from the radar imaging techniques based on the range-Doppler (RD) principle, where scattering centers are resolved based on the analysis of time-delay and Doppler frequency [3].

Because of the different imaging principles, the radar coincidence imaging technique has two advantages over most of the RD imaging methods. Firstly, radar coincidence imaging does not depend on the aspect-angle integration or the Doppler gradient to achieve high azimuth resolution. Thus, it does not require relative motion between radars and targets and can obtain images of the targets which remain stationary with respect to radars. Furthermore, radar coincidence imaging can achieve high-resolution using a single pulse. The extremely short imaging time, which is shorter than a pulse width, considerably decreases the impact of the noncooperative motion to imagery quality. Therefore, radar coincidence imaging can obtain high-resolution, fine-quality images either for stationary targets or for the ones in complex maneuvers.

In radar coincidence imaging, the target area needs to be discretized to a grid and target-scattering centers are assumed to be located at the grid points. If scattering centers are located off the grid points, then the grid mismatch yields. There are two main image-reconstruction methods used in radar coincidence imaging. One is the correlation method with lower resolution. The other is the parameterized method

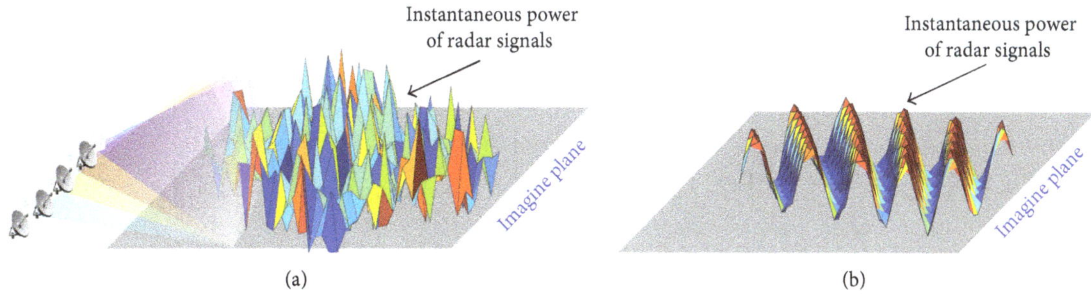

FIGURE 1: (a) The spatial distribution of time-space detecting signals. (b) The spatial distribution produced by coherent transmitted signals.

which can produce much higher resolution but is too sensitive under grid mismatch to give recognizable target images. Furthermore, current algorithms that are applicable to solve the sensitivity to basis mismatch are unfortunately ineffective for the grid mismatch in radar coincidence imaging [4, 5]. This limitation seriously affects the applicability of the imaging method in actual scenarios where the grid mismatch generally exists. This paper proposes a joint correlation-parameterization method for image reconstruction which can achieve both high resolution and good imagery quality under grid mismatch.

The paper is organized as follows. Section 2 is devoted to the basic principles of radar coincidence imaging. Section 3 analyzes the impact of grid mismatch and proposes the joint correlation-parameterization method for image reconstruction. Section 4 concludes the work.

2. Basic Principles

Coherent signals, which are widely employed by most of the imaging radars, generally produce detecting signals that show significant spatial correlation, as shown in Figure 1(b). However, the essence of radar coincidence imaging is to produce time-space independent radar signals in the detecting area. In other words, the detecting signals at different instants or in different positions are independent of each other. As shown in Figure 1(a), at an arbitrary instant, the time-space independent detecting signals fluctuate sharply in the imaging plane and its instantaneous spatial distribution presents remarkable variety.

Generally, the target location can be firstly estimated based on the detection and localization techniques [6, 7], which is defined as the center of the target area. Then a local coordinate system is established in the target area center, as shown in Figure 2. The target area is discretized to a grid consisting of L small rectangles of uniform size and shape. Each small rectangle is the grid cell and is approximated by its own center. Thus, target-scattering centers are actually assumed to be initially located at the grid-cell centers. The discrete target area is expressed as $I = \{\mathbf{r}_1, \mathbf{r}_2, \ldots, \mathbf{r}_L\}$, where $\mathbf{r}_l$ is the position vector of the lth grid cell center. There are N transmitters and a receiver, whose position vectors are $\mathbf{R}_n$ and $\mathbf{R}_r$, respectively. The transmitted signal of the nth transmitter is denoted as $St_n(t)$ and the receiving signal is $Sr(t)$. The detecting signals distributed in the target area are labeled as

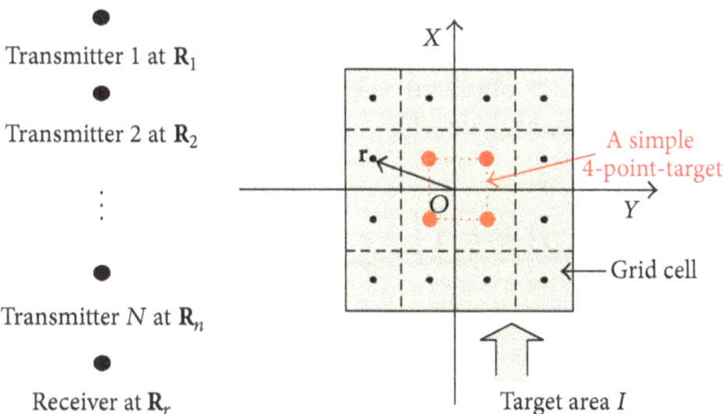

FIGURE 2: Geometry of the target area grid.

$S_I(\mathbf{r}, t)$, where $\mathbf{r}$ is the position vector of an arbitrary grid cell within I.

The ideal transmitted signals for radar coincidence imaging are supposed to be group-orthogonal and time-independent as denoted in (1), which can produce the time-space independent $S_I(\mathbf{r}, t)$ via single transmitting [1]. Consider

$$\begin{aligned} R_T(n_1, n_2; t_1, t_2) &= \int St_{n_1}(t - t_1)\, St_{n_2}(t - t_2)\, dt \\ &= \delta(n_1 - n_2, t_1 - t_2), \end{aligned} \tag{1}$$

where $St_n(t) = \text{rect}(t/T_p) \cdot st_n(t)$, $st_n(t)$ is the envelope function and T_p is the pulse width. If the antenna number is more than 2, for example, $N = 5$, the detecting signal produced by such transmitted signals will show approximate time-space independence [1], expressed as follows:

$$\begin{aligned} R_I(\mathbf{r}, \mathbf{r}'; \tau, \tau') &= \int S_I(\mathbf{r}, t - \tau)\, S_I^*(\mathbf{r}', t - \tau')\, dt \\ &\sim N\delta(\mathbf{r} - \mathbf{r}', \tau' - \tau). \end{aligned} \tag{2}$$

Then, we express the receiving signal as the superposition of the detecting signals. Consider

$$Sr(t) = \sum_{l=1}^{L} \sigma_l S_I\left(\mathbf{r}_l, t - \frac{|\mathbf{r}_l - \mathbf{R}_r|}{c}\right), \tag{3}$$

where σ_l is the scattering coefficient of the scattering center located at the lth grid cell and for the grid cell without target-scattering center $\sigma_l = 0$. For the sake of simplicity, the coincidence imaging formulism needs a reference signal [1], which herein can be simply structured using $S_I(\mathbf{r}, t)$ as follows:

$$S(\mathbf{r}, t) = S_I\left(\mathbf{r}, t - \frac{|\mathbf{r} - \mathbf{R}_r|}{c}\right). \tag{4}$$

Consequently, (3) becomes

$$Sr(t) = \sum_{l=1}^{L} \sigma_l S(\mathbf{r}_l, t). \tag{5}$$

Then, the scattering coefficient of an arbitrary grid cell $\mathbf{r}_x$ can be explicitly obtained via the correlation between the receiving signal and $S(\mathbf{r}_x, t)$,

$$\begin{aligned}\int Sr(t) S^*(\mathbf{r}_x, t)\, dt &= \int \left(\sum_{l=1}^{L} \sigma_l \cdot S_I\left(\mathbf{r}_l, t - \frac{|\mathbf{r}_l - \mathbf{R}_r|}{c}\right)\right) \\ &\quad \cdot S_I^*\left(\mathbf{r}_x, t - \frac{|\mathbf{r}_x - \mathbf{R}_r|}{c}\right) dt \\ &= \sum_{l=1}^{L} \sigma_l \cdot R_I\left(\mathbf{r}_l, \mathbf{r}_x; \frac{|\mathbf{r}_l - \mathbf{R}_r|}{c}, \frac{|\mathbf{r}_x - \mathbf{R}_r|}{c}\right) \\ &\sim \sum_{l=1}^{L} \sigma_l \cdot N\delta(\mathbf{r}_l - \mathbf{r}_x) \\ &= N \cdot \sigma_x.\end{aligned} \tag{6}$$

That is,

$$\sigma_x \sim \frac{1}{N} \int Sr(t) S^*(\mathbf{r}_x, t)\, dt. \tag{7}$$

Note that the detecting signal $S_I(\mathbf{r}, t)$ or $S(\mathbf{r}, t)$ can be computed based on the known transmitted waveforms. Thus, the scattering coefficient of each grid cell can be derived via the correlation between the receiving signal and the reference signal. Therefore, the target scene can be recovered via conducting the correlation in (7) for the whole target area grid. This method to reconstruct target images is defined as the correlation method.

The excellent point-to-point relationship in (7) means a high resolution, but it requires that the transmitted signals have the perfect time-independence as presented in (1). Unfortunately, microwave transmitted signals are far from this requirement and their time-independent degree is inadequate to achieve the high resolution represented in (7) via the correlation method [1]. Therefore, to improve the resolution of radar coincidence imaging, the parameterized method is employed which is less constrained by the signal time-independence. This method uses the relationship between the receiving signal and the reference signal, as shown in (5), to structure a radar coincidence imaging equation, given as follows:

$$\mathbf{Sr} = \mathbf{S} \cdot \boldsymbol{\sigma},$$

$$\begin{bmatrix} Sr(t_1) \\ Sr(t_2) \\ \vdots \\ Sr(t_K) \end{bmatrix} = \begin{bmatrix} S(\mathbf{r}_1, t_1) & S(\mathbf{r}_2, t_1) & \cdots & S(\mathbf{r}_L, t_1) \\ S(\mathbf{r}_1, t_2) & S(\mathbf{r}_2, t_2) & \cdots & S(\mathbf{r}_L, t_2) \\ \vdots & \vdots & \cdots & \vdots \\ S(\mathbf{r}_1, t_K) & S(\mathbf{r}_2, t_K) & \cdots & S(\mathbf{r}_L, t_K) \end{bmatrix} \cdot \begin{bmatrix} \sigma_1 \\ \sigma_2 \\ \vdots \\ \sigma_L \end{bmatrix}, \tag{8}$$

where $K = L$, $\mathbf{S}$ is the reference signal matrix, $\mathbf{Sr}$ is the vector of the receiving signal, and $\boldsymbol{\sigma}$ is the unknown vector of the scattering coefficient. The columns and the rows of $\mathbf{S}$ basically represent the detecting signals in different positions and at different instants, respectively. Hence, the incoherent property of $\mathbf{S}$ is basically determined by the time-space independence of the detecting signals. Therefore, the independent characteristic of the detecting signal will ensure a full-rank $\mathbf{S}$. Consequently, the target scattering-coefficient vector can be uniquely recovered as $\boldsymbol{\sigma} = \mathbf{S}^{-1} \cdot \mathbf{Sr}$.

3. Impact of the Grid Mismatch and Image Reconstruction

The imaging equation presented in (8) is derived under the grid-match condition. Generally, regardless of how finely the target area is gridded, target-scattering centers may not lie in the grid-cell centers. Assume that there exists the grid mismatch. The scattering-center position vector actually is $\mathbf{r}_l + \Delta\mathbf{r}_l$ instead of the assumed $\mathbf{r}_l$ for the lth grid cell. Thus, the actual receiving signal is $Sr(t) = \sum_{l=1}^{L} \sigma_l S(\mathbf{r}_l + \Delta\mathbf{r}_l, t)$. Consequently, the imaging equation should be $\mathbf{Sr} = (\mathbf{S} + \boldsymbol{\varepsilon}_s) \cdot \boldsymbol{\sigma}$, where $\boldsymbol{\varepsilon}_s$ is caused by the difference between $S(\mathbf{r}_l, t)$ and $S(\mathbf{r}_l + \Delta\mathbf{r}_l, t)$. As $\boldsymbol{\varepsilon}_s$ and $\boldsymbol{\sigma}$ are both unknown variables, the above imaging equation can be simplified as

$$\mathbf{Sr} = \mathbf{S} \cdot \boldsymbol{\sigma} + \boldsymbol{\varepsilon}_m, \tag{9}$$

where $\boldsymbol{\varepsilon}_m = \boldsymbol{\varepsilon}_s \cdot \boldsymbol{\sigma}$ and denotes the modeling error caused by grid mismatch. Then (9) represents the well-known linear signal model. Thus, the actual scattering-coefficient vector should be derived as $\boldsymbol{\sigma} = \mathbf{S}^{-1} \cdot (\mathbf{Sr} - \boldsymbol{\varepsilon}_m)$. The previous solution derived from (8) is relabeled as $\widehat{\boldsymbol{\sigma}}$. That is, $\mathbf{Sr} = \mathbf{S} \cdot \widehat{\boldsymbol{\sigma}}$, where $\boldsymbol{\varepsilon}_m$ is neglected because of the grid-match assumption. Certainly, the imaging quality of the parameterized method will be decreased due to the modeling error.

To illustrate the grid-mismatch impact, we give an example to show how the position bias affects the modeling error and how the modeling error affects the imaging quality. In the quantitative manner, the imaging quality discussed here is firstly indicated by the relative imaging error, expressed as $\|\widehat{\boldsymbol{\sigma}} - \boldsymbol{\sigma}\| / \|\boldsymbol{\sigma}\|$. Then the modeling error is normalized as the signal-error-ratio, denoted as $\zeta = 20 \log_{10} \|\mathbf{Sr}\| / \|\boldsymbol{\varepsilon}_m\|$. The position bias is defined as $\sum_{l=1}^{L} |\Delta\mathbf{r}_l| / l_c$, where l_c is the grid-cell perimeter.

The example employs an N-transmitter 1-receiver array, consisting of 5 antennas. The arrangement of the antennas and the target is shown in Figure 3. The target area is

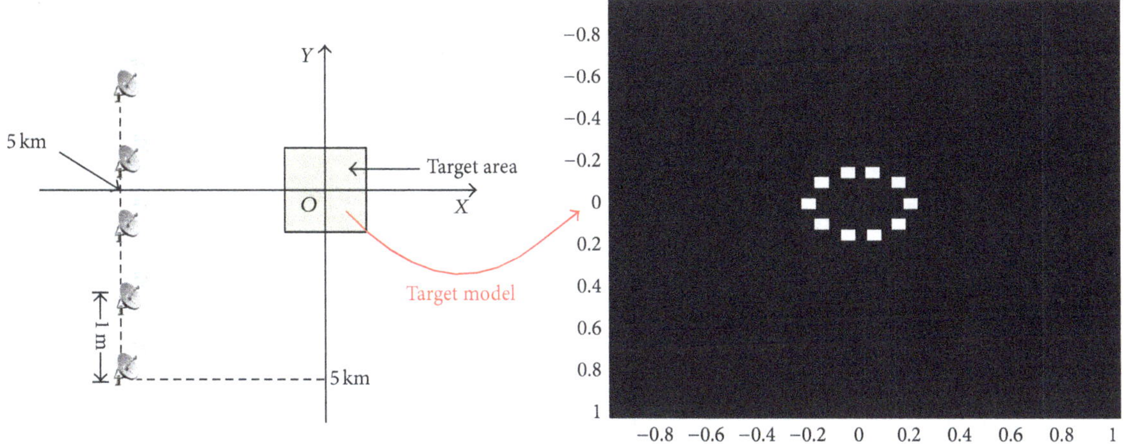

FIGURE 3: The arrangement of antennas and the target model.

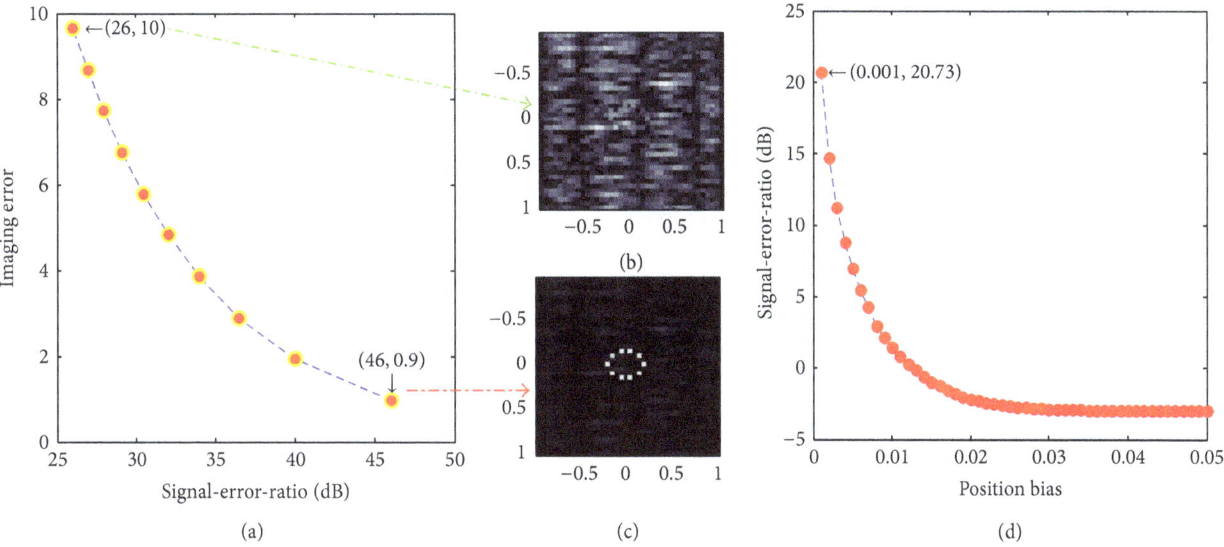

FIGURE 4: The grid-mismatch impact to imaging quality of the parameterized method. (a) The imaging error versus ζ. (b) The imaging result when $\zeta = 26$ dB. (c) The imaging result when $\zeta = 46$ dB. (d) The signal-error-ratio versus the position bias.

2 m × 2 m and is discretized to 40 × 40 grid cells, each of which is 5 cm × 5 cm. The group-orthogonal and time-independent transmitted signals herein are generated by modulating the sinusoid signals on amplitude via mutually independent Gaussian noises. The carrier frequency is 9.5 GHz, the pulse width is 50 μs, the bandwidth is 1 GHz, and the sampling frequency is 2 GHz. The results are shown in Figure 4.

In this example, the target image in Figure 4(b) is badly blurred beyond reorganization when $\zeta = 26$ dB. The target image in Figure 4(c) is recognizable when the imaging error is 0.9, which requires that ζ should not be lower than 46 dB. It demonstrates that the parameterized method has a low tolerance to the modeling error. A recognizable image with a small imaging error requires a small modeling error or a high signal-error-ratio. Then the question is "will the position bias generally produce a high-level signal-error-ratio to guarantee low imaging error?" Unfortunately, as shown in Figure 3(d), even a very small position bias of 0.001 in this example can cause a large modeling error, which only provides 20 dB single-error-ratio. Moreover, the position bias less than 0.02 will generate a negative ζ. It shows that the modeling error is seriously sensitive to the position bias. The position bias of 0.04 can cause −4 dB signal-error-ratio in radar coincidence imaging equation, with no mention of a much larger bias in practice. In this case, modeling error almost overwhelms the receiving signal in the imaging equation. The experiment indicates that the recovery, simply depending on the imaging equation, might impossibly provide a good estimation under such a poor signal-error-ratio condition.

Consider another example to compare the correlation method with the parameterized method under grid mismatch. In the example, target images will be reconstructed via the two methods with and without the grid mismatch. Herein, the grid mismatch leads to $\zeta = -4$ dB. Imaging results are shown in Figure 5.

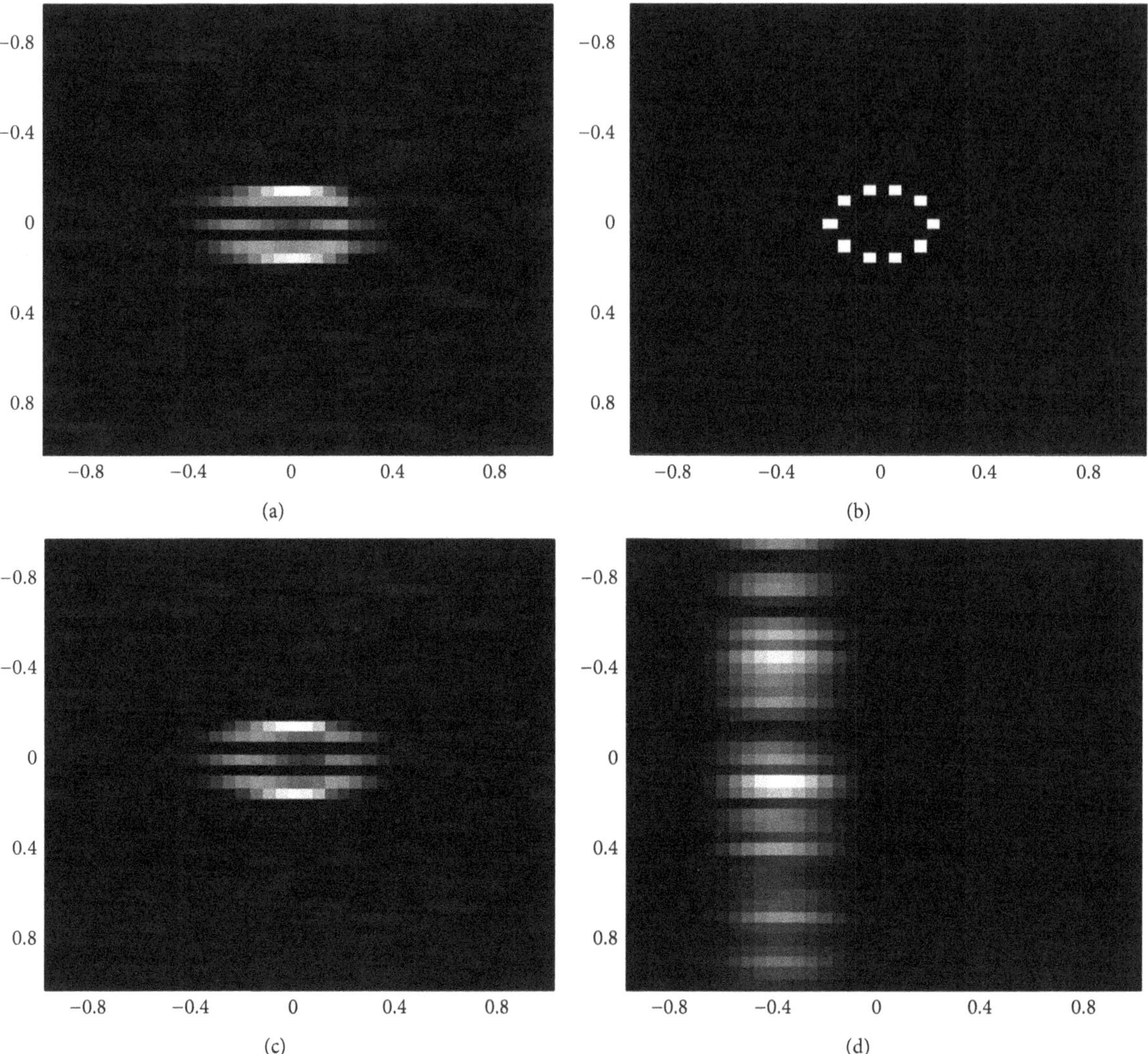

FIGURE 5: The imaging results recovered by different methods. (a) The correlation-method result without grid mismatch. (b) The parameterized-method result without grid mismatch. (c) The correlation-method result when $\zeta = -4$ dB. (d) The parameterized-method result when $\zeta = -4$ dB.

As shown in Figure 5, the parameterized method provides higher resolution than the correlation method in the case of no grid mismatch. On the other side, when $\zeta = -4$ dB, the imaging quality of the correlation method is almost unchanged but the parameterized-method result gets degraded and almost gives no information on the target shape. It is obvious that the correlation method is less sensitive to the modeling error than the parameterized method but has much lower resolution.

In conclusion, neither of the two image-reconstruction methods can achieve both high resolution and good imaging quality in the presence of grid mismatch. According to their respective limitations, there are two ways to solve the image reconstruction under grid mismatch. One way is to improve the resolution of the correlation method. The other one is to decrease the grid-mismatch impact in the parameterized method. As stated previously, however, microwave signal does not have adequate time-independence in nature, resulting in a limited resolution. Thus, the latter might be a possible way to solve the problem of grid mismatch.

A direct idea to decrease grid-mismatch impact in the parameterized method is to estimate the modeling error. Note that the modeling error is relevant to the scattering center number, scattering intensity, and the position-bias level of all scattering centers. If there is no prior information of the target shape, the modeling error will be quite difficult to estimate. Then for the targets that are unknown in advance, the obtainable shape information could be sought from the imaging results of the two reconstruction methods. Thus, a possible source to provide the desired knowledge would be the correlation-method result, considering the seriously distorted images of the parameterized method caused by grid mismatch.

The knowledge that can be used by the imaging equation should be quantitative and precise. However, the result of the correlation method, as observed in Figure 5, presents a

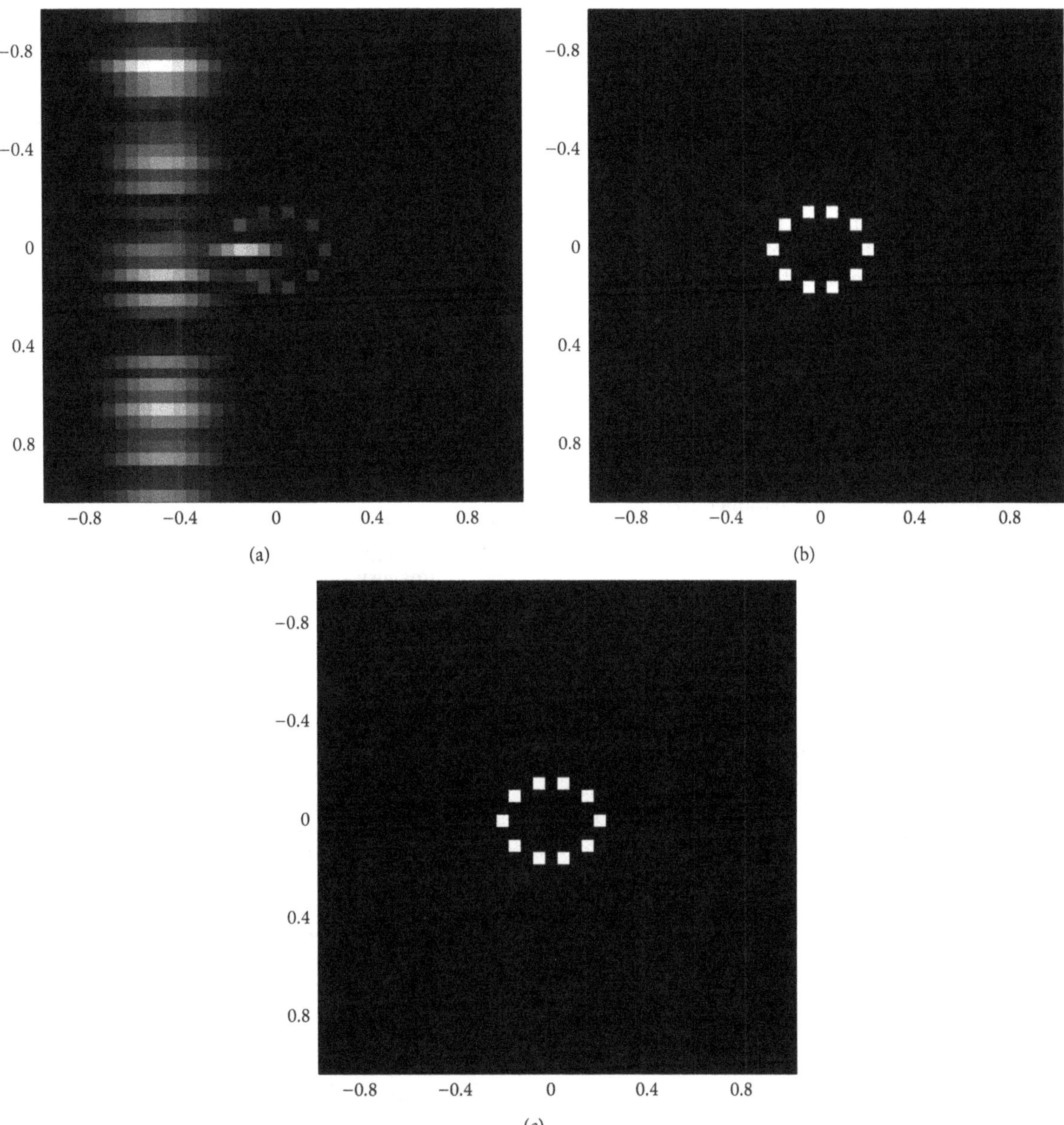

FIGURE 6: The imaging results of the joint correlation-parameter method. (a) The imaging result of the first iteration. (b) The imaging result of the second iteration. (c) The imaging result of the third iteration.

TABLE 1: The flow of the joint correlation-parameterization method.

Step 1	Set the iteration step maximum $T_{\max}$ and the stop limitation η, $k = 0$.
Step 2	Discretize the target area and compute the reference signal.
Step 3	Obtain the scattering-coefficient vector $\widehat{\boldsymbol{\sigma}}_c$ using the correlation method.
Step 4	Set a threshold σ_T to determine $\Lambda = \{l_1, l_2, \ldots, l_M\}$, where $\widehat{\boldsymbol{\sigma}}_c(l_m) < \sigma_T$.
Step 5	Compose the coincidence imaging equation. Obtain the initial estimation $\widehat{\boldsymbol{\sigma}}^{(k)}$ using the parameterized method. Set $\mathbf{Sr}^{(k)} = \mathbf{Sr}$.
Step 6	Compute the inverse matrix of the reference matrix $\mathbf{S}$, and then according to Λ obtain $\mathbf{S}_{\Lambda}^{-1} = [\boldsymbol{\alpha}_{l_1}, \boldsymbol{\alpha}_{l_2}, \ldots, \boldsymbol{\alpha}_{l_M}]^T$.
Step 7	Derive $\widehat{\boldsymbol{\sigma}}_{\Lambda}^{(k)} = [\widehat{\sigma}_{l_1}^{(k)}, \widehat{\sigma}_{l_2}^{(k)}, \ldots, \widehat{\sigma}_{l_M}^{(k)}]^T$.
Step 8	Compute the modeling noise $\widehat{\boldsymbol{\varepsilon}}_m^{(k)} = \mathbf{S}_{\Lambda}^{\dagger}\widehat{\boldsymbol{\sigma}}_{\Lambda}^{(k)}$, $k = k + 1$.
Step 9	Update the scattering-coefficient vector as $\widehat{\boldsymbol{\sigma}}^{(k)} = \mathbf{S}^{-1}(\mathbf{Sr}^{(k-1)} - \widehat{\boldsymbol{\varepsilon}}_m^{(k-1)})$.
Step 10	Stop if $\left\|\widehat{\boldsymbol{\sigma}}^{(k)} - \widehat{\boldsymbol{\sigma}}^{(k-1)}\right\| \le \eta$ or $k \ge T_{\max}$. Otherwise $\mathbf{Sr}^{(k)} = \mathbf{Sr}^{(k-1)} - \widehat{\boldsymbol{\varepsilon}}_m^{(k)}$, and go to Step 7.

blurry target shape. The image cannot precisely indicate the target scattering-center location or provide the exact value of the scattering coefficient. Nonetheless, it indeed provides valuable information. That is, the correlation-method result can denote part of the grid cells which do not contain target-scattering centers. In Figures 5(a) and 5(c), only a minority of image patches, which constitute the target contour, have large values. By contrast, a majority of other image patches have far too small values, which implies the corresponding grid cells have no scattering centers. Hence, these corresponding elements in the scattering-coefficient vector are estimated to be zero. Such quantitative knowledge will be helpful for the estimation of the modeling error. Detailed implementation of image reconstruction using the idea will be described in the following paragraphs.

Firstly, still under the grid-match assumption, we estimate the target scattering-coefficient vector via the two reconstruction methods: $\widehat{\boldsymbol{\sigma}}_c$ is derived via the correlation method and $\widehat{\boldsymbol{\sigma}}$ is obtained via the parameterized method. According to $\mathbf{Sr} = \mathbf{S} \cdot \widehat{\boldsymbol{\sigma}}$ and (9), we have the following relationship:

$$\mathbf{S} \cdot \widehat{\boldsymbol{\sigma}} - \mathbf{S} \cdot \boldsymbol{\sigma} = \boldsymbol{\varepsilon}_m \longrightarrow \widehat{\boldsymbol{\sigma}} - \boldsymbol{\sigma} = \mathbf{S}^{-1} \boldsymbol{\varepsilon}_m, \tag{10}$$

where $\mathbf{S}^{-1}$ is written as $[\boldsymbol{\alpha}_1, \boldsymbol{\alpha}_2, \ldots, \boldsymbol{\alpha}_L]^T$, $\boldsymbol{\alpha}_l$ being the lth row-vector of $\mathbf{S}^{-1}$. The next step is to select the grid cells without scattering centers according to $\widehat{\boldsymbol{\sigma}}_c$. In comparison with the average value of $\widehat{\boldsymbol{\sigma}}_c$, part of the elements in $\widehat{\boldsymbol{\sigma}}_c$, which have far too small values, are determined to be 0. The index of these zero-elements is denoted as $\Lambda = \{l_1, l_2, \ldots, l_M\}$; that is, $\sigma_{l_m} = 0, l_m \in \Lambda$. According to Λ, we define $\mathbf{S}_\Lambda^{-1} = [\boldsymbol{\alpha}_{l_1}, \boldsymbol{\alpha}_{l_2}, \ldots, \boldsymbol{\alpha}_{l_M}]^T$, $\widehat{\boldsymbol{\sigma}}_\Lambda = [\widehat{\sigma}_{l_1}, \widehat{\sigma}_{l_2}, \ldots, \widehat{\sigma}_{l_M}]^T$, and $\boldsymbol{\sigma}_\Lambda = [\sigma_{l_1}, \sigma_{l_2}, \ldots, \sigma_{l_M}]^T$. Based on (10), we have

$$\widehat{\boldsymbol{\sigma}}_\Lambda - \boldsymbol{\sigma}_\Lambda = \mathbf{S}_\Lambda^{-1} \cdot \boldsymbol{\varepsilon}_m. \tag{11}$$

As $\boldsymbol{\sigma}_\Lambda = \mathbf{0}$, finally $\widehat{\boldsymbol{\sigma}}_\Lambda = \mathbf{S}_\Lambda^{-1} \cdot \boldsymbol{\varepsilon}_p$. Therefore, the modeling noise is estimated as

$$\widehat{\boldsymbol{\varepsilon}}_m = \mathbf{S}_\Lambda^{\dagger} \widehat{\boldsymbol{\sigma}}_\Lambda, \tag{12}$$

where $\mathbf{S}_\Lambda^{\dagger}$ is the pseudoinverse matrix of $\mathbf{S}_\Lambda^{-1}$. By substituting $\widehat{\boldsymbol{\varepsilon}}_m$ to (9), the new target scattering-coefficient vector obtained is

$$\widehat{\boldsymbol{\sigma}}^{(1)} = \mathbf{S}^{-1} \left(\mathbf{Sr} - \widehat{\boldsymbol{\varepsilon}}_m \right). \tag{13}$$

If the result remains unsatisfactory, we can repeat the computation of (12) and (13) using the new $\widehat{\boldsymbol{\sigma}}^{(1)}$ and $\widehat{\boldsymbol{\varepsilon}}_m$. That is, update the receiving signal as $\mathbf{Sr}^{(1)} = \mathbf{Sr} - \widehat{\boldsymbol{\varepsilon}}_m$. Then derive $\widehat{\boldsymbol{\sigma}}_\Lambda^{(1)}$ according to Λ and update the modeling error as $\widehat{\boldsymbol{\varepsilon}}_m^{(1)} = \mathbf{S}_\Lambda^{\dagger} \widehat{\boldsymbol{\sigma}}_\Lambda^{(1)}$. Finally, the estimation is updated as $\widehat{\boldsymbol{\sigma}}^{(2)} = \mathbf{S}^{-1}(\mathbf{Sr}^{(1)} - \widehat{\boldsymbol{\varepsilon}}_m^{(1)})$. The update can be performed iteratively until the result is satisfactory. This image-reconstruction method using both the correlation method and the parameterized method is defined as the joint correlation-parameterization method. The detailed steps are given in Table 1.

To examine the correlation-parameterization method, target images are reconstructed according to the iteration process of Table 1, where the grid mismatch is the same as that in Figure 5(d). The imaging results are given in Figure 6.

As shown in Figure 6, a high-quality target image can be obtained through three iterations. The result of the first iteration has presented the target contour, though the contour is light and the imaging error remains remarkable. After the second iteration, the imaging quality is largely improved. By contrast, the imaging quality of Figure 6(b) is much better than the result in Figure 5(d). The third iteration maintains the good imagery quality. Therefore, two iterations actually give a high-resolution and fine-quality target image. It indicates that the joint correlation-parameterization method satisfies the requirement for both the resolution and the robustness in the presence of the modeling noise caused by grid mismatch.

4. Conclusions

Radar coincidence imaging can achieve excellent high-resolution target images on the condition of grid match, but the imagery quality gets degraded beyond recognition in the presence of the modeling error caused by grid mismatch. Therefore, the paper proposes the joint correlation-parameterization method for image reconstruction. The proposed algorithm iteratively modifies the parameterized-method result with the estimated modeling error, which is obtained based on the correlation-method result. Consequently, the grid-mismatch impact on the imaging quality is considerably reduced. The example shows that the joint correlation-parameterization method can achieve high resolution and maintain the robustness under grid mismatch.

Conflict of Interests

The authors declare that there is no conflict of interests regarding the publication of this paper.

References

[1] D. Li, X. Li, Y. Cheng, Y. Qin, and H. Wang, "Radar coincidence imaging: an instantaneous imaging technique with stochastic signals," *IEEE Transactions on Geoscience and Remote Sensing*, vol. 52, no. 4, 2014.

[2] A. Gatti, E. Brambilla, M. Bache, and L. A. Lugiato, "Ghost imaging with thermal light: comparing entanglement and classical correlation," *Physical Review Letters*, vol. 93, no. 9, Article ID 093602, 2004.

[3] Y. Shih, "Quantum imaging," *IEEE Journal on Selected Topics in Quantum Electronics*, vol. 13, no. 4, pp. 1016–1030, 2007.

[4] Y. Chi, L. L. Scharf, A. Pezeshki, and A. R. Calderbank, "Sensitivity to basis mismatch in compressed sensing," *IEEE Transactions on Signal Processing*, vol. 59, no. 5, pp. 2182–2195, 2011.

[5] M. A. Herman and T. Strohmer, "General deviants: an analysis of perturbations in compressed sensing," *IEEE Journal on Selected Topics in Signal Processing*, vol. 4, no. 2, pp. 342–349, 2010.

[6] P. Z. Peebles, *Radar Principles*, John Wiley & Sons, New York, NY, USA, 1998.

[7] J. S. Thorp, "Optimal tracking of maneuvering targets," *IEEE Transactions on Aerospace and Electronic Systems*, vol. 9, no. 4, pp. 512–519, 1973.

Permissions

The contributors of this book come from diverse backgrounds, making this book a truly international effort. This book will bring forth new frontiers with its revolutionizing research information and detailed analysis of the nascent developments around the world.

We would like to thank all the contributing authors for lending their expertise to make the book truly unique. They have played a crucial role in the development of this book. Without their invaluable contributions this book wouldn't have been possible. They have made vital efforts to compile up to date information on the varied aspects of this subject to make this book a valuable addition to the collection of many professionals and students.

This book was conceptualized with the vision of imparting up-to-date information and advanced data in this field. To ensure the same, a matchless editorial board was set up. Every individual on the board went through rigorous rounds of assessment to prove their worth. After which they invested a large part of their time researching and compiling the most relevant data for our readers. Conferences and sessions were held from time to time between the editorial board and the contributing authors to present the data in the most comprehensible form. The editorial team has worked tirelessly to provide valuable and valid information to help people across the globe.

Every chapter published in this book has been scrutinized by our experts. Their significance has been extensively debated. The topics covered herein carry significant findings which will fuel the growth of the discipline. They may even be implemented as practical applications or may be referred to as a beginning point for another development. Chapters in this book were first published by Hindawi Publishing Corporation; hereby published with permission under the Creative Commons Attribution License or equivalent.

The editorial board has been involved in producing this book since its inception. They have spent rigorous hours researching and exploring the diverse topics which have resulted in the successful publishing of this book. They have passed on their knowledge of decades through this book. To expedite this challenging task, the publisher supported the team at every step. A small team of assistant editors was also appointed to further simplify the editing procedure and attain best results for the readers.

Our editorial team has been hand-picked from every corner of the world. Their multi-ethnicity adds dynamic inputs to the discussions which result in innovative outcomes. These outcomes are then further discussed with the researchers and contributors who give their valuable feedback and opinion regarding the same. The feedback is then collaborated with the researches and they are edited in a comprehensive manner to aid the understanding of the subject.

Apart from the editorial board, the designing team has also invested a significant amount of their time in understanding the subject and creating the most relevant covers. They scrutinized every image to scout for the most suitable representation of the subject and create an appropriate cover for the book.

The publishing team has been involved in this book since its early stages. They were actively engaged in every process, be it collecting the data, connecting with the contributors or procuring relevant information. The team has been an ardent support to the editorial, designing and production team. Their endless efforts to recruit the best for this project, has resulted in the accomplishment of this book. They are a veteran in the field of academics and their pool of knowledge is as vast as their experience in printing. Their expertise and guidance has proved useful at every step. Their uncompromising quality standards have made this book an exceptional effort. Their encouragement from time to time has been an inspiration for everyone.

The publisher and the editorial board hope that this book will prove to be a valuable piece of knowledge for researchers, students, practitioners and scholars across the globe.

List of Contributors

Sang Min Yoon
School of Computer Science, Kookmin University, 77 Jeongneung-ro, Sungbuk-gu, Seoul 136-702, Republic of Korea

Gang-Joon Yoon
National Institute for Mathematical Science, KT Daeduk 2 Research Center, 463-1 Jeonmin-dong, Yuseong-gu, Daejeon 305-811, Republic of Korea

Santiago-Omar Caballero-Morales
Technological University of the Mixteca, Road to Acatlima K.m. 2.5, 69000 Huajuapan de Leon, OAX, Mexico

Kristjan Pilt, Rain Ferenets, Kalju Meigas, Kristina Temitski and Margus Viigimaa
Department of Biomedical Engineering, Technomedicum, Tallinn University of Technology, Ehitajate tee 5, 19086 Tallinn, Estonia

Lars-Göran Lindberg
Department of Biomedical Engineering, Linkoping Univeristy, 581 85 Linkoping, Sweden

Qiang Zhang, Xianglian Xue and Xiaopeng Wei
Key Laboratory of Advanced Design and Intelligent Computing, Ministry of Education of Dalian University, Dalian 116622, China

Laia Núñez-Casillas and Manuel Arbelo
Grupo de Observacion de la Tierra y la Atmosfera (GOTA), Universidad de la Laguna, 38200 San Crist´obal de la Laguna, Spain

José Rafael García Lázaro and José Andrés Moreno-Ruiz
Departamento de Informatica, Universidad de Almerıa, 04120 Almerıa, Spain

I. Bosch, A. Serrano and L. Vergara
Signal Processing Group, Institute of Telecommunications and Multimedia Applications (iTEAM), Universitat Politecnica de Valencia, Camino de Vera, S/N, 46022 Valencia, Spain

Lotfi Salhi and Adnane Cherif
Signal Processing Laboratory, Physics Department, Sciences Faculty of Tunis, University of Tunis El Manar, 1060 Tunis, Tunisia

Mohamed Maalej, Sofiane Cherif and Hichem Besbes
Engineering School of Telecommunications of Tunis (SupCom), City of Communications Technologies, El Ghazala 2083, Ariana, Tunisia

Naveed Ishtiaq Chaudhary
Department of Electronic Engineering, International Islamic University, Islamabad 44000, Pakistan

Muhammad Asif Zahoor Raja and Junaid Ali Khan
Department of Electrical Engineering, COMSATS Institute of Information Technology, Attock Campus, Attock 43600, Pakistan

Muhammad Saeed Aslam
Pakistan Institute of Engineering and Applied Sciences, Nilore, Islamabad 45650, Pakistan

Vijay S. Chourasia
Manoharbhai Patel Institute of Engineering & Technology, Kudwa, Gondia, Maharashtra 441 614, India

Anil Kumar Tiwari
Indian Institute of Technology, Rajasthan MBM College Campus, Old Residency Road, Ratanada, Jodhpur, Rajasthan 342 011, India

Erqian Dong, Dengyi Zhang and Zhiyong Yuan
School of Computer, Wuhan University, Wuhan, Hubei 430072, China

Jianhui Zhao
School of Computer, Wuhan University, Wuhan, Hubei 430072, China
Suzhou Institute of Wuhan University, Suzhou, Jiangsu 215123, China

Mingui Sun and Wenyan Jia
Department of Neurosurgery, University of Pittsburgh, Pittsburgh, PA 15213, USA

Noppadon Jatupaiboon and Setha Pan-ngum
Department of Computer Engineering, Faculty of Engineering, Chulalongkorn University, Bangkok 10330, Thailand

Pasin Israsena
National Electronics and Computer Technology Center, Pathumthani 12120, Thailand

Mohamed A. Hankal, Islam A. Eshrah and Hazim Tawfik
Electronics and Electrical Communications Engineering Department, Faculty of Engineering, Cairo University, Giza 12613, Egypt

Jagannath Nirmal, Suprava Patnaik and Pramod Kachare
Department of Electronics Engineering, SVNIT, Surat, India

Mukesh Zaveri
Department of Computer Engineering, SVNIT, Surat, India

Harbinder Singh
Department of Electronics and Communication Engineering, Baddi University of Emerging Sciences and Technology, District Solan, Himachal Pradesh 173205, India

Vinay Kumar
Grupo de ProcesadoMultimedia, Departamento de Teoría de la Senal y Comunicaciones, Universidad Carlos III deMadrid Leganes, Madrid 28911, Spain

Sunil Bhooshan
Department of Electronics and Communication Engineering, Jaypee University of Information Technology, Waknaghat, District Solan, Himachal Pradesh 173215, India

Mithilesh Kumar Jha, Brejesh Lall and Sumantra Dutta Roy
Department of Electrical Engineering, Indian Institute of Technology Delhi, Hauz Khas, New Delhi 110016, India

Dongze Li, Xiang Li, Yongqiang Cheng, Yuliang Qin and Hongqiang Wang
School of Electronic Science and Engineering, National University of Defense Technology, Changsha 410073, China

www.ingramcontent.com/pod-product-compliance
Lightning Source LLC
Chambersburg PA
CBHW080258230326
41458CB00097B/5112

* 9 7 8 1 6 3 2 4 0 2 2 5 7 *